Data Driven Methods for Civil Structural Health Monitoring and Resilience

Data Driven Methods for Civil Structural Health Monitoring and Resilience: Latest Developments and Applications provides a comprehensive overview of data-driven methods for structural health monitoring (SHM) and resilience of civil engineering structures, mostly based on artificial intelligence or other advanced data science techniques. This allows existing structures to be turned into smart structures, thereby allowing them to provide intelligible information about their state of health and performance on a continuous, relatively real-time basis. Artificial intelligence-based methodologies are becoming increasingly more attractive for civil engineering and SHM applications; machine learning and deep learning methods can be applied and further developed to transform the available data into valuable information for engineers and decision makers.

Taylor and Francis Series in Resilience and Sustainability in Civil, Mechanical, Aerospace and Manufacturing Engineering Systems

Series Editor
Mohammad Noori
Cal Poly San Luis Obispo

PUBLISHED TITLES

Reliability-Based Analysis and Design of Structures and Infrastructure
Ehsan Noroozinejad Farsangi, Mohammad Noori, Paolo Gardoni, Izuru Takewaki, Humberto Varum, and Aleksandra Bogdanovic

Seismic Analysis and Design using the Endurance Time Method
Homayoon E. Estekanchi and Hassan A. Vafai

Thermal and Structural Electronic Packaging Analysis for Space and Extreme Environments
Juan Cepeda-Rizo, Jeremiah Gayle, and Joshua Ravich

Sustainable Development for the Americas
Science, Health, and Engineering Policy and Diplomacy
Honorary Editor: E. William Colglazier
Hassan A. Vafai and, Kevin E. Lansey
With the administrative assistance of Molli D. Bryson

Endurance Time Excitation Functions
Intensifying Dynamic Loads for Seismic Analysis and Design
Homayoon E. Estekanchi and Hassan A. Vafai

Mohammad Noori is a Professor of mechanical engineering at California Polytechnic State University, San Luis Obispo. He received his BS (1977), his MS (1980) and his PhD (1984) from the University of Illinois, Oklahoma State University and the University of Virginia respectively; all degrees in Civil Engineering with a focus on Applied Mechanics. His research interests are in stochastic mechanics, non-linear random vibrations, earthquake engineering and structural health monitoring, AI-based techniques for damage detection, stochastic mechanics, and seismic isolation. He serves as the executive editor, associate editor, the technical editor or a member of editorial boards of 8 international journals. He has published over 250 refereed papers, has been an invited guest editor of over 20 technical books, has authored/co-authored 6 books, and has presented over 100 keynote and invited presentations. He is a Fellow of ASME, and has received the Japan Society for Promotion of Science Fellowship.

For more information about this series, please visit: https://www.routledge.com/Resilience-and-Sustainability-in-Civil-Mechanical-Aerospace-and-Manufacturing/book-series/ENG

Data Driven Methods for Civil Structural Health Monitoring and Resilience

Latest Developments and Applications

Edited by
Mohammad Noori
Carlo Rainieri
Marco Domaneschi
and
Vasilis Sarhosis

CRC Press
Taylor & Francis Group
Boca Raton London New York

CRC Press is an imprint of the
Taylor & Francis Group, an **informa** business

Designed cover image: Shutterstock

First edition published 2024
by CRC Press
6000 Broken Sound Parkway NW, Suite 300, Boca Raton, FL 33487-2742

and by CRC Press
4 Park Square, Milton Park, Abingdon, Oxon, OX14 4RN

CRC Press is an imprint of Taylor & Francis Group, LLC

ISBN: 978-1-032-30837-1 (hbk)
ISBN: 978-1-032-30838-8 (pbk)
ISBN: 978-1-003-30692-4 (ebk)

DOI: 10.1201/9781003306924

Typeset in Times LT Std
by KnowledgeWorks Global Ltd.

Contents

Foreword vii
A Brief Overview of the Book by the Editors ix
About the Editors x
List of Contributors xiii

Chapter 1 Structural Resilience through Structural Health Monitoring: A Critical Review 1

Marco Domaneschi and Raffaele Cucuzza

Chapter 2 The Differential Evolution Algorithm: An Analysis of More than Two Decades of Application in Structural Damage Detection (2001–2022) 14

Parsa Ghannadi, Seyed Sina Kourehli, and Andy Nguyen

Chapter 3 Fatigue Assessment and Structural Health Monitoring of Steel Truss Bridges 58

Manuel Buitrago, Elisa Bertolesi, Pedro A. Calderón, and José M. Adam

Chapter 4 Sensor-Based Structural Assessment of Aging Bridges........ 76

Haris Alexakis, Sam Cocking, Nikolaos I. Tziavos, F. Din-Houn Lau, Jennifer Schooling, and Matthew DeJong

Chapter 5 Pile Integrity Assessment through a Staged Data Interpretation Framework........ 98

Qianchen Sun and Mohammed Z. E. B. Elshafie

Chapter 6 Data-Centric Monitoring of Wind Farms: Combining Sources of Information........ 120

Lawrence A. Bull, Imad Abdallah, Charilaos Mylonas, Luis David Avendaño-Valencia, Konstantinos Tatsis, Paul Gardner, Timothy J. Rogers, Daniel S. Brennan, Elizabeth J. Cross, Keith Worden, Andrew B. Duncan, Nikolaos Dervilis, Mark Girolami, and Eleni Chatzi

Chapter 7 From Structural Health Monitoring to Finite Element Modeling of Heritage Structures: The Medieval Towers of Lucca.................. 181

Riccardo Mario Azzara, Maria Girardi, Cristina Padovani, and Daniele Pellegrini

Chapter 8 Development of an Adaptive Linear Quadratic Gaussian (LQG) Controller for Structural Control Using Particle Swarm Optimization 201

Gaurav Kumar, Wei Zhao, M. Noori, and Roshan Kumar

Chapter 9 Application of AI Tools in Creating Datasets from a Real Data Component for Structural Health Monitoring............ 223

Minh Q. Tran, Hélder S. Sousa, and José C. Matos

Chapter 10 Ambient Vibration Prediction of a Cable-Stayed Bridge by an Artificial Neural Network 242

Melissa De Iuliis, Cecilia Rinaldi, Francesco Potenza, Vincenzo Gattulli, Thibaud Toullier, and Jean Dumoulin

Chapter 11 Modeling Uncertainties by Data-Driven Bayesian Updating for Structural and Damage Detection 258

Chiara Pepi, Massimiliano Gioffrè, and Mircea D. Grigoriu

Chapter 12 Image Processing for Structural Health Monitoring: The Resilience of Computer Vision–Based Monitoring Systems and Their Measurement 279

Nisrine Makhoul, Dimitra V. Achillopoulou, Nikoleta K. Stamataki, and Rolands Kromanis

Chapter 13 Automatic Structural Health Monitoring of Road Surfaces Using Artificial Intelligence and Deep Learning........................ 297

Andrea Ranieri, Elia Moscoso Thompson, and Silvia Biasotti

Chapter 14 Computer Vision–Based Intelligent Disaster Mitigation from Two Aspects of Structural System Identification and Local Damage Detection.. 312

Ying Zhou, Shiqiao Meng, Shengyun Peng, and Abouzar Jafari

Index.. 335

Foreword

Traditional structural health monitoring (SHM) strategies rely on updating a finite element model of the structure under study. However, this approach is not very reliable regarding complex systems. Data-driven approaches, instead, rely on a surrogate data-based model constructed based on data obtained from a complex structure to be substituted for an actual model. Although such models do not necessarily capture all the physics of a system, they can sufficiently serve the purpose of damage detection.

Generally, two techniques are used in the literature for SHM. These are (i) physics-based and (ii) data-based methods. The former uses the physical principle governing the structural mechanics to monitor structures for damage based on their response to natural or synthesized excitation forces. In contrast, the second category exploits data manipulation techniques to detect any structural response change and refer it to damage. While physics-based techniques rely on the physics of the problem, data-based methods seek to find an anomaly in systems based on pure mathematical principles governing the nature of data recorded from the structure under study.

There are two main categories of data-based techniques widely used in the context of SHM: (i) signal-processing and (ii) machine learning techniques. Signal-processing techniques seek to indicate symptoms in signals that can be related to damage. These symptoms are generally known as damage-sensitive features (DSFs), which need to be extracted from signals recorded on the structure subjected to some natural or synthesized excitation force. As such, signal-processing techniques consist of (i) a hardware device, such as an ultrasonic, electromagnetic, and, in general, advanced sensing device, to record data on the structures; and (ii) an algorithm to extract damage features from the data and relate them to damage. Change detection in signals is a conventional way of detecting damage using time–frequency domain representation of data through signal decomposition algorithms. These algorithms include wavelet transformation (WT), empirical mode decomposition (EMD), or one of their successors, such as ensemble empirical mode decomposition (EEMD) or variational mode decomposition (VMD). Another field of signal-processing approaches has roots in time series analysis and exploits several theories such as hypothesis testing and change detection algorithms. These algorithms generally consist of (i) data preprocessing, (ii) feature extraction and selection, (iii) pattern recognition, and (iv) data and information fusion. One challenge with signal-processing techniques is optimal sensor placement (OSP), which relates to optimization procedures and information theory.

There are, however, some challenges involved with conventional data analysis, such as (i) that these methods are not robust enough to deal with noisy information, (ii) that they are not well-suited to tackle the environmental and operational variations (EOV) effect, (iii) that they cannot be used to analyze large volumes of data, (iv) that they often suffer from a lack of baseline data recorded from the intact version of the structure under study, and (v) that they cannot deal with errors stemming from missing data that often have roots in sensor failure. Several advancements in data-processing units and deep-learning-based algorithms have facilitated the analysis of massive amounts of data recorded from SHM systems. Hence, deep-learning-based methods are currently in the perspective of researchers for addressing all of the above challenges involved with SHM.

Professor Amir Gandomi
Professor of Data Science
University of Technology Sydney
Sydney, Australia

A Brief Overview of the Book by the Editors

The materials presented in this book are organized in three themes:

- The focus of the first five chapters is on structural characterization and damage identification (detection–localization–quantification–remaining life) to improve structural resilience.
- Chapters 6 through 8 address important research topics related to distributed sensors and big data in SHM applications, data-driven methods, and AI methods.
- The last two chapters address image processing and non-contact methods for SHM.

Mohammad Noori
Carlo Rainieri
Marco Domaneschi
Vasilis Sarhosis

About the Editors

Mohammad Noori is a professor of mechanical engineering at California Polytechnic State University, San Luis Obispo; a fellow and life member of the American Society of Mechanical Engineering (ASME); and a recipient of the Japan Society for Promotion of Science Fellowship. His work in nonlinear random vibrations, especially hysteretic systems, in seismic isolation and application of artificial intelligence methods for structural health monitoring is widely cited. He has authored over 300 refereed papers, including over 150 journal articles; has published 15 scientific books, and 31 book chapters in archival volumes; has edited 15 technical books; and has been the guest editor of 15 journal volumes and proceedings. Noori was a co-founder of the National Institute of Aerospace, established through a $379 million 15-year NASA contract in partnership with NASA Langley Research Center. He has also received over $14 million in support of his research from the National Science Foundation (NSF), the Office of Naval Research (ONR), the National Sea Grant, and industry. He has supervised 24 postdoctoral, 26 PhD, and 53 MS projects. He has given over 20 keynote and 76 invited talks and lectures. He is the founding executive editor of a scientific journal, serves on the editorial board or as the associate editor of over 15 other journals, and has been a member of the scientific committee of numerous conferences. He directed the Sensors Program at the NSF in 2014, has been a distinguished visiting professor at several highly ranked universities in Europe and Asia, and serves as the scientific advisor for several organizations and technical firms. He was the dean of engineering at Cal Poly, served as a chaired professor and department head at North Carolina State University and Worcester Polytechnic Institute, and served as the chair of the national committee of mechanical engineering department heads. Noori has developed a unique online course, How to Write an Effective Research Paper, offered by Udemy.com and taken by over 9,000 students worldwide. Noori is an elected member of the Sigma Xi, Pi Tau Sigma, Chi-Epsilon, and Sigma Mu Epsilon honorary research societies. In 1996, Noori was invited by President Clinton's Special Commission on Critical Infrastructure Protection and presented a testimony as a national expert on that topic. Noori is the founding editor of the Resilience and Sustainability in Civil, Mechanical, Aerospace and Manufacturing Engineering Systems Series of CRC Press/Taylor & Francis Group.

Carlo Rainieri is currently a research scientist at the National Research Council of Italy. His research interests are in the fields of civil structural health monitoring, operational modal analysis, and smart materials. He has been a number of national as well as international research projects focused on civil SHM. He is member of the editorial board of a number of scientific journals, such as *Shock and Vibration*, *Infrastructures*, *Mathematical Problems in Engineering*, and *Advances in Civil Engineering*, and he serves as guest editor for the *Journal of Civil Structural Health Monitoring*. Moreover, he was lead editor of the special issue on "Automated Operational Modal Analysis and Its Applications in Structural Health Monitoring" published in *Shock and Vibration*. In 2019, he received the International Operational Modal Analysis Conference (IOMAC) scientific award for his contribution to the development of operational modal analysis. He was chair of the 8th Civil Structural Health Monitoring Workshop (2021), and he was member of the scientific committee of a number of international conferences in the fields of operational modal analysis and SHM.

He is author of the first book on operational modal analysis that appeared in the literature (*Operational Modal Analysis of Civil Engineering Structure: An Introduction and Guide for Applications*, Springer) and of about 170 papers published in international peer-reviewed journals and national and international conference proceedings. His main achievements in the field of civil SHM have been the development of data-processing methods for vibration-based SHM applications, including a number of original automated operational modal analysis procedures and novel methods for compensation of environmental and operational influences on modal properties. Dr. Rainieri is also the founder and former CEO of S2X s.r.l. (www.s2x.it), a spin-off company of the University of Molise aimed at providing highly qualified solutions and services in the fields of civil SHM and output-only modal analysis of civil engineering structures.

Marco Domaneschi is currently an assistant professor at the Department of Structural, Geotechnical and Building Engineering of Politecnico di Torino, where he teaches earthquake engineering and structural design courses. Formerly, he was a research associate and appointed professor of structural engineering at Politecnico di Milano. He is a professional structural engineer for special structures and serves as an R&D consultant in industrial manufacturing and mechanical engineering. He received his PhD from the University of Pavia (2006) and was a visiting researcher at several global universities. He currently serves as an associate editor and editorial board member for several international journals such as the *Journal of Vibration and Control* (SAGE) and *Bridge Engineering* (Institution of Civil Engineers, UK). He is also a member of several research associations such as the International Society of Structural Health Monitoring of Intelligent Infrastructure (ISHMII) and the International Association for Bridge Maintenance and Safety (IABMAS). He is also a reviewer for more than 40 international journals. He has been a speaker, sessions chair, editorial board member, and organizer in several international conferences. He received numerous awards for best presentations at conferences, and for research papers and activities. He supports/supported the coordination of several research projects and has/had scientific responsibility in numerous research projects. He has authored over 70 journal articles and 130 international conference papers. His research interests and activities include structural control and health monitoring, resilience and robustness of structures and communities, earthquake engineering and seismic risk, special structures, small- and large-scale simulations, emergency evacuation, and structural collapse analysis.

Vasilis Sarhosis is an associate professor of structural engineering at the School of Civil Engineering, University of Leeds, UK. He holds both undergraduate and postgraduate degrees in civil engineering from the University of Leeds and worked as a consultant civil engineer in the UK. His main expertise lies in the development of advanced high-fidelity models of nonlinear response to quantify degradation and understand the long-term behavior of existing masonry infrastructure stock (e.g., bridges, tunnels, historic structures, and monuments) subjected to extreme loading conditions. Recently, he pioneered the development of the Cloud2DEM procedure to transform three-dimensional (3D) point clouds of complex structures obtained from photogrammetry and laser scanning to 3D discrete element models. The approach enables the realistic structural analysis of "as-is" masonry infrastructure in an accurate and computationally efficient manner. Dr. Sarhosis is a Chartered Engineer (CEng), Fellow of the Institute of Mechanical Engineering (FIMEchE), and Fellow of the Higher Education Academy (FHEA) in the UK. He is currently chairing the National Scientific Committee on the Analysis and Restoration of Structures of Architectural Heritage (ISCARSAH-UK), which is part of the International Council on Monuments and Sites (ICOMOS). He has edited a book on computational modeling of masonry structures using the discrete element method and published more than 100 peer-reviewed journal articles. His research has been cited more than 1,500 times, and his h-index is 22 (Scopus). For more information, please visit his website (https://eps.leeds.ac.uk/civil-engineering/staff/1798/professor-vasilis-sarhosis).

Contributors

Imad Abdallah
ETH Zurich
Zurich, Switzerland

Dimitra V. Achillopoulou
University of Glasgow
Glasgow, United Kingdom

José M. Adam
Universitat Politècnica de València
Valencia, Spain

Haris Alexakis
Aston University
Birmingham, United Kingdom
and
University of Cambridge
Cambridge, United Kingdom

Luis David Avendaño-Valencia
University of Southern Denmark
Odense M, Denmark

Riccardo Mario Azzara
Istituto di Scienza e Tecnologie dell'Informazione "Alessandro Faedo"
Consiglio Nazionale delle Ricerche
Pisa, Italy

Elisa Bertolesi
Brunel University
London, United Kingdom

Silvia Biasotti
Istituto per le Tecnologie della Costruzione
Consiglio Nazionale delle Ricerche
Naples, Italy

Daniel S. Brennan
University of Sheffield
Sheffield, United Kingdom

Manuel Buitrago
Universitat Politècnica de València
Valencia, Spain

Lawrence A. Bull
The British Library
London, United Kingdom
University of Cambridge
Cambridge, United Kingdom

Pedro A. Calderón
Universitat Politècnica de València
Valencia, Spain

Eleni Chatzi
ETH Zurich
Zurich, Switzerland

Sam Cocking
University of Cambridge
Cambridge, United Kingdom

Elizabeth J. Cross
University of Sheffield
Sheffield, United Kingdom

Raffaele Cucuzza
Politecnico di Torino
Turin, Italy

Melissa De Iuliis
Sapienza University of Rome
Rome, Italy

Matthew DeJong
University of California
Berkeley, California

Nikolaos Dervilis
University of Sheffield
Sheffield, United Kingdom

Marco Domaneschi
Politecnico di Torino
Turin, Italy

Jean Dumoulin
Université Gustave Eiffel
INRISA, COSYS-SII
Bouguenais, France

Andrew B. Duncan
The British Library
London, United Kingdom
and
Imperial College
London, United Kingdom

Mohammed Z. E. B. Elshafie
Qatar University
Doha, Qatar

Paul Gardner
University of Sheffield
Sheffield, United Kingdom

Vincenzo Gattulli
Sapienza University of Rome
Rome, Italy

Parsa Ghannadi
Islamic Azad University
Ahar, Iran

Massimiliano Gioffrè
University of Perugia
Perugia, Italy

Maria Girardi
Istituto di Scienza e Tecnologie dell'Informazione "Alessandro Faedo"
Consiglio Nazionale delle Ricerche
Pisa, Italy

Mark Girolami
The British Library
London, United Kingdom
and
University of Cambridge
Cambridge, United Kingdom

Mircea Grigoriu
Cornell University
Ithaca, New York

Abouzar Jafari
Tongji University
Shanghai, China

Seyed Sina Kourehli
Azarbaijan Shahid Madani University
Tabriz, Iran

Rolands Kromanis
University of Twente
Enschede, The Netherlands

Gaurav Kumar
Alliance University
Bengaluru, India

Roshan Kumar
Miami College of Henan University
Kaifeng, China

F. Din-Houn Lau
Costain Ltd
Maidenhead, United Kingdom

Nisrine Makhoul
Politecnico di Milano
Milano, Italy

José C. Matos
University of Minho, ISISE
Guimarães, Portugal

Shiqiao Meng
Tongji University
Shanghai, China

Charilaos Mylonas
Deloitte AG
Zurich, Switzerland

Andy Nguyen
University of Southern Queensland
Springfield, Australia

Mohammed Noori
California Polytechnic State University
San Luis Obispo, California

Cristina Padovani
Istituto di Scienza e Tecnologie dell'Informazione "Alessandro Faedo"
Consiglio Nazionale delle Ricerche
Pisa, Italy

Daniele Pellegrini
Istituto di Scienza e Tecnologie dell'Informazione "Alessandro Faedo"
Consiglio Nazionale delle Ricerche
Pisa, Italy

Shengyun Peng
Tongji University
Shanghai, China

Chiara Pepi
University of Perugia
Perugia, Italy

Francesco Potenza
University of Chieti–Pescara
Pescara, Italy

Andrea Ranieri
Istituto per le Tecnologie della Costruzione
Consiglio Nazionale delle Ricerche
Naples, Italy

Cecilia Rinaldi
Sapienza University of Rome
Rome, Italy

Timothy J. Rogers
University of Sheffield
Sheffield, United Kingdom

Jennifer Schooling
University of Cambridge
Cambridge, United Kingdom

Hélder S. Sousa
University of Minho, ISISE
Guimarães, Portugal

Nikoleta K. Stamataki
Democritus University of Thrace
Xanthi, Greece

Qianchen Sun
University of Cambridge
Cambridge, United Kingdom

Konstantinos Tatsis
ETH Zurich
Zurich, Switzerland

Elia Moscoso Thompson
Istituto per le Tecnologie della Costruzione
Consiglio Nazionale delle Ricerche
Naples, Italy

Thibaud Toullier
Université Gustave Eiffel
INRISA, COSYS-SII
Bouguenais, France

Minh Quang Tran
University of Minho, ISISE
Guimarães, Portugal

Nikolaos I. Tziavos
Aston University
Birmingham, United Kingdom
and
University of Cambridge
Cambridge, United Kingdom

Keith Worden
University of Sheffield
Sheffield, United Kingdom

Wei Zhao
Miami College of Henan University
Kaifeng, China

Ying Zhou
Tongji University
Shanghai, China

1 Structural Resilience through Structural Health Monitoring

A Critical Review

Marco Domaneschi and Raffaele Cucuzza

1.1 OVERVIEW

This chapter aims to evaluate structural resilience improvements based on structural health monitoring (SHM). The entire chapter is organized into two main parts. In the first one, the key stages of the monitoring method are described at varying degrees of difficulty and knowledge levels. A brief digression of established and cutting-edge approaches that may be traced back to the discipline is also included. Next, structural resilience and its dimensions are described by referring to the literature. At the end of this general discussion concerning the theoretical aspect of these disciplines, a bibliographical and bibliometric study of the references is conducted, and the relationship between monitoring and resilience is identified. In this way, any conceptual and practical links present in the existing literature will be highlighted.

The second part of the chapter is devoted to the analysis of the conceptual relations between the four dimensions of structural resilience and the four stages of the SHM discipline. Schemes and examples are provided in this chapter to demonstrate how information at each monitoring stage can be employed for the evaluation of the corresponding resilience dimension.

1.2 STRUCTURAL HEALTH MONITORING AND RESILIENCE: THE STATE OF THE ART ON DAMAGE DIAGNOSIS AND PROGNOSIS

Different types of degradation, such as long-term corrosion or fatigue, as well as transient occurrences like earthquakes, can cause damage to civil engineering structures. Damage diagnosis allows informing decision-makers about the appropriate response action to hazardous structural situations. The foundation phases of SHM for damage diagnosis are damage detection, localization, and quantification. The ultimate phase, or the final SHM objective, is the damage prognosis, or the estimation of the system's remaining life, given the measurement and evaluation of its current structural status (Cheung et al. 2008; Doebling et al. 1998;

DOI: 10.1201/9781003306924-1

Domaneschi et al. 2013, 2017; Faravelli & Casciati 2004; Morgese et al. 2021). In detail, the four phases can be characterized as follows (Doebling et al. 1998):

- **Phase 1**: Establishing whether there is damage to the structure
- **Phase 2**: Phase 1 plus locating the damage's geometric position
- **Phase 3**: Includes Phase 2 as well as a measurement of the extent of the damage
- **Phase 4**: Phase 3 plus an estimation of the structure's remaining useful life

The first two phases represent the main supports offered by the most common vibration-based damage identification techniques (without the use of structural models). In some circumstances, Phase 3 damage identification can be achieved by combining vibration-based techniques with a structural model. Phase 4 remains a challenging problem for engineers that requires multidisciplinary and predictive modeling capabilities, and it could provide invaluable safety and economic benefits in structural and infrastructural management.

Doebling (1996) provided a survey of the literature on the various techniques for identifying damage and monitoring the health of structures based on changes in their measured dynamic parameters. They are methods based on (i) the modification of the modal characteristics, (ii) the changing of dynamic flexibility, (iii) the updating of structural matrices under constrained optimization, but also (iv) nonlinear approaches and (v) methods based on neural networks. As a result, they are divided into groups according to the type of measured data that was used and/or to the technique for determining the damage from the measured data that is implemented. A dataset of the undamaged structure is assumed by many methods, while others involve a thorough finite element method (FEM) of the structure. Frequently, the unavailability of this kind of data can make a method impractical for some applications. Although there is no guarantee that dependency on numerical models and historical data can be eliminated, steps can and should be taken to reduce it.

Focusing on the more recent developments in the field, SHM systems have undergone substantial developments due to emerging computing power and sensing technology, which have thus enabled numerous installations on various examples of civil infrastructure for structural health assessment, to detect damage or anomalies, by exploiting continuous monitoring of environmental conditions, loads, and, most importantly, structural responses. As a result, the number of sensors in an SHM system has increased, as has, of course, the amount of data collected, which has reached enormous quantities. To handle this large amount of data, advanced processing methods had to be employed to convert the collected data (heterogeneous and multisource) into different types of specific useful indicators for making effective inspection, maintenance, and management decisions. Consequently, the use of automatic algorithms for data management, computation, and structural (e.g., damage) identification has become increasingly popular in the scientific community. With elegant performance and frequently rigorous accuracy, machine learning algorithms, especially deep learning algorithms, have become increasingly practical and widely employed in vibration-based structural damage diagnosis. Unlike the second (deep learning algorithms), which can identify a direct mapping from the raw inputs to the final outputs without the requirement for feature extraction, the first ones (machine learning algorithms) require preprocessing data samples to extract certain features or attributes that reflect the most distinctive pieces of information. When it is difficult to manually select a new good group of features to be used for training the system, deep learning algorithms are of utmost relevance. As a result, deep learning is regarded as a branch of machine learning, and numerous platforms have successfully used it to deal with large amounts of data (i.e., big data). Fortunately, both deep learning and developments in parallel computing have been successfully used in several applications in a wide range of research areas, such as SHM (Avci et al. 2021; Azimi et al. 2020; Sun et al. 2020; Ye et al. 2019).

1.3 THE FOUR DIMENSIONS OF STRUCTURAL RESILIENCE

When a structure is hit by a severe event, it can develop a certain level of damage: its ability to recover its original functionality in the undamaged configuration in the shortest possible time is an important aspect and of great interest to the scientific community. This positive property is related to the concept of *structural resilience*, which has been defined as "the ability of a system to reduce the chances of a shock, to absorb such a shock if it occurs, and to recover quickly after the shock" (Bruneau et al. 2003, p. 735).

Researchers at MCEER (Multidisciplinary Center for Earthquake Engineering Research) have identified four important factors that contribute to resilience: robustness, resourcefulness, redundancy, and rapidity. These factors have also been referred to as the four dimensions of resilience (Bruneau et al. 2003; Bruneau & Reinhorn 2006, 2007):

1. **Robustness**: The characteristics of a structure, or one of its elements, to withstand a given level of stress or demand (e.g., damage) while maintaining its characteristic level of functionality; can also be thought of as the concept of damage tolerance.
2. **Redundancy** (e.g., of load-bearing elements): The ability to develop alternative load-supporting pathways once degradation of the main elements occurs (i.e., original elements that are substitutable).
3. **Resourcefulness**: The ability to identify issues, set priorities, and gather resources when circumstances threaten to upset the structure or one of its elements (apply monetary, physical, technological, informational, and human resources in the process of recovery to meet established priorities and achieve goals).
4. **Rapidity**: The ability to prioritize tasks and complete tasks quickly to limit losses, restore functionality, and prevent further interruption.

A review of the scientific literature shows that many studies focused on the structural resilience of buildings. For example, Cimellaro et al. (2010) pointed out a framework for evaluation of health care facilities subjected to earthquake. In this work, the evaluation of disaster resilience was based on dimensionless analytical functions related to the variation of functionality during a period of interest, taking into account the losses in disaster and the recovery path. Structural resilience has also been developed in different directions; for example, Domaneschi and Martinelli (2016) introduced the concept of *immediate resilience* as associated with the automatic function of specific elements to compensate for local out-of-services. For example, if one component don't work properly, another nearby component, capable of changing its characteristics through an automatic external command, also fulfills the function of the damaged component.

1.4 BIBLIOGRAPHIC AND BIBLIOMETRIC ANALYSIS

State-of-the-art papers on the recent developments of monitoring techniques can be found in the literature (Salar et al. 2022; Yuan 2016), but none of these dealt specifically with the relationship between SHM and resilience topics. The authors believe that an understanding of how SHM techniques can be useful to provide information regarding structural resilience will play a crucial role in future years.

To this end, bibliometric analysis has been conducted to organize available publications based on the above-mentioned relationship. This analysis was performed by achieving two main purposes:

- Comprehensively describe the state of the art; and
- Identify and organize links between main research topics related to this field and/or connections between the most active authors.

In this study, the relevant works of literature were collected by adopting Elsevier's abstract and citation database, Scopus. Because of the specificity of the topic, conference papers and book chapters have been included too. In the first step, we found a total of 97 scientific documents by searching keywords such as *structural health monitoring* and *resilience* within the title, abstract, and authors components of each paper.

The total number of papers dropped to 40 once scientific contributions belonging to only the structural engineering subject area were selected. At this stage, the database was populated by 40 scientific articles published between 2007 and 2021, belonging to 33 different indexed scientific journals and written by 136 authors. Once the database was completely defined, a bibliometric analysis was conducted and graphical outcomes were obtained by using the Bibliometrix R-package (Aria & Cuccurullo 2017). Figure 1.1 shows the total number of publications per year. The trend reveals that the highest annual growth rate occurred between 2015 and 2019. A descendent trend is recognized between 2019 and 2021, when the peak of the curve passed from 10 articles to about 7.5 articles per year, and currently (September 2022) it is about two.

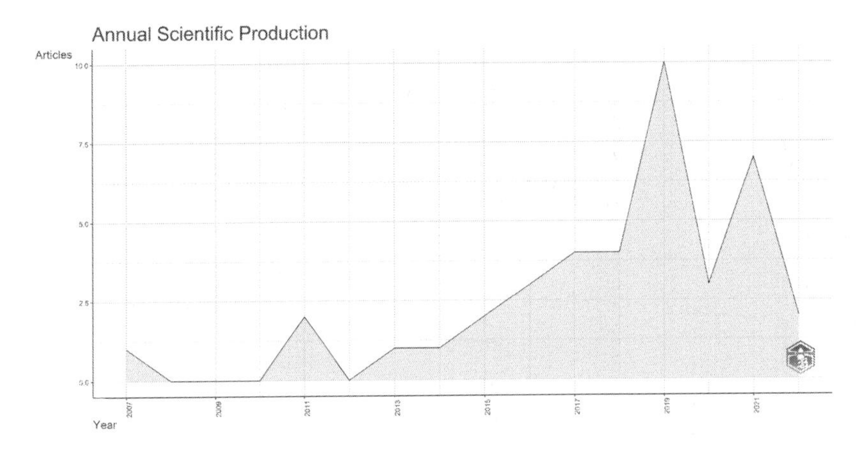

FIGURE 1.1 Total number of publications obtained by reviewing 40 papers.

The statistics of publication per journal are provided in Figure 1.2, where the most relevant sources have been shown and the first 10 journals have been plotted.

To help us analyze the keyword trends in recent years, VOSviewer software was adopted for a graphical representation of the data. It is a software tool for constructing and visualizing bibliometric networks. Figure 1.3 depicts network visualization co-occurrence analysis, while, in Figure 1.4, an overlay visualization is shown with the keyword trends in recent years. Maps created, visualized, and explored using VOSviewer include items that are the authors' keywords.

In the network visualization (see Figure 1.3), items are represented by their label and (by default) by a circle. The size of the label and the circle of an item is determined by the weight of the item. The higher the weight of an item, the

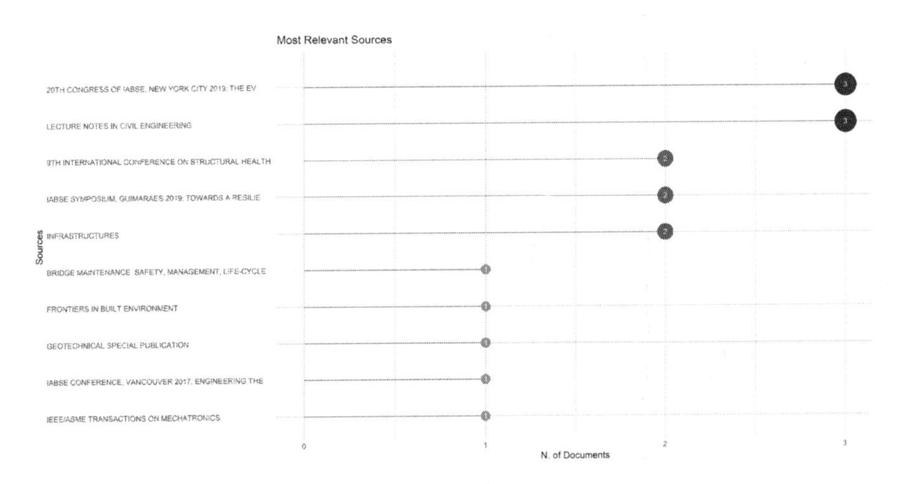

FIGURE 1.2 The most relevant sources carried out from the bibliometric analysis.

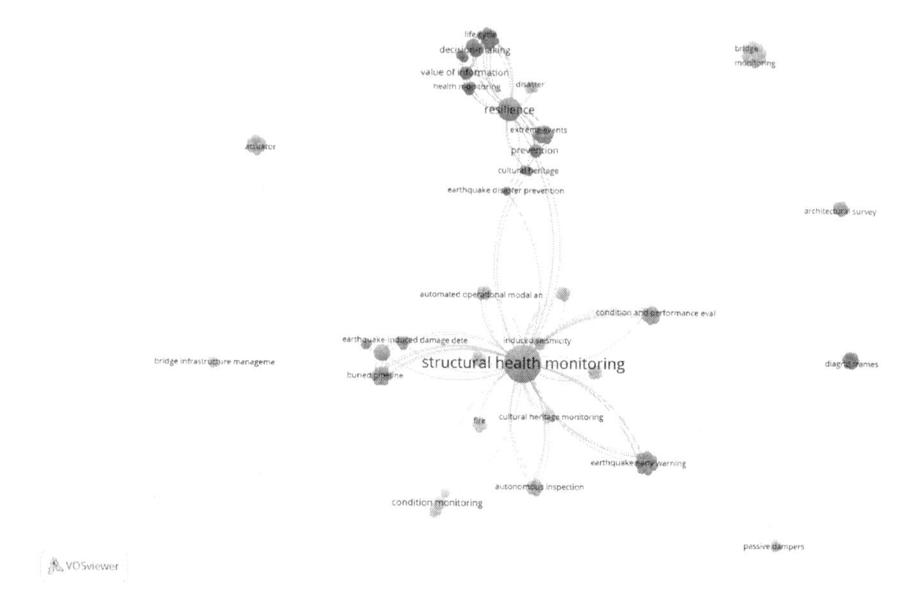

FIGURE 1.3 Network visualization of the co-occurrence of authors' keywords.

larger the label and the circle of the item itself. Moreover, between any pair of items, there can be only one link. A *link* is a connection or a relation between items. In this case, the focus is on detecting the keywords that occur together in the reviewed documents and, hence, finding the research topics that have a sort of relationship with each other.

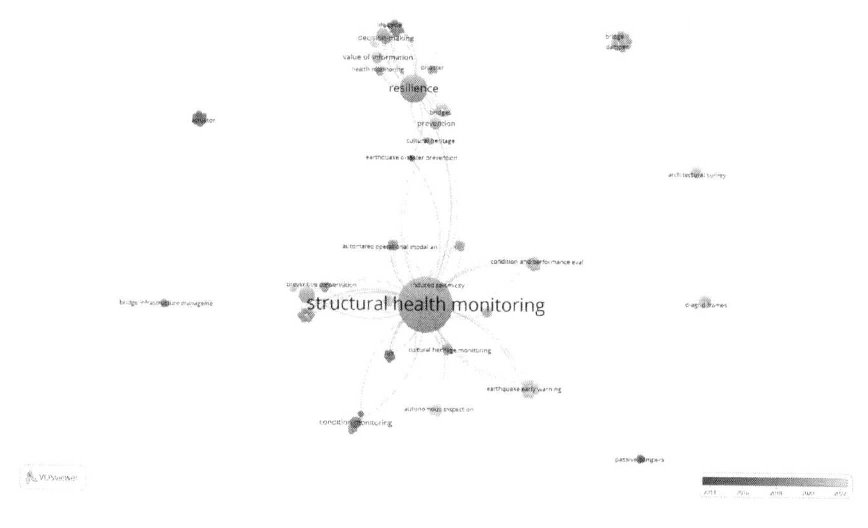

FIGURE 1.4 Overlay visualization of the co-occurrence of authors' keywords in recent years.

Each link is characterized by *strength*, represented by a positive numerical value that varies in a range between 0 (minimum strength) and 1 (maximum strength). The higher this value, the stronger the link.

Moreover, if the link attribute indicates the number of links to an item with other items, the *total strength attribute* indicates the total strength of the links of an item with other items. Hence, a network is a set of items together with the links between the items.

A set of items can be grouped into clusters. Clusters are non-overlapping in VOSviewer; in other words, an item may belong to only one cluster. Clusters are represented by colors; hence, the items having the same color belong to the same cluster. The distance between two items in the visualization approximately indicates the relatedness of the journals in terms of co-citation links. In general, the closer two items are located to each other, the stronger their relatedness.

As could be expected, in the network visualization depicted in Figure 1.3, it is quite evident that among the 131 keywords contained within the overall reviewed documents, *structural health monitoring*, *resilience*, and *value of information* are the most common. The strength and the total link that connect these three main keywords with the other ones are not only the thickest but also the most numerous. In this way, it is proved that the most significant keywords, pointed out by the bibliometric analysis, exhibit the most robust connections with the other secondary branches of the same research field. Therefore, the link between *structural health monitoring* and *resilience* passes through intermediate keywords such as *earthquake disaster prevention*, *cultural heritage*, and *extreme events*. This means that, actually, these keywords don't appear as the only keywords in the same document but are always in conjunction with other keywords.

Moreover, other secondary topics are depicted as *bridge monitoring* and *bridge infrastructure management*. These don't exhibit any type of connection with both *structural health monitoring* and *resilience*, and the distance, in terms of research topic, between the former and the latter is emphasized by the graphical distance between the items of each cluster.

In Figure 1.4, a so-called overlay visualization is shown. The overlay visualization is identical to the network visualization except that items are colored differently. The color of an item is assigned depending on the scores of the item. The score, adopted in this analysis, corresponds to the year of the appearance of each item (keywords).

It becomes clear how the most used keywords, such as *structural health monitoring*, *resilience*, and *value of information*, belong to documents that were published between 2016 and 2020. It proves that the above-mentioned topics received great attention from the scientific community and their growth took place in a few years.

Another interesting map representation of the data involved in the analysis is reported in Figure 1.5, where the network of the 136 authors among collaborating researchers is shown. At the center of the figure are placed all the authors who are the most active in this field. Each item in the network displays an author/coauthor, and a link between items illustrates the co-occurrence of the knowledge channels.

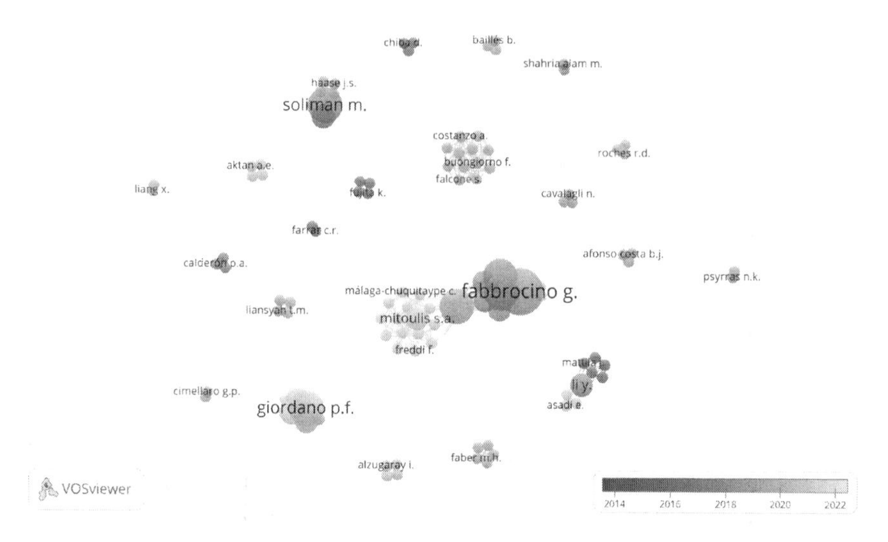

FIGURE 1.5 Overlay visualization of the coauthorship and collaborations within the scientific community.

It is rather evident that there is not a connected collaborating scientific community interested in some topics. Conversely, several independent groups can be seen, composed of three or four authors, which have no links with other research groups.

1.5 RELATIONSHIP BETWEEN THE FOUR LEVELS OF SHM AND THE FOUR DIMENSIONS OF STRUCTURAL RESILIENCE

Once the four different levels of SHM and the resilience dimensions are identified, a sort of link between these apparently independent categories can be provided. However, based on the bibliometric analysis conducted in this study, few works pointed out the possible benefits in terms of resilience related to the adoption of SHM systems. Therefore, the question has not been completely settled: how can the four SHM phases be employed as efficient tools and strategies for the evaluation and improvement of the resilience dimensions?

Limongelli et al. (2019) and Iannacone et al. (2022) demonstrated how SHM may enhance resilience by providing information that can support decision-making, aiming to minimize the impact of disturbances. The research described here is entirely focused on cultural heritage constructions, which are particularly important for local communities due to the provided functions (i.e., strategic, cultural, and social). The authors offered a different perspective on the resilience concept concerning the investigated case studies, intending it as "the capability to overcome a disturbance with the minimum total loss of functionality in time" (Limongelli et al. 2019, p. 1). Therefore, the authors described possible prospects of structural resilience based on SHM aiming at three different decision situations: before, during, and after the disturbance.

Other authors (e.g., Cimellaro and Kammouh 2019; Freddi et al. 2021) presented new dynamic data-driven frameworks for effectively selecting and executing SHM systems in critical infrastructures. The introduced framework was developed to help decision-makers and infrastructure operators efficiently use resources for SHM, for the prediction of the best rehabilitation strategies or the refining of the risk assessment and infrastructure prioritization plan.

Achieving more inclusive resilient infrastructure frameworks represented a priority for many researchers, such as Achillopoulou et al. (2021). They developed resilient strategic monitoring for the estimation of the risk level of assets with no disruption to traffic in reinforced concrete bridges. Haria et al. (2019) proposed an innovative approach that employs satellite measurements of displacement with a 3D building information model (BIM). On the same topic, Zhang et al. (2019) assessed bridge resilience considering the combined effects of reinforcement corrosion and foundation scour under extreme loading using SHM techniques. Concerning the bridge application field, Ngamkhanong et al. (2018) realized a state-of-the-art review of railway track resilience monitoring in which the most promising SHM strategies have been collected.

The influence of SHM on resilience was also researched by Celebi (2015) concerning different tall buildings in San Francisco and by Asadi et al. (2021) concerning health monitoring for seismic resilience quantification and safety evaluation. A few authors, such as Aloisio et al. (2022), addressed the problem of pinching in reinforced concrete and timber structures for the prediction of structural resilience.

However, despite the critical survey conducted among all the contributions concerning the treated research field, a clear theoretical or analytical relation between resilience and the four levels of SHM has not yet been achieved.

If *damage detection* as the first level of SHM is considered, the provided information is the presence (or not) of structural damage. Any information regarding the position or intensity of the damage cannot be provided. In this case, the damaged state of the structure is strictly related to the redundancy dimension of the same structure. Indeed, a redundant structure, independent of the type of damage, results in being safer than an isostatic (statically determinate) one. In Italy, the tragic event of the Polcevera Bridge, which collapsed in 2018, proves that the problem of corrosion of the steel strands inside the concrete stays reasonably represented the main cause of the failure because of the intrinsic lack of redundancy of the structure (Domaneschi et al. 2020; Morgese et al. 2020). For this reason, the simplest information on the presence of pathologies in the structures, determined through SHM strategies, could be correlated to the grade of redundancy of the structure to take the necessary countermeasures, such as emergency measures (e.g., evacuation, traffic reduction, and shutting of critical facilities).

The second level of SHM, *damage localization*, can be linked to the rapidity dimension of resilience. Once the pathology is detected and the local or diffuse damage is localized, the rapidity of the intervention and recovery phases plays a crucial role. In this context, the SHM system may support decision-making that aims to accelerate the interventions following a priority plan, and efficiently managing the emergency.

A relationship between the resourcefulness dimension and the third SHM level of *damage intensity* also can be identified. Indeed, information about the damage intensity subsequent to disturbances effectively supports the organizational and recovery phases, allowing the accurate estimation of the necessary economic and social resources to be made available.

Finally, the fourth level of SHM, *prognosis*, is linked to the robustness dimension of resilience. It is focused on the estimation of the structure's remaining life given the assessment of its current conditions. This SHM level indirectly gives a measure of structural robustness, providing the attempt to forecast the structural performance to tolerate a specific amount of damage while retaining its usual degree of functionality.

Figure 1.6 illustrates the links between structural resilience dimensions and the four SHM phases. In the circle representation, at each slice corresponds a direct link between the selected resilience dimension and the related damage level (e.g., damage detection with redundancy and damage localization with rapidity). Reading this clockwise is key to understanding the graphical representation: the external circle describes the increasing intensity level of the damage investigation, while the internal one shows the appropriate level of resilience knowledge that can be derived from those specific monitoring activities.

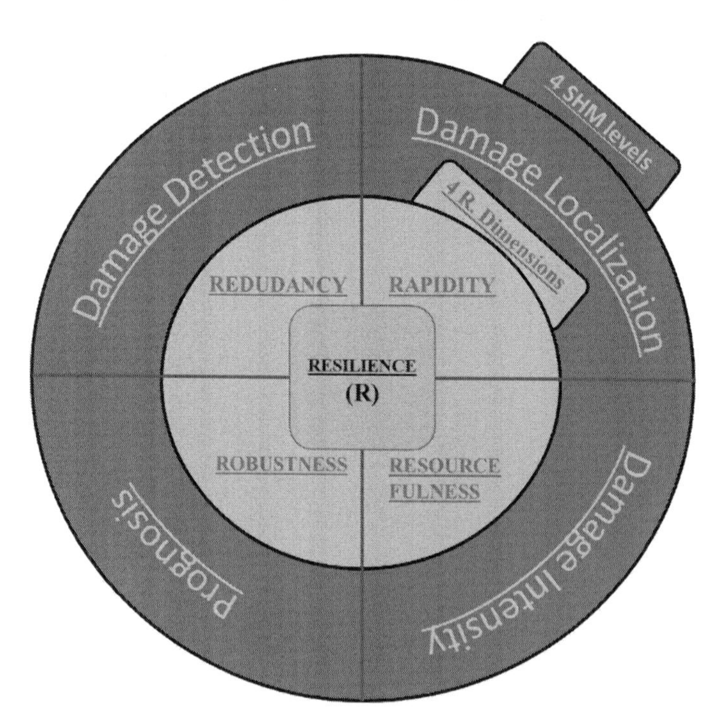

FIGURE 1.6 The conceptual relation between SHM and resilience.

1.6 CONCLUSIONS

The subject of this chapter is the study of the connection between SHM and resilience, from their characterizations in components and phases. Subsequently, a bibliographic and bibliometric study of the state of the art of their interaction was conducted to highlight the conceptual and practical links in the existing literature. It was observed how: (i) such research topics have attracted the interest of researchers, specifically in recent years, highlighting how they are topical issues; (ii) concerning the keywords of the contributions analyzed, the main ones (i.e., resilience and SHM) are often connected to intermediate keywords related to infrastructure, its management and safety, and the value of information; and (iii) although there are a good number of research groups active on these topics, (iv) the connections between these groups are very low, highlighting a scenario in which there are many active but essentially independent groups in which (v) interactions are active only between members of the same individual group.

It is therefore possible to conclude from the state-of-the-art analysis that the issue of possible benefits to structural resilience brought by monitoring is emerging but is still in an embryonic state. So, the second part of this chapter is focused on determining the conceptual relationships between the four dimensions of structural resilience and the four phases of the discipline of SHM. It looked at the direct links by which the different phases of SHM can determine, influence, and improve the dimensions of structural resilience, and thus resilience itself.

This purely conceptual dissertation is intended to represent a proposal that can also be a cornerstone for fostering and determining further conceptual and application developments.

REFERENCES

Achillopoulou, D. V., Mitoulis, S. A., & Stamataki, N. K. 2021. Resilient monitoring of the structural performance of reinforced concrete bridges using guided waves. In Bridge Maintenance, Safety, Management, Life-Cycle Sustainability and Innovations (pp. 3275–3281). CRC Press.

Aloisio, A., Pelliciari, M., Bergami, A. V., Alaggio, R., Briseghella, B., & Fragiacomo, M. 2022. Effect of pinching on structural resilience: Performance of reinforced concrete and timber structures under repeated cycles. *Structure and Infrastructure Engineering* 1–17.

Aria, M., & Cuccurullo, C. 2017. Bibliometrix: An R-tool for comprehensive science mapping analysis. *Journal of Informetrics 11*(4): 959–975.

Asadi, E., Salman, A. M., Li, Y., & Yu, X. 2021. Localized health monitoring for seismic resilience quantification and safety evaluation of smart structures. *Structural Safety 93*: 102127.

Avci, O. et al. 2021. A review of vibration-based damage detection in civil structures: From traditional methods to machine learning and deep learning applications. *Mechanical Systems and Signal Processing* 147: 107077.

Azimi, M. et al. 2020. Data-driven structural health monitoring and damage detection through deep learning: State-of-the-art review. *Sensors (Switzerland) 20*(10): 2778.

Bruneau, M. et al. 2003. A framework to quantitatively assess and enhance the seismic resilience of communities. *Earthquake Spectra* 19(4): 733–752.

Bruneau, M., & Reinhorn, A. 2006. Overview of the resilience concept. In Proceedings of the 8th U.S. National Conference on Earthquake Engineering, Earthquake Engineering Research Institute (EERI), Oakland, CA.

Bruneau, M., & Reinhorn, A. 2007. Exploring the concept of seismic resilience for acute care facilities. *Earthquake Spectra 23*(1): 41–62.

Celebi, M. 2015. Rapid, feasible and technically sound structural health monitoring to facilitate resiliency of tall buildings. In SHMII 2015—7th International Conference on Structural Health Monitoring of Intelligent Infrastructure, Torino, Italy 1–3 July 2015.

Cheung, A. et al. 2008. The application of statistical pattern recognition methods for damage detection to field data. *Smart Materials and Structures 17*(6): 065023.

Cimellaro, G. P., & Kammouh, O. 2019. SHM role in the framework of infrastructure resilience. In 9th International Conference on Structural Health Monitoring of Intelligent Infrastructure: Transferring Research into Practice, SHMII 2019, St. Louis, United States,4–9 August 2019. (pp. 1595–1601).

Cimellaro, G. P., Reinhorn, A. M., & Bruneau, M. 2010. Framework for analytical quantification of disaster resilience. *Science China Technological Sciences 55*(11): 3081–3089.

Doebling, S. W. 1996. Los Alamos National Laboratory Report LA-13070-MS. Los Alamos National Laboratory.

Doebling, S. W. et al. 1998. A summary review of vibration-based damage identification methods. *Shock and Vibration Digest 30*(2): 91–105.

Domaneschi, M. et al. 2017. Damage detection on output-only monitoring of dynamic curvature in composite decks. *Structural Monitoring and Maintenance 4*(1): 1–15.

Domaneschi, M., Limongelli, M. P., & Martinelli, L. 2013. Vibration based damage localization using MEMS on a suspension bridge model. *Smart Structures and Systems 12*(6): 679–694.

Domaneschi, M., & Martinelli, L. 2016. Earthquake resilience-based control solutions for the extended benchmark cable-stayed bridge. *Journal of Structural Engineering, ASCE 142*(8): 4015009.

Domaneschi, M., Pellecchia, C., De Iuliis, E., Cimellaro, G. P., Morgese, M., Khalil, A. A., & Ansari, F. 2020. Collapse analysis of the Polcevera Viaduct by the applied element method. *Engineering Structures 214*: 110659.

Faravelli, L., & Casciati, S. 2004. Structural damage detection and localization by response change diagnosis. *Progress in Structural Engineering and Materials 6*(2): 104–115.

Freddi, F., Galasso, C., Cremen, G., Dall'Asta, A., Di Sarno, L., Giaralis, A., & Woo, G. 2021. Innovations in earthquake risk reduction for resilience: Recent advances and challenges. *International Journal of Disaster Risk Reduction 60*: 102267.

Haria, K., de Farago, M., Dawood, T., & Bush, M. 2019. Integration of regional and asset satellite observations for assessment of infrastructure resilience. In International Conference on Smart Infrastructure and Construction 2019 (ICSIC) Driving Data-Informed Decision-Making (pp. 29–34). ICE Publishing.

Iannacone, L., Francesco Giordano, P., Gardoni, P., & Pina Limongelli, M. 2022. Quantifying the value of information from inspecting and monitoring engineering systems subject to gradual and shock deterioration. *Structural Health Monitoring 21*(1): 72–89.

Limongelli, M. G., Turksezer, Z. I., & Giordano, P. F. 2019. Structural health monitoring for cultural heritage constructions: A resilience perspective. In IABSE Symposium Towards a Resilient Built Environment – Risk and Asset Management, 27–29 March 2019, Guimarães, Portugal. (pp. 1552–1559).

Morgese, M. et al. 2021. Improving distributed FOS measures by DIC: A two stages SHM. *ACI Structural Journal 18*(6): 91–102.

Morgese, M., Ansari, F., Domaneschi, M., & Cimellaro, G. P. 2020. Post-collapse analysis of Morandi's Polcevera Viaduct in Genoa Italy. *Journal of Civil Structural Health Monitoring 10*: 69–85.

Ngamkhanong, C., Kaewunruen, S., & Costa, B. J. A. 2018. State-of-the-art review of railway track resilience monitoring. *Infrastructures 3*(1): 3.

Salar, M., Entezami, A., Sarmadi, H., Behkamal, B., De Michele, C., & Martinelli, L. 2022. Vibration-based structural health monitoring of bridges based on a new unsupervised machine learning technique under varying environmental conditions. In Current Perspectives and New Directions in Mechanics, Modelling and Design of Structural Systems (pp. 1748–1753). CRC Press.

Sun, L. et al. 2020. Review of bridge structural health monitoring aided by big data and artificial intelligence: From condition assessment to damage detection. *Journal of Structural Engineering (United States) 146*(5): 04020073.

Ye, X. W. et al. 2019. A review on deep learning-based structural health monitoring of civil infrastructures. Smart Structures and Systems *24*(5): 567–585.

Yuan, F. G. (Ed.). 2016. Structural Health Monitoring (SHM) in Aerospace Structures. Woodhead Publishing.

Zhang, Y., DesRoches, R., & Tien, I. 2019. Updating Bridge Resilience Assessment Based on Corrosion and Foundation Scour Inspection Data. 9th International Conference on Structural Health Monitoring of Intelligent Infrastructure: Transferring Research into Practice, SHMII 2019, St. Louis, United States, 4–7 August 2019.

2 The Differential Evolution Algorithm

An Analysis of More than Two Decades of Application in Structural Damage Detection (2001–2022)

Parsa Ghannadi, Seyed Sina Kourehli, and Andy Nguyen

2.1 OVERVIEW

Vibration-based structural damage detection methods using optimization techniques have increased extensively in recent decades due to the rapid development of swarm intelligence and the introduction of robust and computationally efficient optimizers. The differential evolution algorithm (DEA) is a widely used optimization algorithm that has been successfully implemented for different engineering problems since Storn and Price released it in 1997. This study analyzes more than two decades of application of the DEA in structural damage detection problems between 2001 and 2022. The main contribution of the present chapter is to provide detailed and tabulated reviews of the methodologies, objectives, and main findings of about 50 publications. This study also presents statistical analysis to investigate the contribution of objective functions, types of structures, number of publications per year, and percentage of utilized single-step, two-step and multiple-step methods within the past two decades.

2.2 INTRODUCTION

As a result of earthquakes, fatigue, overloading, joint loosening, and other human-induced or nature-induced factors, structural damage may progressively expand [1, 2]. Monitoring structural conditions and making timely decisions to repair damaged elements can prevent human disasters and reduce maintenance costs [3, 4]. There are several existing techniques for localizing damaged elements and identifying their severities [5]. Some damage detection

DOI: 10.1201/9781003306924-2

methods, such as acoustic emission [6, 7], guided wave [8–11], and electromechanical impedance [12, 13], are classified as local nondestructive testing techniques and enable us to evaluate the condition of certain elements close to the sensors [14]. However, the global vibration characteristics of the structure, such as natural frequencies [15, 16], mode shapes [17–19], and modal flexibility [20–22], are analyzed by vibration-based methods to assess the structural health condition [14]. Vibration-based methods relying on frequency-domain and time-domain responses have attracted remarkable attention [23] due to the availability of measuring signals by single or multiple accelerometers without the need to have a sensor adjacent to the damaged element [14]. In vibration-based methods, the inverse problem of structural damage identification can be mathematically formulated as an optimization process by defining an objective function [24, 25]. The objective function describes the discrepancy between the measured vibration characteristics and those calculated from the finite element model (FEM) [26, 27]. The optimization algorithms attempt to minimize the objective function by finding design variables, including a vector of structural elements whose damage severities are rated between 0 and 1 [28]. The healthy and fully damaged elements are represented by 0 and 1, respectively [29]. Many optimization methods have been developed over the past several decades due to technological advancements to address challenging engineering problems [30–40]. Many researchers have evaluated the performance of classic and novel optimization algorithms using different objective functions to solve the inverse damage detection problem in structural engineering [41]. Ghannadi et al. analyzed the previously published papers on the application of different variants of particle swarm optimization (PSO) and frequently used objective functions [42]. Alkayem et al. investigated the capability of the social swarm algorithm and a novel hybrid objective function based on modal strain energy and mode shape curvature [43]. Jahangiri et al. developed a robust const function, namely the holistic objective function, and employed it for damage identification of large-scale structures [44, 45]. Beheshti Aval and Mohebian [46] proposed an efficient two-step approach for joint damage detection in frame structures. In the first step, the residual moment-based joint damage index is applied to recognize the possibly damaged connections. For identifying the severity of damaged connections in the second step, the equilibrium optimizer is utilized to minimize a hybrid objective function based on natural frequencies and modal assurance criteria (MAC). In another study, Beheshti Aval and Mohebian introduced a methodology for simultaneously detecting damaged joints and elements in skeletal structures. They have utilized the improved biology migration algorithm and a weighted hybrid objective function in their method [47]. In terms of the applicability of novel optimization algorithms in damage detection problems, several scholars have reported successful applications of different optimizers, such as the slime mold algorithm [48–51], modal force information-based optimization [52], the ant lion optimization algorithm [53], the visible particle series search algorithm [54], the improved cuckoo search algorithm [55], the YUKI algorithm [56], the multiverse optimizer [57], the guided water strider algorithm [58], the grey wolf optimizer (GWO) [59, 60], the bat algorithm [61], teaching–learning-based optimization (TLBO) [62], and modified TLBO [63].

In real-world damage detection problems, measuring all degrees of freedom (DOFs) is impossible because of the limited number of sensors [64]. Therefore, several researchers have proposed practical approaches when solving a vibration-based damage detection problem as an optimization scheme. A frequently used method to tackle incomplete modal data is condensing the FEM in the size of measured DOFs [65]. Ghannadi and Kourehli [66] compared the efficiency of different FEM reduction techniques, such as Guyan's reduction method, the improved reduced system (IRS), the iterated improved reduced system (IIRS), and the system equivalent reduction expansion process (SEREP). Kahya et al. [67] and Şimşek et al. [68] adopted Guyan's reduction method to deal with incomplete modal data for damage identification of laminated composite beams. Several optimization-based damage detection procedures have also been established based on the IRS [69], IIRS [70], and SEREP [71].

A well-known evolutionary algorithm inspired by Charles Darwin's theory of evolution is called the differential evolution algorithm (DEA), which has been widely implemented to address various engineering problems since it was first released in 1997 [72]. This study is divided into five sections to review more than two decades of application of DEA in structural damage detection from 2001 to 2022. The current section, the introduction, briefly reviewsrecent vibration-based damage detection methods formulated as an optimization problem. Section 2.3 presents an introduction to DEA and its related mathematical definition. Section 2.4 analyzes nearly 50 published papers to investigate methodologies, objectives, types of structures, and their findings. The most important results of this study are demonstrated graphically in Section 2.5 to provide a discussion. Finally, conclusions are provided in Section 2.6 to highlight the key points.

2.3 THE DIFFERENTIAL EVOLUTION ALGORITHM (DEA)

The differential evolution algorithm (DEA) is a straightforward but efficient heuristic method that Storn and Price [73] originally introduced for handling global optimization in continuous space. The DEA is easy to employ, requires few control variables, performs exceptionally well in parallel computation, and provides reliable results. Therefore, the DEA has grown in popularity and has been used to solve various optimization problems in practical applications [72]. The DEA is an evolutionary algorithm that includes three types of operators: mutation, crossover, and selection [74]. Figure 2.1 demonstrates how DEA attempts to minimize the objective function and solve the optimization problem. As depicted in Figure 2.1, the optimization procedure begins with the initialization phase. Then, the DEA undergoes a loop that contains the processes of mutation, crossover, and

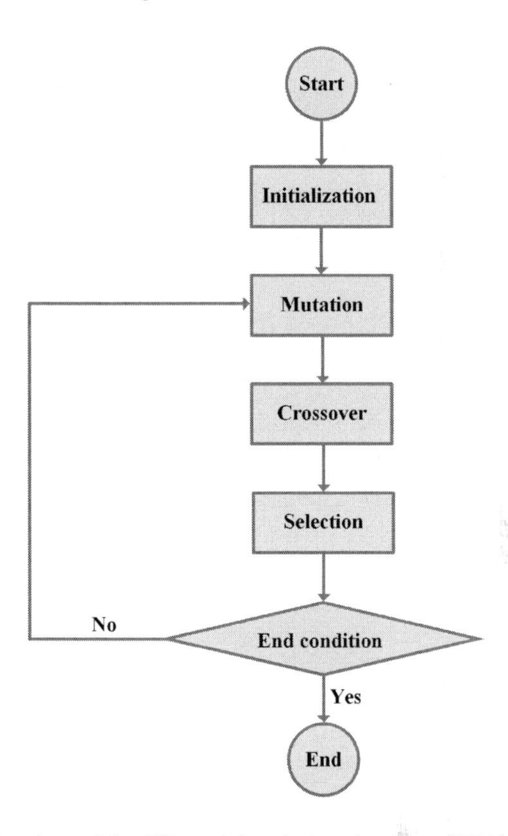

FIGURE 2.1 Flowchart of the differential evolution algorithm (DEA) [74].

selection, and this loop continues while the stop condition is satisfied [74]. The remainder of this section is a description of the initialization phase and three operators [74, 75].

2.3.1 Initialization

The initial population, together with the control parameters, are generated during the initialization phase. The initial population consists of NP solutions (vectors) with D variables. The following definition applies [74] to a solution (individual) of the population at generation G:

$$X_i^G = \left(X_{i,1}^G, X_{i,2}^G, \ldots, X_{i,D}^G\right), \qquad i = 1, 2, \ldots, NP \tag{2.1}$$

where D is the search space's dimension, and NP is the population's size.

By uniformly randomizing individuals inside the search space while keeping the search space constrained by the specified minimum and maximum parameter ranges, the initial population should sufficiently cover the whole search space as far as possible [75]. Individual solutions are frequently initialized using the following equation:

$$X_{i,j}^{0} = X_{j}^{\min} + rand(0,1).\left(X_{j}^{\max} - X_{j}^{\min}\right) \tag{2.2}$$

where *rand* (0, 1) is a randomly generated number with uniform distribution that ranges between 0 and 1. The maximum and minimum bounds of the jth dimension of the search space are $X_j^{\max}$ and $X_j^{\min}$, respectively.

2.3.2 Mutation

The DEA uses the mutation procedure to generate a mutant vector V_i^G for each individual X_i^G during every generation *G*. The following list includes some of the most widespread DEA mutation techniques [74, 75]:

$$DEA/rand/1: \quad V_i^G = X_{r1}^G + F.\left(X_{r2}^G - X_{r3}^G\right) \tag{2.3}$$

$$DEA/rand/2: \quad V_i^G = X_{r1}^G + F.\left(X_{r2}^G - X_{r3}^G\right) + F.\left(X_{r4}^G - X_{r5}^G\right) \tag{2.4}$$

$$DEA/best/1: \quad V_i^G = X_{best}^G + F.\left(X_{r1}^G - X_{r2}^G\right) \tag{2.5}$$

$$DEL/current—to—rand/1: \quad v_i^G = X_i^G + F.\left(X_{r1}^G - X_i^G\right) + F.\left(X_{r2}^G - X_{r3}^G\right) \tag{2.6}$$

where *r1*, *r2*, *r3*, *r4*, and *r5* are random numbers generated from 1 to *NP*, neither of which are equal to index *i*. In the population at generation *G*, X_{best}^G is the best individual vector with the best fitness value. To control the differential variation's amplification [73, 74], *F* is a positive scaling factor typically ranging between [0, 1] or ranging between [0, 2].

2.3.3 Crossover

Following the mutation step [73–75], a trial vector $U_i^G = \left(U_{i,1}^G, U_{i,2}^G, ..., U_{i,D}^G\right)$ is produced for every individual using the binomial crossover operator on V_i^G and X_i^G, as given in Eq. (2.7):

$$U_{i,j}^G = \begin{cases} V_{i,j}^G & if\left(rand_{i,j}(0,1) \le CR \quad or \quad j = j_{rand}\right) \\ X_{i,j}^G & otherwise \end{cases}, \tag{2.7}$$

$$j = 1, 2, ..., D.$$

where j_{rand} is a uniform random number in the range of [1, D] that should be computed for each individual in Eq. (2.7). CR is the crossover rate [73] and can be defined by users between 0 and 1.

The jth variable $U_{i,j}^{G}$ of the trial vector U_i^G will be updated as follows if it exceeds the boundary constraints [74]:

$$U_{i,j}^{G} = X_j^{\min} + rand(0,1).\left(X_j^{\max} - X_j^{\min}\right) \tag{2.8}$$

2.3.4 Selection

The fitness values of the target and trial vectors are assessed by the selection operator to decide which will survive and enter the following generation. During the minimization procedure, the decision vector that has the lowest fitness value would enter the upcoming generation [74], defined as follows:

$$X_i^{G+1} = \begin{cases} U_i^G & if\left(fit\left(U_i^G\right) \leq fit\left(X_i^G\right)\right) \\ X_i^G & otherwise \end{cases} \tag{2.9}$$

2.4 A Tabulated Review of Structural Damage Detection Using the DEA (2001–2022)

This section provides a tabular approach to critically discuss different variants of DEA, characteristics of various utilized objective functions, and types of structures. The main results of the previously released publications between 2001 and 2022 are also summarized to highlight the key points. Table 2.1 has been organized to fulfill the role of a detailed review according to the following categories:

Reference and Year: These columns indicate the names of the authors and the year that the work was published, respectively.

Objective: This column explains the primary contribution of the work and its motivation for presentation.

Methodology: This column outlines the algorithms and methods used to address damage detection problems.

Structure: The kinds of structures that are used to accomplish the structural damage detection approach are addressed in this column.

Result and Finding: The key findings of the publications are abstracted in this column.

TABLE 2.1
A Review of the Application of the Differential Evolution Algorithm (DEA) for Structural Damage Identification

Reference	Year	Objective	Methodology	Structure(s)	Result and Finding
Manson and Worden [76]	2001	Several studies based on Lamb-wave propagation have promising results for damage localization in composite plates. Lamb-waves are generated by piezoceramic actuators, and the ensuing signals are measured by piezoceramic sensors located throughout the structure. The Lamb-wave will be altered if damage is applied to the structure. The Lamb-wave modification depends on the distance between the damaged area and the sensor/actuator. Therefore, this study introduced an optimization-based approach for finding the optimal location of sensors and piezoceramic actuators.	The DEA was employed to minimize an objective function relying on angles between the sensor, actuator, and damage location.	Composite plate	The proposed strategy for optimal sensor placement on a simple structure has provided successful results. However, this method can be applied to more complex systems in future studies.
Casciati [77]	2008	In this paper, the inverse problem of damage identification is solved by considering the stiffness of structural elements as optimization variables. Previous efforts mainly were concerned with multi-story shear buildings. In this study, a discretized model of the cantilever beam is adopted for investigation.	The DEA was applied to minimize the discrepancy between the measured and calculated modal parameters (natural frequencies and mode shapes).	Cantilever beam	The results revealed that the presented approach is sufficient for damage localization and identification when the identified stiffness matrix is compared to the initial one.

(Continued)

TABLE 2.1 *(Continued)*
A Review of the Application of the Differential Evolution Algorithm (DEA) for Structural Damage Identification

Reference	Year	Objective	Methodology	Structure(s)	Result and Finding
Kang et al. [78]	2012	This study compared the efficiency of an improved version of particle swarm optimization (PSO) with the DEA, standard PSO, and a real-coded genetic algorithm (RCGA) in structural damage detection problems.	The mode shape and natural frequency changes are used as the cost function.	Simply supported beam Planar truss	The comparison results showed that the performance of improved PSO is more efficient than that of the DEA, standard PSO, and the RCGA.
Rao et al. [79]	2012	This paper provides a damage detection method based on the self-adaptive DEA and proper orthogonal decomposition (POD), considering noisy data and environmental variability.	The three-stage procedure is presented for damage identification, localization, and quantification. In the first and second steps, the exact time instant of damage and the location are identified using POD. In the third step, the constrained optimization problem is solved by employing self-adaptive DEA to determine the damage's severity. During the optimization procedure, the objective function is formulated as the discrepancy between the proper orthogonal value of the damaged state and the calculated values from the finite element model (FEM).	Cantilever beam Concrete slab bridge Plane truss	The numerical investigations demonstrated the robustness of the proposed structural health monitoring (SHM) methodology, even under changing environmental conditions and considering measurement noise. However, experimental validation is still necessary for assessing the performance of POD-based methodology in real-world applications.

(Continued)

TABLE 2.1 *(Continued)*

A Review of the Application of the Differential Evolution Algorithm (DEA) for Structural Damage Identification

Reference	Year	Objective	Methodology	Structure(s)	Result and Finding
Bighamian and Mirdamadi [80]	2012	Most damage assessment studies identify just stiffness reduction and assume no mass decrease. Mass reduction is a critical consideration and inevitable in aircraft composite structures. Therefore, this research outlines a novel method for simultaneously identifying the reduction of stiffness and mass in aerospace structures.	The presented algorithm to find the mass and stiffness is a signal-driven method that minimizes the differences between a system digital pulse response and equivalent virtual damped SDOF (single degree of freedom) using the DEA as an optimizer.	Mass–spring system Bar model Shear frame Plane truss	The performance of the utilized procedure for single and multiple damage detection is satisfactory, even in noise conditions.
Kang et al. [81]	2013	This paper introduces a new variant of PSO to improve the convergence speed and accuracy of the standard PSO. The obtained results are also compared with those obtained from the DEA and PSO.	The optimization algorithms minimize an objective function that has been given the dynamic (natural frequencies) and static (displacements) responses as inputs.	Clamped–clamped beam	Compared to the standard PSO and DEA, improved PSO is more successful in detecting structural damage. However, the accuracy of improved PSO is decreased with noisy inputs.
Reed et al. [82]	2013	The main contribution of this work is to improve the standard DEA to solve structural inverse problems accurately. The proposed variant of the DEA provides a reasonable convergence rate while still properly exploring the parameter space and maximizing the likelihood.	This research focuses on the maximum likelihood technique and employs a cost function based on maximum likelihood estimators (MLEs). Then, an improved version of the DEA is applied to minimize the suggested cost function.	Barrel vault shell	Results are shown that the presented method provides impressive performance in structural parameter estimation. Additionally, the modified DEA can swiftly converge to the global minimum compared to the standard DEA.

(Continued)

TABLE 2.1 *(Continued)*

A Review of the Application of the Differential Evolution Algorithm (DEA) for Structural Damage Identification

Reference	Year	Objective	Methodology	Structure(s)	Result and Finding
Jena et al. [83]	2013	This paper has introduced an optimization-based inverse strategy to find the depth and location of transverse surface cracks in beam-like structures.	The damage parameters such as crack depth and crack location are formulated as a constrained optimization problem. Then, the DEA is utilized to minimize the difference between the first three calculated and measured natural frequencies as an objective function.	Cantilever beam	This study indicates that the proposed approach is robust in determining crack parameters and can be extended to different SHM applications.
Vincenzi et al. [84]	2013	This study compares the damage detection in a cracked beam using the coupled local minimizers (CLM) method and the DEA.	The numerical examples with error (natural frequencies and mode shapes contaminated by some error) and without error are studied to compare the performance of the DEA and the CLM method. During the optimization procedure, the discrepancy between the measured and calculated modal properties (natural frequencies and mode shapes) is the cost function that must be reduced.	Simply supported beam	Statistical analysis of the obtained results by optimization algorithms demonstrated that the CLM method and the DEA can provide good results. However, when the number of optimization parameters is limited, the CLM method performs better in terms of accuracy and speed. When the number of parameters is increased or when pseudo-experimental data (modal properties with error) is used, the DEA becomes more efficient.

(Continued)

TABLE 2.1 *(Continued)*
A Review of the Application of the Differential Evolution Algorithm (DEA) for Structural Damage Identification

Reference	Year	Objective	Methodology	Structure(s)	Result and Finding
Villamizar et al. [85]	2014	This paper proposes an expert system based on self-organizing maps (SOMs) and principal component analysis (PCA) to find the simulated damage (adding a mass on the surface).	In the first step, PCA is employed to reduce the time signals and prepare a database for training SOMs. The DEA is utilized to tune the training parameters in the second step.	Aircraft turbine blade	The identification error is approximately 22% when using default training parameters. The identification error is 20% when implementing the DEA and training the neural network with tuned parameters.
Villalba-Morales and Laier [86]	2014	This paper compares the performance of different objective functions based on natural frequencies, modal flexibilities, mode shapes, modal strain energies, and the residual force vector when applying the adaptive DEA as an optimizer for structural damage localization and quantification.	This paper uses a simple yet efficient adaptation method to avoid utilizing the trial-and-error method to determine the DEA parameters, and users must define only the population size. Then, cost functions based on dynamic parameters such as natural frequencies, modal flexibilities, modal strain energies, mode shapes, and the residual force vector are minimized by using the adaptive DEA.	Plane truss	The objective function based on natural frequencies and mode shapes has provided the most accurate results. The cost function based on modal flexibility produced comparable results to those achieved using natural frequencies and mode shapes as an objective function. The approach does not function well when using an objective function based on natural frequencies and modal strain energies. The findings of the objective function based on the residual force vector were unreliable because several false identifications were observed.

(Continued)

TABLE 2.1 *(Continued)*
A Review of the Application of the Differential Evolution Algorithm (DEA) for Structural Damage Identification

Reference	Year	Objective	Methodology	Structure(s)	Result and Finding
Fu and Yu [87]	2014	The improved version of the adaptive DEA has provided better exploration ability, higher accuracy, and fast convergence. Therefore, this revised optimization algorithm is considered for structural damage identification.	The damage detection problem is converted into a constrained optimization problem. Then, minimizing the differences between measured and calculated modal parameters (natural frequencies and mode shapes) is defined as an objective function.	Space truss	The following are the significant findings of this research: 1. The improved adaptive DEA is accurate for damage detection and can find the damage parameters in single and multiple damage scenarios. 2. The improved adaptive DEA performs well when damage severity is high. 3. The improved adaptive DEA is robust to noise. However, the noise level is related to the convergence rate. 4. There is still an opportunity to improve the premature convergence of the utilized optimizer. 5. It is necessary to investigate how to increase accuracy and overcome the influence of structural symmetry in future studies.

(Continued)

TABLE 2.1 *(Continued)*

A Review of the Application of the Differential Evolution Algorithm (DEA) for Structural Damage Identification

Reference	Year	Objective	Methodology	Structure(s)	Result and Finding
Cavalini Jr. et al. [88]	2015	The efficiency of the self-adaptive DEA is evaluated to reduce the discrepancies between experimental and analytical results through FEM updating. The control parameters of the algorithm, including perturbation rate, population size, crossover, and crossover parameter, are automatically adjusted by the self-adaptive DEA.	The philosophy of the self-adaptive DEA is based on convergence rate concepts and population diversity. This technique decreases the number of objective function evaluations by defining a convergence rate to assess population homogeneity in the evolutionary process. This proposed algorithm is used to minimize the difference between the calculated and measured frequency response functions (FRFs) as an objective function. For comparison, the efficiency of the standard DEA is also evaluated to determine the optimal solution.	Rotating machine	According to the results, the self-adaptive DEA is a potential alternative for solving the inverse problems associated with FEM updating.
Jena and Parhi [89]	2015	This paper introduces a modified version of PSO to accelerate the search strategy while keeping the standard form of PSO. The results of modified PSO are compared with the results obtained by the DEA.	The squeezing approach was introduced into the standard PSO formulation to restrict the search domain in each iteration, resulting in a faster convergence time for reaching the best solution. Natural frequency alterations caused by the existence of a crack are beneficially utilized to identify the crack depth and crack	Cantilever beam	Based on the obtained results, modified PSO can predict the crack parameters more accurately than the DEA.

(Continued)

TABLE 2.1 *(Continued)*
A Review of the Application of the Differential Evolution Algorithm (DEA) for Structural Damage Identification

Reference	Year	Objective	Methodology	Structure(s)	Result and Finding
			location by employing modified PSO and the DEA through minimizing a cost function based on the differences between the measured and the calculated natural frequencies.		
Seyedpoor et al. [90]	2015	In this paper, the performance of the DEA is investigated to handle the optimization-based damage detection problem by minimizing a frequency-based objective function. The obtained results are also compared with PSO.	An objective function is defined using the efficient correlation-based index (ECBI). The ECBI is a hybrid cost function based on a multiple damage location assurance criterion (MDLAC).	Simply supported beam Plane truss Space frame	Numerical results indicate the effectiveness of the DEA and ECBI for accurately finding the location and severity of the damage compared to those results obtained from PSO.
Seyedpoor and Yazdanpanah [91]	2015	The performance comparison of two optimization techniques, including the DEA and PSO, is carried out to identify the robust optimization method that works properly in highly nonlinear problems such as damage detection of structures.	The hybrid objective function (ECBI) is created through the MDLAC and a weighted frequency term.	Cantilever beam Plane truss Portal frame	Compared to PSO, the DEA was able to produce accurate solutions with a significantly lower number of function evaluations.

(Continued)

TABLE 2.1 *(Continued)*
A Review of the Application of the Differential Evolution Algorithm (DEA) for Structural Damage Identification

Reference	Year	Objective	Methodology	Structure(s)	Result and Finding
Vincenzi and Savoia [92]	2015	The surrogate-assisted DEA is presented in this study to provide higher accuracy and faster convergence in model updating procedures and dynamic parameter identification problems.	In the proposed approach based on a surrogate and the DEA, the response surface second-order approximation is introduced in the mutation operation of the DEA. In the presented algorithm, multiple search points are employed simultaneously. Therefore, the robustness of the DEA is preserved for global minimum search. Then, the surrogate-assisted DEA is applied to minimize the weighted objective function relying on natural frequencies and mode shapes.	Pontelagoscuro bridge, Italy	The proposed algorithm decreases the number of objective function evaluations. Therefore, the surrogate-assisted DEA can be efficiently implemented for optimizing problems with computationally expensive objective functions.
Seyedpoor and Montazer [93]	2015	This paper develops a two-step procedure using a flexibility-based damage probability index (FBDPI) and the DEA to find the location and severity of structural damages.	In the first step, potentially damaged elements are identified by the FBDPI. Then, the severity of damaged elements is quantified by minimizing the differences between measured and calculated mode shapes through the DEA.	Plane truss Space truss	The results indicate that the proposed FBDPI can accurately determine potentially damaged elements while only requiring a few modal data. When the damaged elements are recognized in the first step, the DEA could determine the damage severity with a few iterations during the optimization procedure.

(Continued)

TABLE 2.1 *(Continued)*
A Review of the Application of the Differential Evolution Algorithm (DEA) for Structural Damage Identification

Reference	Year	Objective	Methodology	Structure(s)	Result and Finding
Vo-Duy et al. [94]	2016	A two-step technique with a combination of a modal strain energy-based method and an improved version of the DEA is presented in this study. The improved DEA is developed by adjusting the mutation and selection phases of the traditional DEA. Multiple mutation operators are employed adaptively in the mutation phase to maintain the trade-off between global exploration and local exploitation of the optimization algorithm. The elitist scheme replaces the standard section scheme of the DEA in the selection phase.	The possible damaged elements are initially detected by implementing a modal strain energy-based method. In the second step, the improved DEA minimizes the error between measured and calculated mode shapes.	Laminated composite plate	The numerical study revealed that regardless of noise, the modal strain energy-based method successfully identifies damaged elements. The improved DEA and DEA can accurately assess damage severities even if mode shapes are contaminated with 3% random noise. Additionally, the improved DEA needs far fewer structural analyses than the DEA.
Seyedpoor and Montazer [95]	2016	This study introduces a two-stage method with the assistance of the modal residual vector-based indicator (MRVBI) and DEA for correctly identifying the location and severity of damage in truss structures.	In the first stage, the MRVBI is employed to identify elements that may have been damaged. The DEA is implemented to minimize the ECBI as a cost function in the second stage to find the severity of damage for candidate elements.	Plane truss Space truss	The numerical investigations demonstrate that the utilized strategy relying on the MRVBI can efficiently locate possibly damaged elements and significantly reduce the design variables. Furthermore, it has been found that the DEA can successfully handle the optimization problem in order to determine the severity of damaged elements in the narrowed search space.

(Continued)

TABLE 2.1 *(Continued)*
A Review of the Application of the Differential Evolution Algorithm (DEA) for Structural Damage Identification

Reference	Year	Objective	Methodology	Structure(s)	Result and Finding
Ding et al. [96]	2016	The artificial bee colony (ABC) algorithm is a swarm-based optimizer with a simple implementation. However, the ABC algorithm has some drawbacks, such as a slow convergence rate and trapping in the local optimal solutions. In this study, an improved version of the ABC algorithm with the assistance of the DEA is proposed to be a more effective optimization algorithm.	A new mechanism based on the DEA is introduced in the employed bee phase to enhance the exploration ability of the standard ABC algorithm. In the improved ABC algorithm, the tournament selection strategy is used instead of the roulette selection strategy, and simulation of the onlooker bee's behavior is performed by a novel formula. The cost function is defined only based on differences in the first few natural frequencies.	Cantilever beam Fixed–fixed beam	Compared to the standard ABC algorithm, the DEA, genetic algorithm (GA), and PSO, the improved ABC algorithm has produced more accurate damage detection results. Additionally, the improved DEA converges rapidly and produces identifications with minor standard deviations.
Anh [97]	2016	This paper presents a simple but efficient modification of the DEA for reducing the number of fitness evaluations in computationally expensive inverse problems.	The modified DEA employs the nearest-neighbor comparison technique to evaluate a trial vector in the search population. The adopted objective function is based on the natural frequencies and mode shape components.	Two-dimensional beam structure	The results show that the modified DEA has been successfully implemented to reduce the computational cost in structural damage detection problems.

(Continued)

TABLE 2.1 *(Continued)*

A Review of the Application of the Differential Evolution Algorithm (DEA) for Structural Damage Identification

Reference	Year	Objective	Methodology	Structure(s)	Result and Finding
Nguyen-Thoi et al. [98]	2016	This research nominates a two-step method for structural damage detection that combines the damage-locating vector (DLV) method and the DEA.	Damage localization in structures is accomplished in the first step by utilizing the DLV technique with normalized cumulative energy. In the second step, a combined mode shape error function and MDLAC function are suggested to address the limitations of the MDLAC function, including the exclusive solution and the problem of symmetric structures.	Space truss Portal frame	The performance and accuracy of the presented two-step method are assessed numerically, and the following results have been obtained: 1. The proposed two-step method can effectively determine damage locations and their severities. Additionally, robustness to noise is the other advantage of this method. 2. A large number of modes are needed to provide better results.
Georgioudakis and Plevris [99]	2016	In this article, the capability of four objective functions—the modal assurance criterion (MAC), modified total modal assurance criterion (MTMAC), coordinate modal assurance criterion (COMAC), and modal flexibility assurance criterion (MACFLEX)—to determine structural damage in location and severity was evaluated.	The inverse problem of damage detection was formulated as an optimization problem, and the following four objective functions were taken into account: 1. 1-MAC 2. 1-MTMAC 3. 1-MACFLEX 4. 1-COMAC The utilized optimization algorithm for minimizing the above-mentioned objective functions is the DEA.	Simply supported beam	Overall, the objective function based on the MTMAC works appropriately for all damage scenarios, even though limited measurements were available.

(Continued)

TABLE 2.1 *(Continued)*
A Review of the Application of the Differential Evolution Algorithm (DEA) for Structural Damage Identification

Reference	Year	Objective	Methodology	Structure(s)	Result and Finding
Vo-Duy et al. [100]	2016	The research introduces a two-stage technique for damage identification in laminated composite structures (beam and plate), employing the modal strain energy method and an improved version of the DEA. Two improvements in the mutation phase and selection phase of the standard DEA are addressed to adaptively determine the mutant factor and crossover control parameter. An adaptive mutation method with multiple mutation operators is presented in the mutation phase, while a new selection method is proposed in the selection phase.	The modal strain energy-based method is first implemented to detect potential damage elements, as well as to decrease the design variables of the optimization problem for the second stage. In the second step, the improved DEA is employed to identify the severity of damaged elements by minimizing the error function of measured and calculated mode shapes.	Laminated composite beam Laminated composite plate	The numerical studies show that regardless of noise, the modal strain energy-based method provides successful results in locating damaged elements. The improved DEA and DEA can identify damage severities accurately. However, the improved DEA needs far fewer function evaluations than the DEA.
Dinh-Cong et al. [101]	2017	A two-stage technique incorporating the DLV method and the DEA is developed for damage identification of laminated composite beams.	The DLV method is used to locate the damaged elements in the first stage, which uses normalized cumulative energy. The extent of potentially damaged elements is identified in the	Laminated composite beam	Two numerical examples, a symmetric cross-ply (0/90/0) beam and an asymmetric (0/90/0/90) beam, are adopted to demonstrate the potential of the

(Continued)

TABLE 2.1 *(Continued)*
A Review of the Application of the Differential Evolution Algorithm (DEA) for Structural Damage Identification

Reference	Year	Objective	Methodology	Structure(s)	Result and Finding
			second stage through optimizing a cost function by the DEA.		proposed methodology. The following outcomes have been achieved: 1. The two-step procedure can accurately detect the location and severity of multiple damages at individual layers. 2. In the case of a high level of random noise or a limited number of modes, the damage detection results may become erroneous.
Bureerat and Pholdee [102]	2017	This paper proposes the adaptive sine cosine algorithm integrated with DEA (ASCA-DEA) to enhance the performance of the sine cosine algorithm for solving structural damage detection problems.	The ASCA-DEA includes an adaptive strategy and the mutation operator from the DEA, which increase the algorithm's performance. The ASCA-DEA is applied to minimize the root mean square error between measured and computed natural frequencies to solve the damage identification problem.	Space truss	The results show that the ASCA-DEA is a practical and dependable strategy for tackling damage identification problems.

(Continued)

TABLE 2.1 *(Continued)*
A Review of the Application of the Differential Evolution Algorithm (DEA) for Structural Damage Identification

Reference	Year	Objective	Methodology	Structure(s)	Result and Finding
Dinh-Cong et al. [103]	2017	This paper compares three optimization algorithms—Jaya, DEA, and cuckoo search (CS)—for structural damage detection.	The structural damages and their severities are identified by minimizing a hybrid objective function based on the MDLAC and modal flexibility matrix.	Portal frame Plane truss Plane frame	The Jaya algorithm, DEA, and CS can provide the exact solution to damage identification problems, even in the presence of noise. However, the convergence speed of the Jaya algorithm is much faster than that of the DEA and CS.
Dinh-Cong et al. [104]	2017	For damage assessment in plate-like structures, the research provides a multi-stage optimization strategy utilizing the modified version of the DEA. The modified DEA is employed as an optimizer, which can improve the balance of global and local searches.	When using the proposed multi-stage procedure, elements with a damage severity of less than 2% can be regarded as healthy and set to zero to remove them from the design variables and improve the convergence rate. The objective function in this study is the discrepancy between a measured flexibility matrix and the equivalent one from a FEM.	Square isotropic thick plate Laminated composite plate	The numerical examinations demonstrate that the presented two-stage method can accurately recognize the location and extent of damage with low computation cost.
Seyedpoor et al. [105]	2018	This paper introduces a multi-stage method for structural damage detection. First, the inverse damage identification problem is formulated as an optimization problem. Then, an improved version of	The location of identified damaged elements in each optimization step is imposed on the following step. In contrast, the impacts of undamaged elements on the subsequent step are	Cantilever beam Plane truss Space frame	The numerical and experimental results reveal that the presented multi-stage technique is more efficient than the single-step method relying on the DEA.

(Continued)

TABLE 2.1 *(Continued)*
A Review of the Application of the Differential Evolution Algorithm (DEA) for Structural Damage Identification

Reference	Year	Objective	Methodology	Structure(s)	Result and Finding
		the DEA is employed to minimize the cost function. In the improved DEA, to accelerate the convergence rate, a new mutation scheme is introduced instead of the standard mutation phase, and a random variation scheme is operated to modify the mutation constant.	omitted. This method eliminates undamaged elements one by one during certain stages, and the algorithm eventually converges to the actual solution (location and severity of damaged members) with a lower computational cost. This study uses the ECBI as the cost function during the multi-stage optimization procedure.		
Georgioudakis and Plevris [106]	2018	In this study, a hybrid objective function is developed as a sensitive criterion to provide a reliable optimization-based approach for finding the location and severity of structural damage.	The DEA is utilized to minimize a combined cost function, which takes into account the values of the MTMAC and MACFLEX.	Simply supported beam Portal frame	When compared to the MTMAC or MACFLEX alone, it was found that the hybrid objective function based on the MTMAC and MACFLEX performs best.
Alkayem and Cao [107]	2018	In terms of accuracy, consistency, and computational cost, the performance of five optimization methods—PSO, GA, the DEA, the Lévy flight–DEA (LFDEA), and the elitist artificial bee colony–PSO (EABCPSO)—is compared.	Presents a hybrid cost function that combines the residuals of the mode shape and the modal strain energy with weighting factors.	IASC-ASCE benchmark structure	The following is a summary of the findings from this study: 1. GA identified many undamaged elements instead of damaged ones. 2. In comparison to GA and the DEA, PSO provides accurate

(Continued)

TABLE 2.1 *(Continued)*

A Review of the Application of the Differential Evolution Algorithm (DEA) for Structural Damage Identification

Reference	Year	Objective	Methodology	Structure(s)	Result and Finding
					results. However, after several tests, it was discovered that it lacked stability. 3. The EABCPSO and LFDE, respectively, enhanced the accuracy and consistency of the basic version of PSO and the DEA. 4. The EABCPSO outperforms the LFDE in terms of accuracy. Besides, the computational time of the EABCPSO is shorter than that of the LFDE.
Kim et al. [108]	2018	The most significant contribution of this paper is the implementation of the modified DEA as a swift optimizer to solve the damage detection problem. Then, a comparative study is performed to assess the performance of the standard DEA, GA, and PSO with the modified DEA.	The natural frequency and mode shape differences are considered as the cost function. A penalty function is also added to the objective function for more precise detection of damage parameters.	Plane truss Space frame	The numerical results indicate that the modified DEA can identify the location and severity of damaged elements more accurately than the standard DEA, PSO, and GA. Additionally, the convergence rate of the modified DEA is faster than that of the other studied algorithms.

(Continued)

TABLE 2.1 *(Continued)*

A Review of the Application of the Differential Evolution Algorithm (DEA) for Structural Damage Identification

Reference	Year	Objective	Methodology	Structure(s)	Result and Finding
Seyedpoor et al. [109]	2018	This paper presents an effective method for damage identification using the time-dependent acceleration response and the DEA.	The cost function is formulated using measured and calculated acceleration response vectors from a limited number of sensors. Then, the DEA was employed as a global optimization technique to address the optimization problem.	Cantilever beam Plane truss Portal frame Plane frame	According to numerical results, the combination of the acceleration response–based objective function and the DEA can provide a potent tool for structural damage identification, even at a high random noise level (15%).
Fallahian et al. [110]	2018	This study introduces a practical method with the assistance of changes in acceleration response and the DEA to handle the structural damage detection problem.	The proposed objective function utilizes the time-domain acceleration response as a sensitive criterion for damage occurrence.	Plane truss Portal frame	The suggested approach based on the DEA and changes in acceleration response can accurately determine the location and severity of damage when different noise levels are imposed (1%, 2%, and 3%).
Bassoli et al. [111]	2018	A model-updating strategy through minimization of the vibration-based objective function by employing an improved surrogate-assisted DEA (DEA-S) for severely damaged historic masonry structures is provided in this research. The DEA-S significantly decreased the number of objective	The weighted objective function is defined as the difference between measured natural frequencies and mode shapes from ambient vibration testing and corresponding numerical values from the Cloud2FEM.	San Felice sul Panaro Fortress, Italy, a historical masonry structure	Three design parameters are considered to investigate the influence of structural parameters on dynamic behaviors. The results for FEM updating are presented as follows: 1. When considering the homogeneous distribution of

(Continued)

TABLE 2.1 *(Continued)*

A Review of the Application of the Differential Evolution Algorithm (DEA) for Structural Damage Identification

Reference	Year	Objective	Methodology	Structure(s)	Result and Finding
		function evaluations and can be successfully implemented to optimize the highly time-consuming problems.			the masonry elastic properties (E_M), the updated FEM does not represent the actual behavior of the structure. 2. When elastic moduli of the portions of walls connecting the Mastio to the Fortress on the west (E_{CW}) and north (E_{CN}) sides are adopted as design parameters, there is still a poor correlation between numerical and experimental models for the fourth and fifth modes. 3. When masonry elastic properties for undamaged (E_U) and damaged (E_D) states are considered design parameters, the updated E_U and E_D equal 892 and 700 MPa, respectively. The updated model has much better consistency between the experimental and numerical results.

(Continued)

TABLE 2.1 *(Continued)*
A Review of the Application of the Differential Evolution Algorithm (DEA) for Structural Damage Identification

Reference	Year	Objective	Methodology	Structure(s)	Result and Finding
Bureerat and Pholdee [112]	2018	In this paper, the radial basis function is incorporated into the DEA (RBFDEA) to accelerate the convergence rate of the standard DEA in solving inverse damage identification problems.	This research practices a natural frequency–based cost function for minimizing by the RBFDEA.	Space truss	The results obtained from numerical examples clearly demonstrate the advantage of the RBFDEA compared to the standard DEA, the whale optimization algorithm (WOA), the sine cosine algorithm, the moth flame optimization algorithm, real-code ant colony optimization, the charged system search, the league championship algorithm, simulated annealing, evolution strategies, teaching–learning-based optimization, adaptive differential evolution, the evolution strategy with covariance matrix adaptation, PSO, and the ABC algorithm.
Seyedpoor and Nopour [113]	2019	An efficient and swift two-step approach through a machine learning method and the DEA is developed for localizing and quantifying the damaged connections in moment frames.	The possible location of damaged connections is identified in the first stage using a support vector machine (SVM), which reduces the size of the search space. The second stage employs the	Plane frame	The SVM exhibited high accuracy in locating probably damaged connections based on the numerical results. When implementing the DEA in the

(Continued)

TABLE 2.1 *(Continued)*
A Review of the Application of the Differential Evolution Algorithm (DEA) for Structural Damage Identification

Reference	Year	Objective	Methodology	Structure(s)	Result and Finding
			DEA to minimize an objective function relying on the MDLAC in order to accurately detect the severity of damage in connections.		reduced search space, the severity of damaged connections can be swiftly and accurately determined.
Kim et al. [114]	2019	The main contribution of this research is to develop a practical damage detection technique for plane and space truss structures utilizing the DEA and vibration data extracted from the force method.	The natural frequencies, as well as mode shapes, are taken into account to construct the objective function. In the objective function, the force mode vectors are introduced as eigenvectors.	Plane truss Space truss	The combination of the DEA with the force method is significantly more efficient than GA in recognizing damaged elements, according to three numerical examples.
Sobrinho et al. [115]	2020	The structural responses of experimental simply supported beams under various loading were employed in combination with the DEA to identify damaged elements.	The objective function is defined based on the least squares of the difference between experimental and numerical responses in the time domain.	Simply supported beam	The results demonstrate that using the DEA and time-domain responses as an objective function to address damage identification problems has a lot of promise.
Seyedpoor and Pahnabi [116]	2020	This paper identifies structural damage using a sensitive damage indicator based on FRFs and the DEA.	The FRFs are placed instead of natural frequencies in the ECBI formula to form the objective function.	Cantilever beam Portal frame Plane frame	The results show that using the FRF-based objective function in conjunction with the DEA to identify the damaged elements and their severity is highly effective, even when there is a lot of noise (up to 5%). However, a

(Continued)

TABLE 2.1 *(Continued)*

A Review of the Application of the Differential Evolution Algorithm (DEA) for Structural Damage Identification

Reference	Year	Objective	Methodology	Structure(s)	Result and Finding
					sensitivity analysis is necessary to determine the exact number of utilized modes because this is an important parameter that influences the accuracy of the damage identification method and varies from one example to the next.
Lieu et al. [117]	2020	An inverse two-stage technique with the assistance of the modal strain energy-based index and adaptive hybrid evolutionary firefly algorithm (AHEFA) is applied to damage detection of truss structures. The AHEFA algorithm is a mixture of the DEA and the firefly algorithm, and a dynamically adapted parameter is employed to determine an optimal mutation scheme. As a result, global exploration and local exploitation capabilities are appropriately balanced [118].	In the first step, the potentially damaged elements are identified through an efficient criterion called the modal strain energy-based index. The second step aims to find the exact severities of the damaged elements by addressing an optimization procedure based on the AHEFA and ECBI as an objective function.	Plane truss Space truss	The performance of the AHEFA is significantly better than those obtained from the standard DEA and firefly algorithm in terms of accuracy and computational efforts.

(Continued)

TABLE 2.1 *(Continued)*

A Review of the Application of the Differential Evolution Algorithm (DEA) for Structural Damage Identification

Reference	Year	Objective	Methodology	Structure(s)	Result and Finding
Guedria [119]	2020	This research establishes the accelerated DEA, a modified optimizer for detecting damage in large-scale problems with a rapid convergence rate and producing accurate solutions while avoiding being entrapped in the local solutions.	The DEA algorithm is redesigned by making three modifications to the basic DEA. Firstly, a realistic choice is utilized to generate the initial population instead of producing them randomly. This kind of initialization aids the algorithm in achieving a quick convergence. Secondly, an innovative mutation operator depending on the dispersion of individuals through the search space is introduced to ensure the automatic balance between local and global searching capabilities. Lastly, a specialized exchange operator is developed and implemented to prevent premature convergence. By minimizing an objective function created by the flexibility matrix, the location and severity of damage have been determined.	Isotropic plate Laminated composite plate	The following conclusions have been drawn based on the numerical examples: 1. The established objective function successfully determines the location and severity of damaged elements while avoiding false identifications. 2. The accelerated DEA has been shown to be a proper optimizer for recognizing locations and extents of damage while only requiring lower modes. 3. For all examples investigated, the accelerated DEA outperforms its competitors in terms of mean, standard deviation, and computational time.

(Continued)

TABLE 2.1 *(Continued)*
A Review of the Application of the Differential Evolution Algorithm (DEA) for Structural Damage Identification

Reference	Year	Objective	Methodology	Structure(s)	Result and Finding
Su et al. [120]	2021	The main contribution of this paper is developing a modified version of the bat algorithm to overcome shortcomings such as lack of diversity and premature convergence. The optimization capability of the modified bat algorithm is also compared to the DEA, PSO, the shuffled frog-leaping algorithm (SFLA), and different versions of the bat algorithm.	The optimization algorithms attempt to minimize a hybrid objective function (a combination of natural frequency, mode shape, and flexibility matrix).	Simply supported beam Plane truss	The modified bat algorithm has a higher accuracy and convergence rate than the DEA, PSO, the SFLA, and different variants of the bat algorithm.
Wang et al. [121]	2021	This paper proposes a damage localization method incorporating the B-spline wavelet on the interval finite element method and the optimized singular value decomposition method.	The damaged structures are modeled using a B-spline wavelet on an interval finite element method. The attractor trajectory matrix is calculated using mode shape vectors extracted by modal analysis, and the singular value decomposition-based approach is utilized to determine damage locations. For matrix trajectory decomposition, the DEA is employed to explore the optimal parameters adaptively.	Cantilever beam	Numerical and experimental investigations show that the proposed strategy based on the B-spline wavelet on the interval finite element method and optimized singular value decomposition method is robust to damage localization in beam-like structures.

(Continued)

TABLE 2.1 *(Continued)*
A Review of the Application of the Differential Evolution Algorithm (DEA) for Structural Damage Identification

Reference	Year	Objective	Methodology	Structure(s)	Result and Finding
Firouzi et al. [16]	2021	This paper evaluates the computational efficiency of different optimization algorithms for open-edge crack identification in Euler–Bernoulli beams. In this study, eight optimization algorithms—the DEA, WOA, GWO, Harris hawk optimization (HHO), pathfinder algorithm (PFA), electrostatic discharge algorithm (ESDA), Henry gas solubility optimization (HGSO), and covariance matrix adaptation–evolution strategy (CMA-ES)—are taken into account. Then, new hybrid versions are introduced to improve computational efficiency.	Hybridized versions of studied algorithms with the Nelder–Mead algorithm (PFA-NM, ESDA-NM, HHO-NM, DE-NM, and CMA-ES-NM) are proposed to decrease the number of function evaluations. The cost function for determining the crack location and depth is the weighted squared difference between the measured and computed natural frequencies.	Cantilever beam	Regarding the obtained results, the crack location and depth can be accurately predicted with 3500 function evaluations in 150 iterations when using the ESDA. However, the PFA provides the same results with 15,000 function evaluations in 500 iterations. When implementing hybridized versions with the Nelder–Mead algorithm, the PFA-NM algorithm requires 360 function evaluations to determine the crack parameters. The ESDA-NM algorithm needs 400 function evaluations to find reliable results.
Pahnabi and Seyedpoor [122]	2021	This paper introduces a method for detecting joint damage in moment frames using an improved version of the DEA and time-domain responses. When employing the improved DEA, a newmutated vector is operated to produce the new generation for improving the performance of the standard DEA.	The objective function is assembled by substituting acceleration response vectors for natural frequencies in the ECBI equation.	Plane frame	The results reveal the effectiveness of the presented approach for determining the location and extent of joint damage with a limited number of measurements and noise effects. The influence of sensor placement on damage

(Continued)

TABLE 2.1 *(Continued)*
A Review of the Application of the Differential Evolution Algorithm (DEA) for Structural Damage Identification

Reference	Year	Objective	Methodology	Structure(s)	Result and Finding
					identification results is minor in small structures. However, increasing the number of sensors might improve the accuracy of the damage detection methodology for large-scale structures.
Aloisio et al. [123]	2022	In this paper, the indirect estimation of concrete resistance using the modulus of elasticity identified through an optimization-based FEM updating procedure is compared to the resistance directly estimated from concrete samples. To determine the optimal solution in FEM updating, two optimization techniques, the DEA and PSO, were implemented.	For FEM updating with the DEA, the deck (E_d) and girder (E_b) modulus of elasticity are considered design parameters, and a cost function (combination of natural frequencies and mode shapes) based on changes in measured and calculated modal parameters is taken into account. The measured modal parameters are extracted through ambient vibration tests, and the numerical model is developed in Sap2000.	Corvara bridge, Italy	The resistances of the concrete specimens estimated from each span confirm the indirect results (FEM updating) for assessing concrete compressive strength. The percentage difference between the compressive strengths determined by the direct and indirect methods is about 20%. The validation is only limited to girders, and the researchers could not prepare the specimens from the deck. The utilized optimization algorithms (the DEA and PSO) have provided the same results in FEM updating.

2.5 ANALYSIS AND DISCUSSION

This study's central purpose is to attract readers' attention to the DEA and various modified versions for solving structural damage detection problems and other similar optimization-based problems in SHM, such as optimal sensor placement, FEM updating, and crack detection. This chapter makes an effort to give a summary of the DEA's journey over slightly more than 20 years (2001–2022). Figure 2.2 displays the number of reviewed articles released in the last two decades. The DEA was initially introduced in 1997. However, the first paper related to SHM was presented in 2001, according to Figure 2.2. Additionally, the number of published papers between 2016 and 2019 is more significant than in other periods.

The review of improved versions of the DEA revealed that the suggested improvements to the DEA's standard parameters and operators typically included modifications in the initialization mechanism, mutation, crossover, and selection operators. For example, Fu and Yu [87], Vincenzi and Savoia [92], Vo-Duy et al. [94, 100], and Bureerat and Pholdee [102] are some of them to be mentioned.

Figure 2.3 illustrates a pie graph of employed structures, showing that beam-like structures were used more often to validate the effectiveness of different vibration-based methodologies. The minor contribution is related to bridges among the other five categories.

Figure 2.4 depicts the percentage of papers that have been published in the fields of damage detection, FEM updating, crack detection, and optimal sensor placement. It can be seen that a large number of papers present damage detection methodologies and only a few papers focus on optimal sensor placement.

Figure 2.5 provides the statistical analysis of utilized objective functions over the past two decades. It is clear that the hybrid objective function based on natural frequencies and mode shapes and the defined cost function using the ECBI term are also popular objective functions when using the DEA as an optimizer in vibration-based damage detection problems.

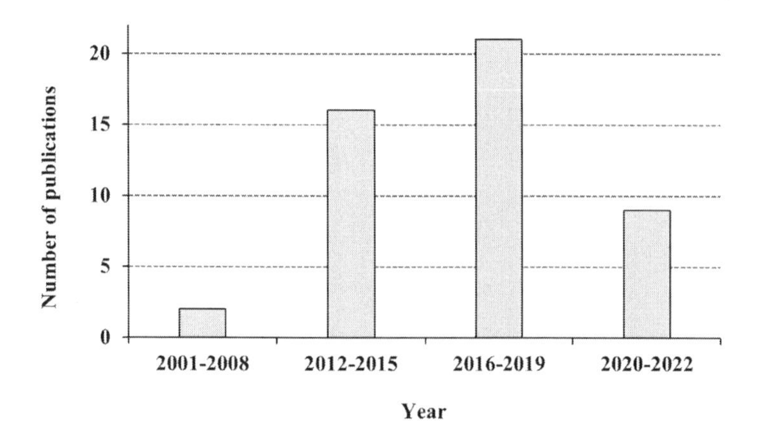

FIGURE 2.2 Number of publications in the field of structural damage detection using the DEA.

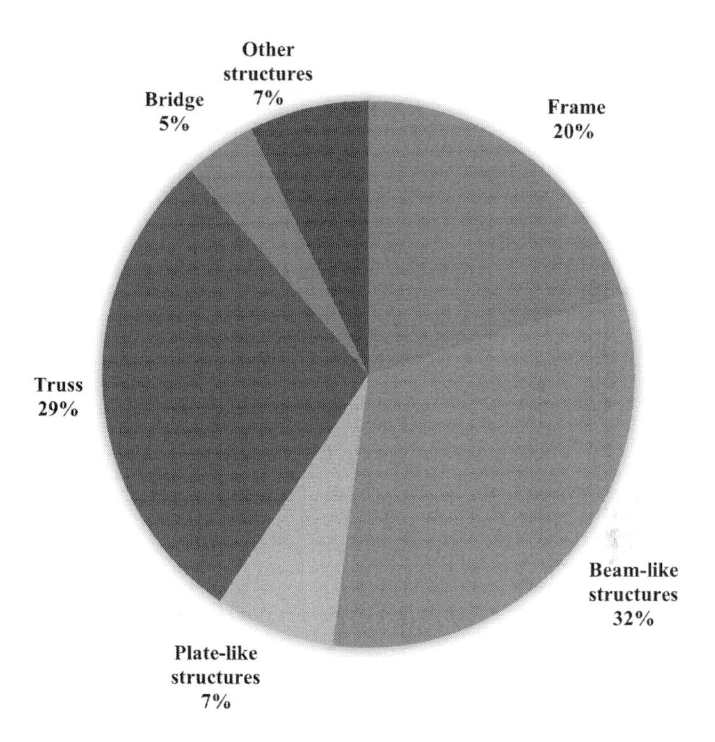

FIGURE 2.3 The distribution of structures employed in publications to show the effectiveness of suggested techniques.

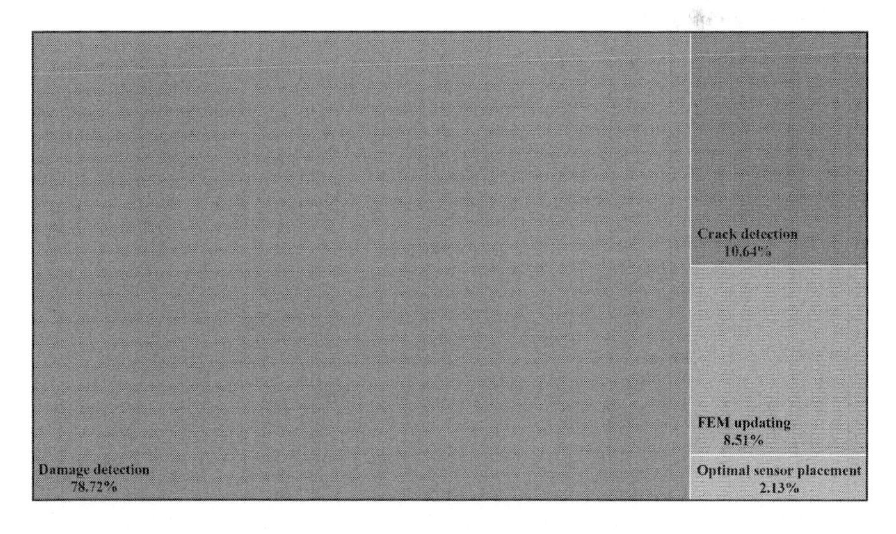

FIGURE 2.4 Contribution of publications in the areas of damage detection, FEM updating, crack detection, and optimal sensor placement.

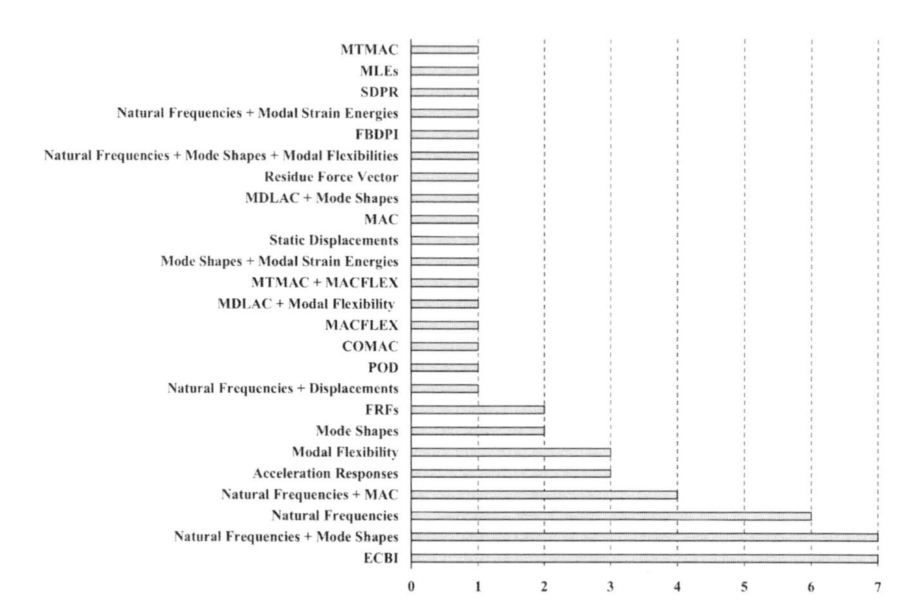

FIGURE 2.5 The categorization of implemented objective functions by the number of publications.

According to reviewed papers, several two-step and multiple-step methods have been proposed to deal with large search areas in high-dimensional optimization problems. The two-step and multiple-step methods generally attempt to reduce the number of design variables by identifying the damaged elements in the first step. Then, the DEA is implemented to minimize an objective function in the narrowed search space with lower design variables. This methodology is efficient in eliminating false alarms that are identified by single-step methods. Figure 2.6 demonstrates the ratio of single-step, two-step, and multiple-step procedures utilized from 2001 to 2022. According to Figure 2.6, 17% of methodologies have been developed based on two-step methods. The categorization of different two-step procedures is provided in Figure 2.7. Most researchers have applied modal strain energy to find the potentially damaged elements in the first step.

2.6 CONCLUSIONS

The present chapter is conducted in two main phases. The first phase reviews the methodologies, objectives, types of structures, and results of published papers from 2001 to 2022 in tabulated form. The organization of the tabular review helps readers find the essential points of each paper and address the critical questions as well as future directions. The second phase statistically analyzes the extracted data from the tabular review and graphically quantifies the number of publications per year, the percentage of different types of structures employed to assess the damage detection methodologies, the ratio of utilized objective functions, and the contributions of single-step, two-step, and multiple-step approaches with the

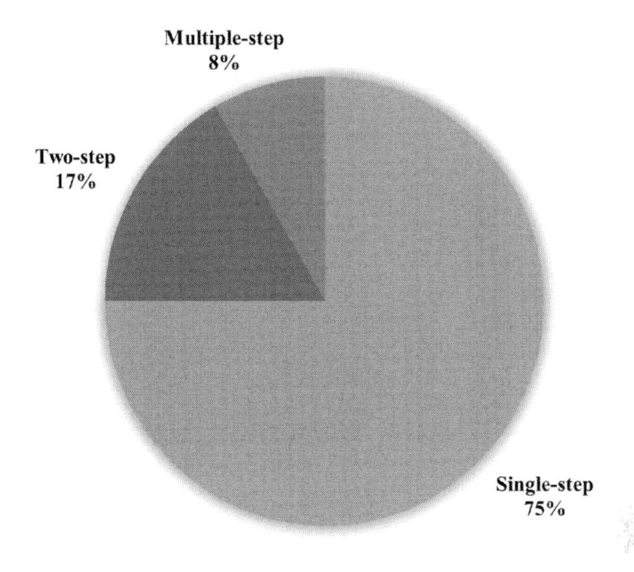

FIGURE 2.6 The ratio of single-step, two-step, and multiple-step techniques used in publications.

assistance of the DEA. The overall results of this review can be summarized as follows:

- The highest number of papers was published between 2016 and 2019.
- The contribution of beam-like structures is more significant than that of other structures. In contrast, only a few papers verify their methodology using bridge structures.
- Considerable articles propose techniques for damage detection (78.72%). The ratio of other SHM problems, such as crack detection, FEM updating, and optimal sensor placement, are 10.64%, 8.51%, and 2.13%, respectively.

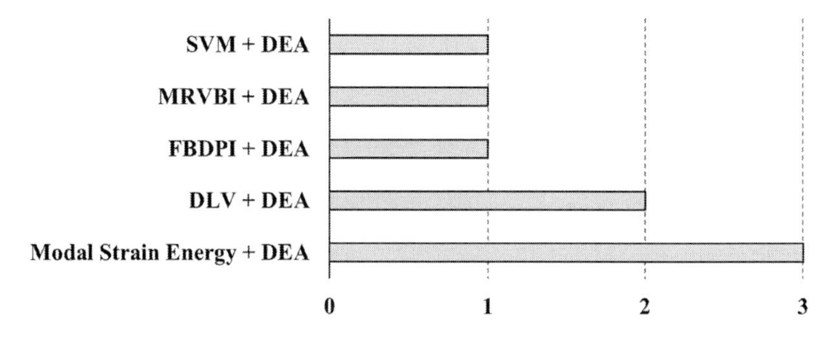

FIGURE 2.7 The categorization of various two-step techniques by the number of publications.

- The ECBI and the combination of natural frequencies and mode shapes have been the most widespread objective functions over the past two decades.
- The percentage of utilized single-step, two-step, and multiple-step methods are 75%, 17%, and 8%, respectively. Additionally, among other methods such as DLV, FBDPI, MRVBI, and SVM, the modal strain energy has often been employed to detect damaged elements in the first step.

REFERENCES

1. Ding, Z., R. Hou, and Y. Xia, *Structural damage identification considering uncertainties based on a Jaya algorithm with a local pattern search strategy and L0. 5 sparse regularization*. Engineering Structures, 2022. **261**: pp. 114312.
2. Le, N.T., et al., *A new method for locating and quantifying damage in beams from static deflection changes*. Engineering Structures, 2019. **180**: pp. 779–792.
3. Ding, Z., J. Li, and H. Hao, *Simultaneous identification of structural damage and nonlinear hysteresis parameters by an evolutionary algorithm-based artificial neural network*. International Journal of Non-Linear Mechanics, 2022. **142**: pp. 103970.
4. Hassani, S., M. Mousavi, and A.H. Gandomi, *Structural health monitoring in composite structures: A comprehensive review*. Sensors, 2021. **22**(1): pp. 153.
5. Gharehbaghi, V.R., et al., *A critical review on structural health monitoring: definitions, methods, and perspectives*. Archives of Computational Methods in Engineering, 2022. 29(4): pp. 2209–2235.
6. Shigeishi, M., et al., *Acoustic emission to assess and monitor the integrity of bridges*. Construction and Building Materials, 2001. **15**(1): pp. 35–49.
7. Nair, A. and C. Cai, *Acoustic emission monitoring of bridges: Review and case studies*. Engineering structures, 2010. **32**(6): pp. 1704–1714.
8. Dafydd, I. and Z. Sharif Khodaei, *Analysis of barely visible impact damage severity with ultrasonic guided Lamb waves*. Structural Health Monitoring, 2020. **19**(4): pp. 1104–1122.
9. Thiene, M., Z. Sharif-Khodaei, and F.M. Aliabadi, Optimal sensor placement for damage detection based on ultrasonic guided wave. Key Engineering Materials. 2016. 665: pp. 269–272. https://doi.org/10.4028/www.scientific.net/KEM.665.269
10. Salmanpour, M., Z. Sharif Khodaei, and M. Aliabadi, *Guided wave temperature correction methods in structural health monitoring*. Journal of Intelligent Material Systems and Structures, 2017. **28**(5): pp. 604–618.
11. De Luca, A., et al., *Damage characterization of composite plates under low velocity impact using ultrasonic guided waves*. Composites Part B: Engineering, 2018. **138**: pp. 168–180.
12. Sharif-Khodaei, Z., M. Ghajari, and M. Aliabadi, *Impact damage detection in composite plates using a self-diagnostic electro-mechanical impedance-based structural health monitoring system*. Journal of Multiscale Modelling, 2015. **6**(04): p. 1550013.
13. Schwankl, M., et al. Electro-mechanical impedance technique for structural health monitoring of composite panels. Key Engineering Materials. 2013. **525–526**: pp. 569–572. https://doi.org/10.4028/www.scientific.net/KEM.525-526.569
14. Ding, Z., et al., *Vibration-based FRP debonding detection using a Q-learning evolutionary algorithm*. Engineering Structures, 2023. **275**: p. 115254.

15. Ghannadi, P. and S.S. Kourehli, *Model updating and damage detection in multi-story shear frames using salp swarm algorithm.* Earthquakes and Structures, 2019. **17**(1): pp. 63–73.
16. Firouzi, B., A. Abbasi, and P. Sendur, *Improvement of the computational efficiency of metaheuristic algorithms for the crack detection of cantilever beams using hybrid methods.* Engineering Optimization, 2022. 54(7): pp. 1236–1257.
17. Ghannadi, P. and S.S. Kourehli, *An effective method for damage assessment based on limited measured locations in skeletal structures.* Advances in Structural Engineering, 2021. **24**(1): pp. 183–195.
18. Ghannadi, P. and S.S. Kourehli, *Data-driven method of damage detection using sparse sensors installation by SEREPa.* Journal of Civil Structural Health Monitoring, 2019. **9**(4): pp. 459–475.
19. Kourehli, S.S., *Prediction of unmeasured mode shapes and structural damage detection using least squares support vector machine.* Structural Monitoring and Maintenance, 2018. **5**(3): pp. 379–390.
20. Ghannadi, P. and S.S. Kourehli, *Structural damage detection based on MAC flexibility and frequency using moth-flame algorithm.* Structural Engineering and Mechanics, 2019. **70**(6): pp. 649–659.
21. Catbas, F.N., D.L. Brown, and A.E. Aktan, *Use of modal flexibility for damage detection and condition assessment: case studies and demonstrations on large structures.* Journal of Structural Engineering, 2006. **132**(11): pp. 1699–1712.
22. Sung, S.-H., K.-Y. Koo, and H.-J. Jung, *Modal flexibility-based damage detection of cantilever beam-type structures using baseline modification.* Journal of Sound and Vibration, 2014. **333**(18): pp. 4123–4138.
23. Fan, W. and P. Qiao, *Vibration-based damage identification methods: A review and comparative study.* Structural Health Monitoring, 2011. **10**(1): pp. 83–111.
24. Ding, Z., J. Li, and H. Hao, *Non-probabilistic method to consider uncertainties in structural damage identification based on Hybrid Jaya and Tree Seeds Algorithm.* Engineering Structures, 2020. **220**: p. 110925.
25. Alkayem, N.F., et al., *A new self-adaptive quasi-oppositional stochastic fractal search for the inverse problem of structural damage assessment.* Alexandria Engineering Journal, 2022. **61**(3): pp. 1922–1936.
26. Alkayem, N.F., et al., *Inverse analysis of structural damage based on the modal kinetic and strain energies with the novel oppositional unified particle swarm gradient-based optimizer.* Applied Sciences, 2022. **12**(22): p. 11689.
27. Alkayem, N.F., et al., *The combined social engineering particle swarm optimization for real-world engineering problems: A case study of model-based structural health monitoring.* Applied Soft Computing, 2022. **123**: p. 108919.
28. Kaveh, A., S.M. Hosseini, and H. Akbari, *Efficiency of plasma generation optimization for structural damage identification of skeletal structures based on a hybrid cost function.* Iranian Journal of Science and Technology, Transactions of Civil Engineering, 2021. **45**(4): pp. 2069–2090.
29. Kaveh, A. and A. Dadras, *Structural damage identification using an enhanced thermal exchange optimization algorithm.* Engineering Optimization, 2018. **50**(3): pp. 430–451.
30. Salgotra, R., et al., *Marine predator inspired naked mole-rat algorithm for global optimization.* Expert Systems with Applications, 2023. **212**: p. 118822.
31. Meng, Z., et al., *An efficient two-stage water cycle algorithm for complex reliability-based design optimization problems.* Neural Computing and Applications, 2022. **34**(23): pp. 20993–21013.

32. Shen, Y., et al., *An improved whale optimization algorithm based on multi-population evolution for global optimization and engineering design problems.* Expert Systems with Applications, 2023. **215**: p. 119269.
33. Mohammadi, D., et al., *Quantum Henry gas solubility optimization algorithm for global optimization.* Engineering with Computers, 2022. **38**(3): pp. 2329–2348.
34. Khodadadi, N., V. Snasel, and S. Mirjalili, *Dynamic arithmetic optimization algorithm for truss optimization under natural frequency constraints.* IEEE Access, 2022. **10**: pp. 16188–16208.
35. Kaveh, A., S. Talatahari, and N. Khodadadi, *Stochastic paint optimizer: Theory and application in civil engineering.* Engineering with Computers, 2022. **38**: pp. 1921–1952.
36. Khodadadi, N., F. Soleimanian Gharehchopogh, and S. Mirjalili, *MOAVOA: A new multi-objective artificial vultures optimization algorithm.* Neural Computing and Applications, 2022. **34**(23): pp. 20791–20829.
37. Talatahari, S., H. Bayzidi, and M. Saraee, *Social network search for global optimization.* IEEE Access, 2021. **9**: pp. 92815–92863.
38. Talatahari, S., et al., *Crystal structure algorithm (CryStAl): A metaheuristic optimization method.* IEEE Access, 2021. **9**: pp. 71244–71261.
39. Talatahari, S. and M. Azizi, *Chaos game optimization: A novel metaheuristic algorithm.* Artificial Intelligence Review, 2021. **54**(2): pp. 917–1004.
40. Ezugwu, A.E., et al., *Prairie dog optimization algorithm.* Neural Computing and Applications, 2022. **34**(22): pp. 20017–20065.
41. Ereiz, S., I. Duvnjak, and J.F. Jiménez-Alonso. Review of finite element model updating methods for structural applications. Structures. 2022. **41**: pp. 684–723.
42. Ghannadi, P., S.S. Kourehli, and S. Mirjalili, *The application of PSO in structural damage detection: an analysis of the previously released publications (2005–2020).* Frattura Ed Integrità Strutturale, 2022. **16**(62): pp. 460–489.
43. Alkayem, N.F., M. Cao, and M. Ragulskis, *Damage localization in irregular shape structures using intelligent FE model updating approach with a new hybrid objective function and social swarm algorithm.* Applied Soft Computing, 2019. **83**: p. 105604.
44. Jahangiri, M., et al., *A reliability-based sieve technique: A novel multistage probabilistic methodology for the damage assessment of structures.* Engineering Structures, 2021. **226**: p. 111359.
45. Jahangiri, M. and M.A. Hadianfard, *Vibration-based structural health monitoring using symbiotic organism search based on an improved objective function.* Journal of Civil Structural Health Monitoring, 2019. **9**(5): pp. 741–755.
46. Beheshti Aval, S.B. and P. Mohebian, *Joint damage identification in frame structures by integrating a new damage index with equilibrium optimizer algorithm.* International Journal of Structural Stability and Dynamics, 2022. **22**(05): p. 2250056.
47. Beheshti Aval, S.B. and P. Mohebian, *Combined joint and member damage identification of skeletal structures by an improved biology migration algorithm.* Journal of Civil Structural Health Monitoring, 2020. **10**(3): pp. 357–375.
48. Ghannadi, P. and S.S. Kourehli, *Efficiency of the slime mold algorithm for damage detection of large-scale structures.* The Structural Design of Tall and Special Buildings, 2022. **31**(14): p. e1967.
49. Ghiasi, R., et al., *Structural assessment under uncertain parameters via the interval optimization method using the slime mold algorithm.* Applied Sciences, 2022. **12**(4): p. 1876.

50. Ghiasi, R. and A. Malekjafarian, *An advanced binary slime mould algorithm for feature subset selection in structural health monitoring data*, in *Civil Engineering Research in Ireland 2022 (CERI2022)*. 2022. pp. 165–170.
51. Tiachacht, S., et al., *Inverse problem for dynamic structural health monitoring based on slime mould algorithm*. Engineering with Computers, 2022. **38**(3): pp. 2205–2228.
52. Beheshti Aval, S.B. and P. Mohebian, *A novel optimization algorithm based on modal force information for structural damage identification*. International Journal of Structural Stability and Dynamics, 2021. **21**(07): p. 2150100.
53. Mishra, M., et al., *Ant lion optimisation algorithm for structural damage detection using vibration data*. Journal of Civil Structural Health Monitoring, 2019. **9**(1): pp. 117–136.
54. Mohebian, P., et al., *Visible particle series search algorithm and its application in structural damage identification*. Sensors, 2022. **22**(3): p. 1275.
55. Minh, H.-L., et al., *Structural damage identification in thin-shell structures using a new technique combining finite element model updating and improved Cuckoo search algorithm*. Advances in Engineering Software, 2022. **173**: p. 103206.
56. Benaissa, B., et al., *YUKI Algorithm and POD-RBF for Elastostatic and dynamic crack identification*. Journal of Computational Science, 2021. **55**: p. 101451.
57. Ghannadi, P. and S.S. Kourehli, *Multiverse optimizer for structural damage detection: Numerical study and experimental validation*. The Structural Design of Tall and Special Buildings, 2020. **29**(13): p. e1777.
58. Kaveh, A., P. Rahmani, and A. Dadras Eslamlou, *Guided water strider algorithm for structural damage detection using incomplete modal data*. Iranian Journal of Science and Technology, Transactions of Civil Engineering, 2022. **46**(2): pp. 771–788.
59. Sang-To, T., et al., *A new movement strategy of grey wolf optimizer for optimization problems and structural damage identification*. Advances in Engineering Software, 2022. **173**: p. 103276.
60. Ghannadi, P., et al., *Efficiency of grey wolf optimization algorithm for damage detection of skeletal structures via expanded mode shapes*. Advances in Structural Engineering, 2020. **23**(13): pp. 2850–2865.
61. Gomes, G.F., J.A.S. Chaves, and F.A. de Almeida, *An inverse damage location problem applied to AS-350 rotor blades using bat optimization algorithm and multiaxial vibration data*. Mechanical Systems and Signal Processing, 2020. **145**: p. 106932.
62. Barman, S.K., et al., *Vibration-based damage detection of structures employing Bayesian data fusion coupled with TLBO optimization algorithm*. Structural and Multidisciplinary Optimization, 2021. **64**(4): pp. 2243–2266.
63. Ahmadi-Nedushan, B. and H. Fathnejat, *A modified teaching–learning optimization algorithm for structural damage detection using a novel damage index based on modal flexibility and strain energy under environmental variations*. Engineering with Computers, 2022. **38**(1): pp. 847–874.
64. Sengupta, P. and S. Chakraborty, *An improved iterative model reduction technique to estimate the unknown responses using limited available responses*. Mechanical Systems and Signal Processing, 2023. **182**: p. 109586.
65. Kourehli, S.S., *Damage identification of structures using second-order approximation of Neumann series expansion*. Journal of Rehabilitation in Civil Engineering, 2020. **8**(2): pp. 81–91.
66. Ghannadi, P. and S.S. Kourehli, *Investigation of the accuracy of different finite element model reduction techniques*. Structural Monitoring and Maintenance, 2018. **5**(3): pp. 417–428.

67. Kahya, V., S. Şimşek, and V. Toğan, *Vibration-based damage detection in anisotropic laminated composite beams by a shear deformable finite element and harmony search optimization*. 2022.
68. Şimşek, S., et al., *Damage detection in anisotropic-laminated composite beams based on incomplete modal data and teaching–learning-based optimization*. Structural and Multidisciplinary Optimization, 2022. **65**(11): pp. 1–17.
69. Dinh-Cong, D., S. Pham-Duy, and T. Nguyen-Thoi, *Damage detection of 2D frame structures using incomplete measurements by optimization procedure and model reduction*. Journal of Advanced Engineering and Computation, 2018. **2**(3): pp. 164–173.
70. Das, S. and N. Dhang, *Damage identification of thin plates using multi-stage PSOGSA and incomplete modal data*. Applied Mathematics in Science and Engineering, 2022. **30**(1): pp. 396–438.
71. Cancelli, A., et al., *Vibration-based damage localization and quantification in a pretensioned concrete girder using stochastic subspace identification and particle swarm model updating*. Structural Health Monitoring, 2020. **19**(2): pp. 587–605.
72. Ahmad, M.F., et al., *Differential evolution: A recent review based on state-of-the-art works*. Alexandria Engineering Journal, 2022. **61**(5): pp. 3831–3872.
73. Storn, R. and K. Price, *Differential evolution–a simple and efficient heuristic for global optimization over continuous spaces*. Journal of global optimization, 1997. **11**(4): pp. 341–359.
74. Cui, L., et al. *Enhance differential evolution algorithm based on novel mutation strategy and parameter control method*. in *International Conference on Neural Information Processing*. 2015. Springer.
75. Qin, A.K., V.L. Huang, and P.N. Suganthan, *Differential evolution algorithm with strategy adaptation for global numerical optimization*. IEEE transactions on Evolutionary Computation, 2008. **13**(2): pp. 398–417.
76. Manson, G. and K. Worden. *Lamb wave sensor optimization using differential evolution*. in *Smart Structures and Materials 2001: Modeling, Signal Processing, and Control in Smart Structures*. 2001. SPIE.
77. Casciati, S., *Stiffness identification and damage localization via differential evolution algorithms*. Structural Control and Health Monitoring, 2008. **15**(3): pp. 436–449.
78. Kang, F., J.-j. Li, and Q. Xu, *Damage detection based on improved particle swarm optimization using vibration data*. Applied Soft Computing, 2012. **12**(8): pp. 2329–2335.
79. Rao, A.R.M., K. Lakshmi, and D. Venkatachalam, *Damage diagnostic technique for structural health monitoring using POD and self adaptive differential evolution algorithm*. Computers & Structures, 2012. **106**: pp. 228–244.
80. Bighamian, R. and H.R. Mirdamadi, *Input/output system identification of simultaneous mass/stiffness damage assessment using discrete-time pulse responses, differential evolution algorithm, and equivalent virtual damped SDOF*. Structural Control and Health Monitoring, 2013. **20**(4): pp. 576–592.
81. Kang, F., J. Li, and S. Liu, *Combined data with particle swarm optimization for structural damage detection*. Mathematical Problems in Engineering, 2013. **2013**: pp. 1–10.
82. Reed, H., J. Nichols, and C. Earls, *A modified differential evolution algorithm for damage identification in submerged shell structures*. Mechanical Systems and Signal Processing, 2013. **39**(1–2): pp. 396–408.
83. Jena, P.K., D.N. Thatoi, and D.R. Parhi, *Differential evolution: An inverse approach for crack detection*. Advances in Acoustics and Vibration, 2013. **2013**: pp. 1–10.

84. Vincenzi, L., G. De Roeck, and M. Savoia, *Comparison between coupled local minimizers method and differential evolution algorithm in dynamic damage detection problems.* Advances in Engineering Software, 2013. **65**: pp. 90–100.
85. Villamizar Mejía, R., et al. *Tuning of expert systems for structural damage detection through differential evolutionary algorithms*, in *6th edition of the World Conference of the International Association for Structural Control and Monitoring.* 2014.
86. Villalba-Morales, J.D. and J.E. Laier, *Assessing the performance of a differential evolution algorithm in structural damage detection by varying the objective function.* Dyna, 2014. **81**(188): pp. 106–115.
87. Fu, Y.M. and L. Yu. A DE-based algorithm for structural damage detection, in Advanced Materials Research. 2014. **919–921**: pp. 303–307.
88. Cavalini Jr, A.A., et al., *Model updating of a rotating machine using the self-adaptive differential evolution algorithm.* Inverse Problems in Science and Engineering, 2016. **24**(3): pp. 504–523.
89. Jena, P.K. and D.R. Parhi, *A modified particle swarm optimization technique for crack detection in Cantilever Beams.* Arabian Journal for Science and Engineering, 2015. **40**(11): pp. 3263–3272.
90. Seyedpoor, S., S. Shahbandeh, and O. Yazdanpanah, *An efficient method for structural damage detection using a differential evolution algorithm-based optimisation approach.* Civil Engineering and Environmental Systems, 2015. **32**(3): pp. 230–250.
91. Seyedpoor, S. and O. Yazdanpanah, *Structural damage detection by differential evolution as a global optimization algorithm.* Iranian Journal of Structural Engineering, 2015. **1**(1): pp. 52–62.
92. Vincenzi, L. and M. Savoia, *Coupling response surface and differential evolution for parameter identification problems.* Computer-Aided Civil and Infrastructure Engineering, 2015. **30**(5): pp. 376–393.
93. Seyedpoor, S. and M. Montazer, *A damage identification method for truss structures using a flexibility-based damage probability index and differential evolution algorithm.* Inverse Problems in Science and Engineering, 2016. **24**(8): pp. 1303–1322.
94. Vo-Duy, T., et al., *Damage detection in laminated composite plates using modal strain energy and improved differential evolution algorithm.* Procedia Engineering, 2016. **142**: pp. 182–189.
95. Seyedpoor, S.M. and M. Montazer, *A two-stage damage detection method for truss structures using a modal residual vector based indicator and differential evolution algorithm.* Smart Structures and Systems, 2016. **17**(2): pp. 347–361.
96. Ding, Z., et al., *Improved artificial bee colony algorithm for crack identification in beam using natural frequencies only.* Inverse Problems in Science and Engineering, 2017. **25**(2): pp. 218–238.
97. Anh, P.H., Solving inverse problem in structural health monitoring by differential evolution algorithm with nearest neighbor comparison method, in *The Fourteenth East Asia-Pacific Conference on Structural Engineering and Construction (EASEC-14)*, Ho Chi Minh City, Vietnam, January 2016.
98. Nguyen-Thoi, T., et al., *A combination of damage locating vector method (DLV) and differential evolution algorithm (DE) for structural damage assessment.* Frontiers of Structural and Civil Engineering, 2018. **12**(1): pp. 92–108.
99. Georgioudakis, M. and V. Plevris, *Investigation of the performance of various modal correlation criteria in structural damage identification.* 2016.
100. Vo-Duy, T., et al., *A two-step approach for damage detection in laminated composite structures using modal strain energy method and an improved differential evolution algorithm.* Composite Structures, 2016. **147**: pp. 42–53.

101. Dinh-Cong, D., et al., *A two-stage assessment method using damage locating vector method and differential evolution algorithm for damage identification of cross-ply laminated composite beams*. Advances in Structural Engineering, 2017. **20**(12): pp. 1807–1827.
102. Bureerat, S. and N. Pholdee. *Adaptive sine cosine algorithm integrated with differential evolution for structural damage detection*. in *International Conference on Computational Science and Its Applications*. 2017. Springer.
103. Dinh-Cong, D., et al., *Efficiency of Jaya algorithm for solving the optimization-based structural damage identification problem based on a hybrid objective function*. Engineering Optimization, 2018. **50**(8): pp. 1233–1251.
104. Dinh-Cong, D., et al., *An efficient multi-stage optimization approach for damage detection in plate structures*. Advances in Engineering Software, 2017. **112**: pp. 76–87.
105. Seyedpoor, S.M., E. Norouzi, and S. Ghasemi, *Structural damage detection using a multi-stage improved differential evolution algorithm (Numerical and experimental)*. Smart Struct. Syst, 2018. **21**(2): pp. 235–248.
106. Georgioudakis, M. and V. Plevris, *A combined modal correlation criterion for structural damage identification with noisy modal data*. Advances in Civil Engineering, 2018. **2018**: pp. 1–20.
107. Alkayem, N.F. and M. Cao, *Damage identification in three-dimensional structures using single-objective evolutionary algorithms and finite element model updating: Evaluation and comparison*. Engineering Optimization, 2018. **50**(10): pp. 1695–1714.
108. Kim, S., et al. *Truss damage detection using modified differential evolution algorithm with comparative studies*. in *Proceedings of the International Conference on Advances in Computational Mechanics 2017*. 2018. Singapore: Springer Singapore.
109. Seyedpoor, S., A. Ahmadi, and N. Pahnabi, *Structural damage detection using time domain responses and an optimization method*. Inverse Problems in Science and Engineering, 2019. **27**(5): pp. 669–688.
110. Fallahian, S., A. Joghataie, and M.T. Kazemi, *Damage identification in structures using time domain responses based on differential evolution algorithm*. International Journal of Optimization in Civil Engineering, 2018. **8**(3): pp. 357–380.
111. Bassoli, E., et al., *Ambient vibration-based finite element model updating of an earthquake-damaged masonry tower*. Structural Control and Health Monitoring, 2018. **25**(5): p. e2150.
112. Bureerat, S. and N. Pholdee, *Inverse problem based differential evolution for efficient structural health monitoring of trusses*. Applied Soft Computing, 2018. **66**: pp. 462–472.
113. Seyedpoor, S.M. and M.H. Nopour, *A two-step method for damage identification in moment frame connections using support vector machine and differential evolution algorithm*. Applied Soft Computing, 2020. **88**: p. 106008.
114. Kim, N.-I., S. Kim, and J. Lee, *Vibration-based damage detection of planar and space trusses using differential evolution algorithm*. Applied Acoustics, 2019. **148**: pp. 308–321.
115. Sobrinho, B.E., et al., *Differential evolution algorithm for identification of structural damage in steel beams*. Frattura ed Integrità Strutturale, 2020. **14**(52): pp. 51–66.
116. Seyedpoor, S. and N. Pahnabi, *Structural damage identification using frequency domain responses and a differential evolution algorithm*. Iranian Journal of Science and Technology, Transactions of Civil Engineering, 2021. **45**(2): pp. 1253–1264.

117. Lieu, Q.X., V.H. Luong, and J. Lee, *Structural damage identification using adaptive hybrid evolutionary firefly algorithm*, in *Applications of Firefly Algorithm and Its Variants*. 2020. Springer. pp. 75–97.
118. Lieu, Q.X., D.T. Do, and J. Lee, *An adaptive hybrid evolutionary firefly algorithm for shape and size optimization of truss structures with frequency constraints.* Computers & Structures, 2018. **195**: pp. 99–112.
119. Guedria, N.B., *An accelerated differential evolution algorithm with new operators for multi-damage detection in plate-like structures.* Applied Mathematical Modelling, 2020. **80**: pp. 366–383.
120. Su, Y., L. Liu, and Y. Lei, *Structural damage identification using a modified directional bat algorithm.* Applied Sciences, 2021. **11**(14): p. 6507.
121. Wang, N., et al., *An adaptive damage detection method based on differential evolutionary algorithm for beam structures.* Measurement, 2021. **178**: p. 109227.
122. Pahnabi, N. and S.M. Seyedpoor, *Damage identification in connections of moment frames using time domain responses and an optimization method.* Frontiers of Structural and Civil Engineering, 2021. **15**(4): pp. 851–866.
123. Aloisio, A., et al. Indirect assessment of concrete resistance from FE model updating and Young's modulus estimation of a multi-span PSC viaduct: Experimental tests and validation. Structures. 2022. **37**: pp. 686–697.

3 Fatigue Assessment and Structural Health Monitoring of Steel Truss Bridges

Manuel Buitrago, Elisa Bertolesi, Pedro A. Calderón, and José M. Adam

3.1 OVERVIEW

This chapter describes a double experimental and analytical fatigue residual life prediction of a steel truss-type railway bridge constructed between 1913 and 1915. The experimental part of the study involved full-scale fatigue testing of (i) a full-scale bridge span and (ii) an upper crossbeam. Both structures belong to a twin bridge of that under study. Both tests considered an extensive monitoring system to capture the possible nucleation and propagation of fatigue cracks. The analytical part of the study consisted of applying the fracture mechanics theory in advanced numerical models, and this approach confirmed the crack nucleation and propagation obtained in the experimental part. These studies were also used to define a monitoring method to help in decision making in case of possible fatigue failures. Even though other researchers had previously carried out fatigue tests on full-scale riveted bridge elements, this study is unique in that it is the first time a full-scale bridge has been subjected to fatigue tests.

3.2 INTRODUCTION

The Quebec Bridge (Canada, 1907), the I-35 Bridge in Minneapolis (USA, 2007), and the Chauras Bridge (India, 2012) are some historic and recent examples of disproportionate collapse of truss-like railway and road bridges that involved casualties and economic losses [1–4]. A study carried out in the USA [5] identified the failure of more than 500 bridges over a period of 11 years (1989–2000), 21.3% of which were steel truss-like bridges, thus demonstrating the vulnerability of this type of structure (an average of 9.7 bridges of this type collapsed every year in the USA). Natural disasters, impacts, overloads, structural and design deficiencies, construction and supervision mistakes, or the lack of maintenance or inspection were the main causes of these collapses.

DOI: 10.1201/9781003306924-3

Bridge structures are expected to withstand loads defined in the codes (e.g., gravity, wind, snow). However, these structures may be subjected to extreme events (also called low-probability/high-consequence events), such as hurricanes, tsunamis, explosions, vehicle impacts, fires, human errors, or terrorist attacks [6, 7]; or be exposed to several degradation actions, such as corrosion [8] and fatigue [9]. The latter failure mechanism is especially important due to the damage accumulated over time in the different components. Aircraft wings are a well-known example of this due to their continual flexing under aerodynamic pressure, as are ships' hulls reacting to heavy seas. The case under study in this chapter deals with steel truss railway bridges with riveted joints. Supporting the passage of trains over many years, these structures experience repeated cyclical loads together with cumulative damage and fatigue that can compromise their structural safety without ever reaching the material's yield strength. Some researchers have already studied this phenomenon in this type of structure, the work of Pipinato et al. [10] being among the leaders in this field.

This chapter describes the research team's unique opportunity to test and analyze a full-scale steel-riveted truss bridge under fatigue loads at the ICITECH laboratories of the Universitat Politècnica de València (ICITECH-UPV). The study was both ambitious and novel and provided a considerable advance in the field of structural fatigue assessment, with the final objective of establishing practical recommendations for structural health monitoring (SHM).

After this introduction, the rest of the chapter contains the following elements: (i) a description of the bridge; (ii) the method used; (iii) a summary of the work carried out in the different phases: a preliminary evaluation and in-depth analysis; (iv) some practical recommendations for SHM; and (v) the conclusions drawn from the study.

3.3 DESCRIPTION OF THE BRIDGE

The railway bridge studied in this chapter was built between 1913 and 1915 and belongs to the Alicante–Denia line of the local railway company Ferrocarrils de la Generalitat Valenciana (FGV). In addition to its obvious usefulness, it is considered to be a remarkable piece of the historical heritage of civil engineering in Spain. Its structure was formed by a series of Pratt-type trusses connected by riveted joints. It also had a series of horizontal and vertical braces in the form of St. Andrew's crosses, and longitudinal and transverse beams to locally distribute train loads to the Pratt trusses. The heights of the metal piers varied up to 23.6 m, and the bridge had two isostatic spans at each end with a continuous beam in the two central spans. All the supports were hinged with free rotations. One support in each span also had free longitudinal displacement as a roller (A1 for span 1, P1 for span 2, P2 for span 3, P4 for span 4, P5 for span 5, and A2 for span 6). Figure 3.1 gives the bridge dimensions and a view of a train passing over it. A more in-depth description can be found in Refs. [11–13].

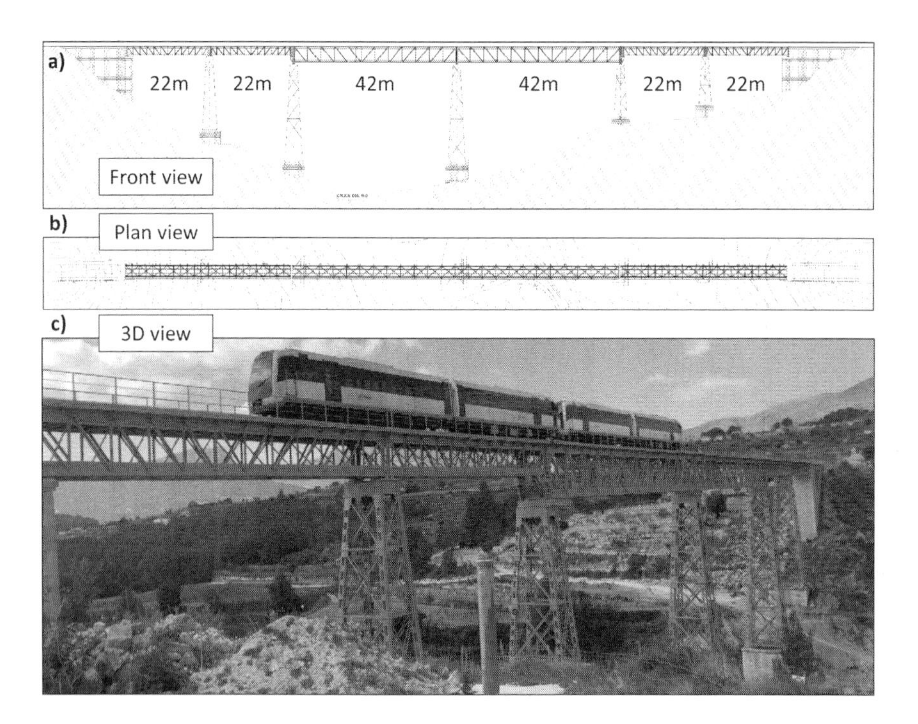

FIGURE 3.1 Geometry of the bridge: (a) front and (b) plan views, and (c) general aerial view.

3.4 METHODOLOGY

This section describes the method used to estimate the bridge's remaining fatigue life. The recommendations adopted were those of the Report of the European Convention for Constructional Steelwork (ECCS) and the European Commission's *Assessment of Existing Steel Structures: Recommendations for Estimation of Remaining Fatigue Life* [14], which proposes a method of estimating structural fatigue behavior and remaining useful life. The work and analyses carried out in the course of this study followed the proposed method. Advantage was taken of this unique opportunity to examine the Quisi Bridge, a twin specimen of the Ferrandet Bridge, which has almost identical spans and was built around the same time, was situated only a few kilometers away on the same line and thus subjected to similar loads. The original Ferrandet Bridge, which had been replaced by a new structure, had been dismantled and was available for tests. This unique case allowed us to carry out more complex tests on the Ferrandet Bridge to obtain valid conclusions for the analysis of the Quisi Bridge without compromising its structure. The tests carried out, which will be described in the subsequent sections of this chapter, were as follows:

- Phase I, preliminary evaluation:
    - Inspections
    - Mechanical characterization
    - Static and dynamic load tests

- Phase II, detailed and in-depth analysis:
  - Definition of the fatigue strength curve test until failure on a local element of the structure
  - Test on a real-scale span of the bridge
  - Remaining life based on experimental evidence
  - Fracture mechanics and failure patterns

Finally, extra studies were performed and summarized in this chapter in order to lay down general SHM recommendations for this type of structure.

3.5 ANALYSIS

This section contains the works carried out for the preliminary evaluation and the detailed and in-depth analysis.

3.5.1 Preliminary Evaluation

The bridge's structural behavior was analyzed in detail by means of: a preliminary inspection following the principles laid down in the ITPF-05 [15]; a detailed characterization of the material's mechanical behavior (as a summary: elastic modulus E = 210 GPa; yield strength F_y = 270 MPa); and a loading test, in which strains were measured by FBG strain sensors in the bottom and top chords and deflections were measured in the center of spans. Shear demand was assessed with strain FBG sensors measurements in diagonal and vertical elements near supports. This load test considered four different load hypotheses (see Figure 3.2). The dynamic response of the structure was also analyzed with trains passing over it at 20 and 50 km/h. In addition to the previously described sensors, accelerations were also monitored on the bridge. The results obtained showed that the bridge was structurally safe, although, due to being more than 100 years old, its components could compromise its behavior and cause an unexpected failure or accident due to fatigue, so that the work method entered on the next stage of detail described in Section 3.5.2.

3.5.2 Detailed and In-Depth Analysis

This section contains two full-scale tests: a test on an element of the local structural system and a test on a complete isostatic span of the bridge to control the overall structural behavior under fatigue loads. It also contains an estimation of its remaining life based on experimental evidence and the advanced computational modeling study, applying the fracture mechanics theory.

3.5.2.1 Definition of the Fatigue Strength Curve Test until Failure on a Local Element of the Structure

A fatigue test was carried out on one of the most critical components subjected to cyclic bending stress increments to assess the bridge's local behavior. As will be seen later in greater detail, this test assesses the fatigue behavior of these elements

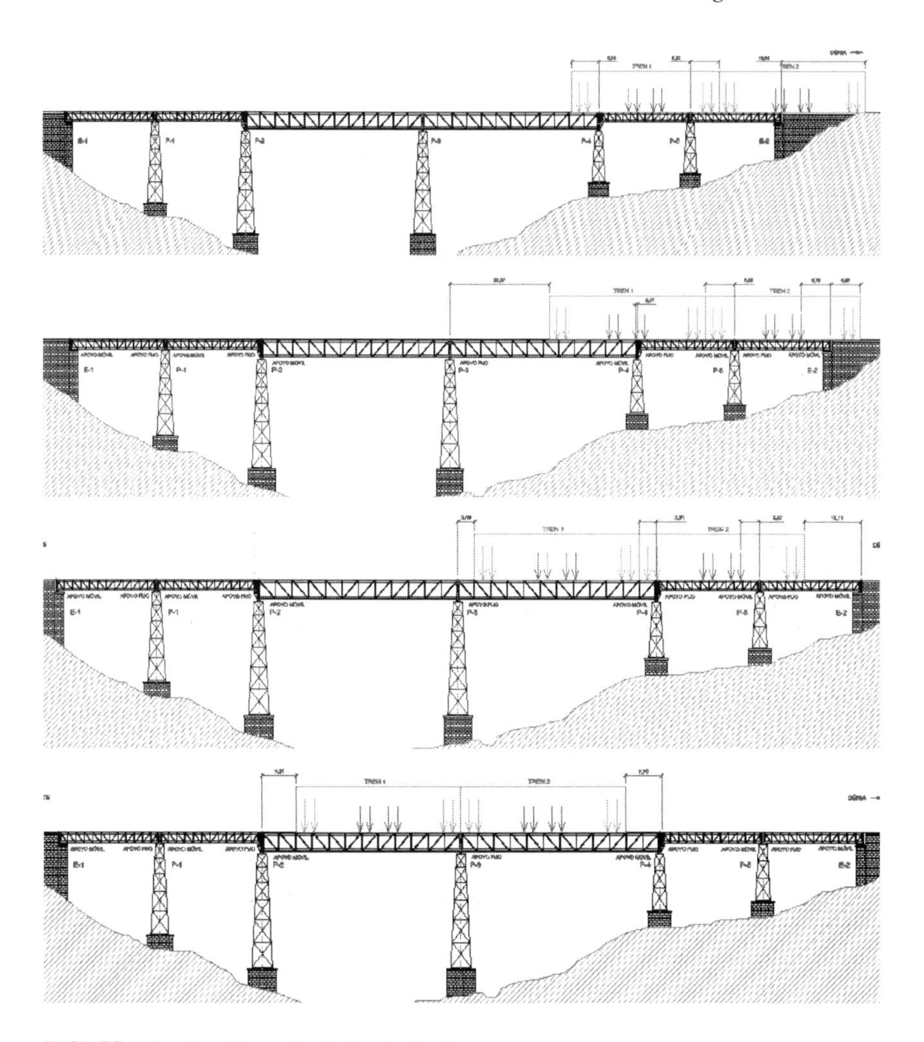

FIGURE 3.2 Load hypothesis for the load tests.

under actual conditions and takes them to their breaking point. The idea was to obtain the category of structural detail for fatigue strength characterization of the Quisi Bridge. For this, we also needed to carry out a historical analysis of the stress history of the elements that had been subjected to such stresses from the beginning of their service life.

This subsection contains (i) a definition of the test, (ii) test results, and (iii) a historical analysis of the stresses to which the item under study had been subjected to define its fatigue strength curve.

3.5.2.1.1 Definition of the Test

The test was performed on a beam-type element that conserved its lateral connections with the top chord of the main truss of the bridge (see Figure 3.3).

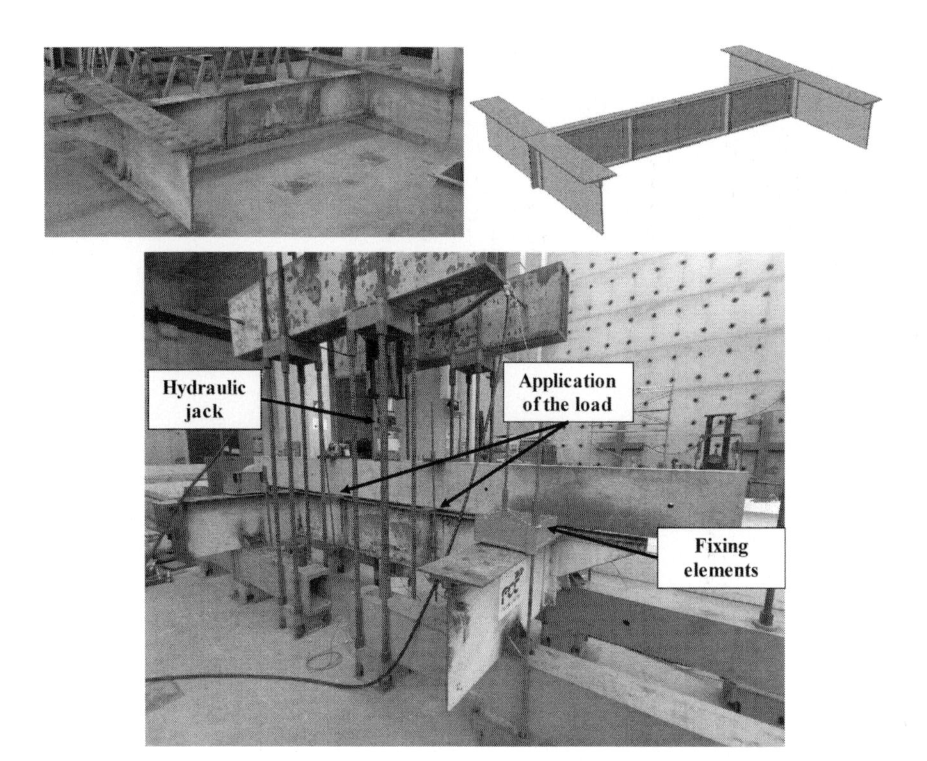

FIGURE 3.3 (Top) Selected element for testing and (bottom) test setup.

Figure 3.3 shows the test setup. This beam-type element is the one that absorbs the load of the train axles that are actually applied at the same points as the ones prepared in the test.

Under these conditions, the item was subjected to loads at two points that reproduced similar bending and shear stresses to those applied on the element in the bridge when the train passed over it. The definition of the load applied during the test was performed by computational models, as shown in Figure 3.4. The model simulated different situations at different loading levels. The load increment finally selected was 600 kN. The expected maximum Von Mises stresses on the element were 212 MPa, considerably lower than the material's yielding point.

The test was continuously monitored by four strain gauges (SG_1 at the bottom part of the beam web; SG_2 and SG_3 at the top and bottom of the beam, respectively; and SG_4 at the L-profile located at the bottom of the beam) and two displacement transducers measuring vertical deflection (LVDT_1) and possible slippage (LVDT_2) between the rivets and steel profiles in the zone where the shear stresses are highest. A scheme of the monitoring system can be seen in Figure 3.4.

3.5.2.1.2 Test Results

Figure 3.5 shows the readings registered by a strain gauge in a short period of time. Figure 3.6 shows the evolution of deformation and displacement in each of

FIGURE 3.4 Position of sensors.

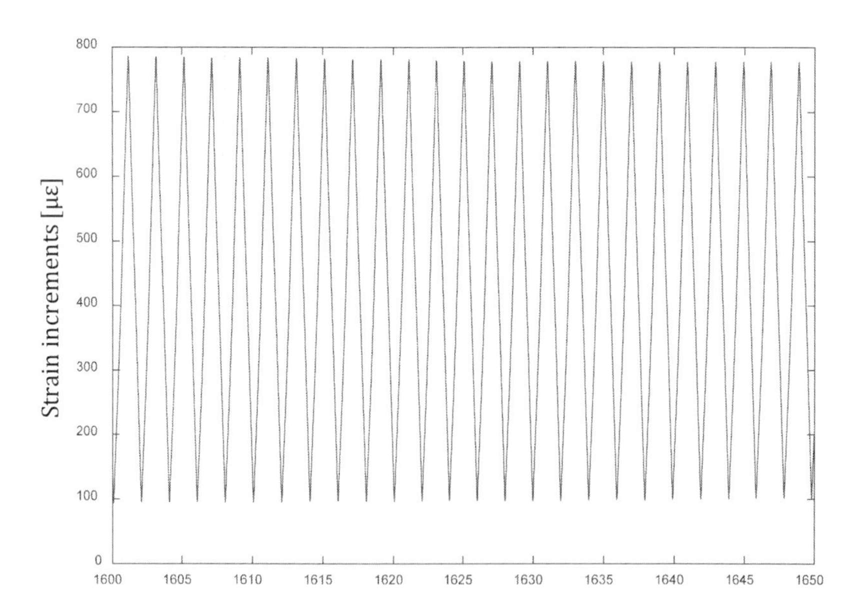

FIGURE 3.5 Example of strain measurements.

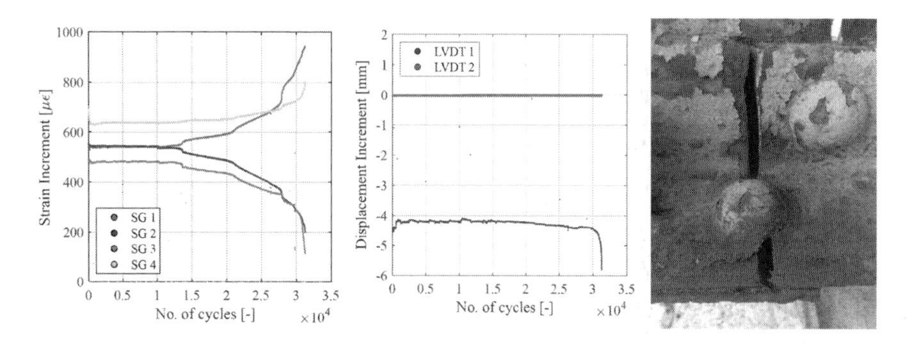

FIGURE 3.6 (Left and middle) Strain increment evolution of strain gauges (SGs) and linear voltage displacement transducers (LVDTs), and (right) fatigue crack at the end of the test.

the cycles from the beginning until failure occurred. The item under study failed under normal stresses in cycle number 31,377. Figure 3.6 shows the fatigue crack that started from the bottom of the element (zone in tension) and propagated to the top of the beam element.

3.5.2.1.3 Historic Analysis of the Stresses Borne by the Elements, and Definition of the Fatigue Strength Curve

The failure of the beam gave us the opportunity to determine the category of the detail to which this type of element with riveted joints belongs. The EU recommendations [14] stated that this type belongs to detail category 71. In our case, to precisely determine this category, we had to carry out a historical analysis of the loads to which this element had been subjected, with which, together with the results of the test, we were able to estimate the real detail category.

The historical data of the railway traffic provided by the FGV railway administration showed that 11 different types of trains had used the bridge since it was put into service in 1915. The only doubtful information associated with this data was related to the weight of the different trains, since there was no information available on how many were empty and how many were fully loaded. Figure 3.7 gives

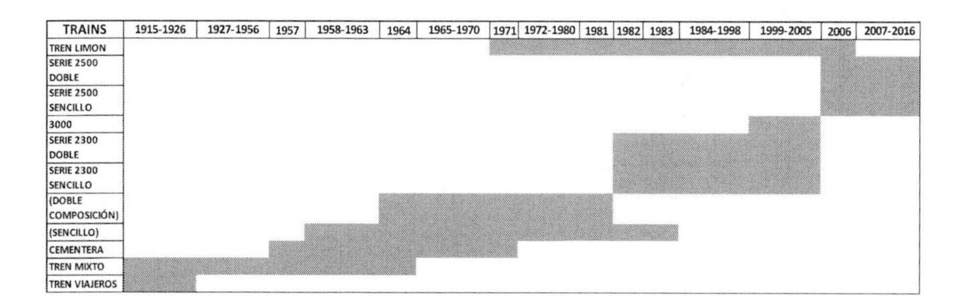

FIGURE 3.7 Running periods of the different trains in the years between 1915 and 2016.

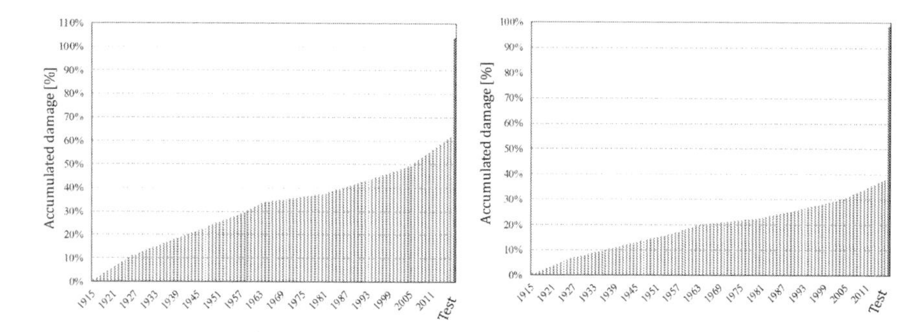

FIGURE 3.8 Hypotheses of (left) fully loaded and (right) completely empty trains.

a summary of the data provided by the FGV on the trains that passed over the bridge from 1915 to 2016.

Adopting the cumulative damage method used in different codes (e.g., Eurocode 3 Part 1-9, "Fatigue" [16]), the detail category to which the element that failed due to fatigue belongs can be determined. Considering that the total cumulative damage at the end of the lab test was 100%, the detail category obtained was:

- Considering fully loaded trains: Detail category 71
- Considering empty trains: Detail category 63

Figure 3.8 gives a summary of the calculations carried out to obtain the detail category. It can be seen that, in both cases, the category obtained is for 100% cumulative damage.

3.5.2.2 Test on a Real-Scale Span of the Bridge

An isostatic span of the Ferrandet Bridge was selected and transported from the FGV installations in Campello (Alicante) to the ICITECH-UPV laboratories. Figure 3.9 shows a photo of the transport and the arrival of the isostatic span at the test center.

The same conditions were reproduced in the laboratory as in the bridge itself: (i) hinges that allowed movement around an axis at each end, and (ii) metal boxes containing steel ball bearings to allow longitudinal movement at one end only. Figure 3.10 shows the devices used in the test to support the bridge, while Figure 3.11 shows a general view of the bridge at the ICITECH-UPV laboratories.

3.5.2.2.1 Description of the Test

The test was defined with the help of computational models that faithfully reproduced the geometry and behavior of the bridge span. To reduce the test time, load increments of 1250 kN, the maximum load capacity of the hydraulic jack, were applied in the computational models. Figure 3.12 shows the results of the finite element models.

FIGURE 3.9 Transport from the FGV depot in Campello (Alicante) to the ICITECH-UPV (Valencia).

The entire test was continuously monitored by 40 strain gauges and eight LVDTs. Dynamic tests were also carried out (which consisted of placing accelerometers at the center of the span and applying an impact at the center of the load distribution system) at the beginning and end to determine whether the first vibrational mode of the structure had changed due to the hypothetical damage sustained. Figure 3.13 shows the positions of the 40 strain gauges (in red) and those of the eight LVTDs.

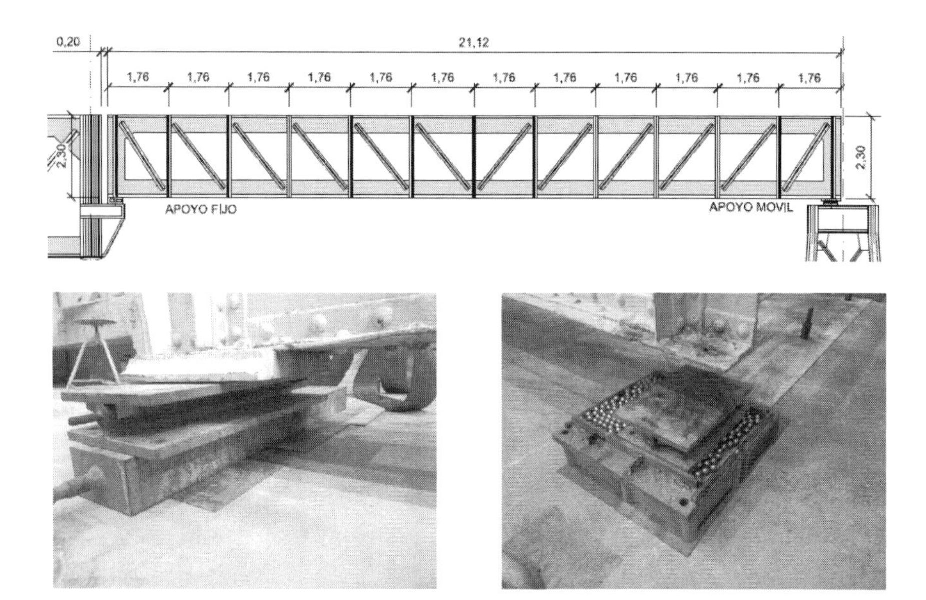

FIGURE 3.10 Bridge support devices used in the tests.

In the defined test conditions, 16,517 cycles were estimated to cause damage equivalent to 10 years of use, and since a total of 45,000 cycles were completed, the damage sustained in the test was equivalent to the damage foreseen for the next 27.2 years.

3.5.2.2.2 Test Results

Figure 3.14 shows an example of the stress increments in the four strain gauges for the first and last test cycles. It can be clearly seen that the structural

FIGURE 3.11 General view of the bridge at the ICITECH-UPV.

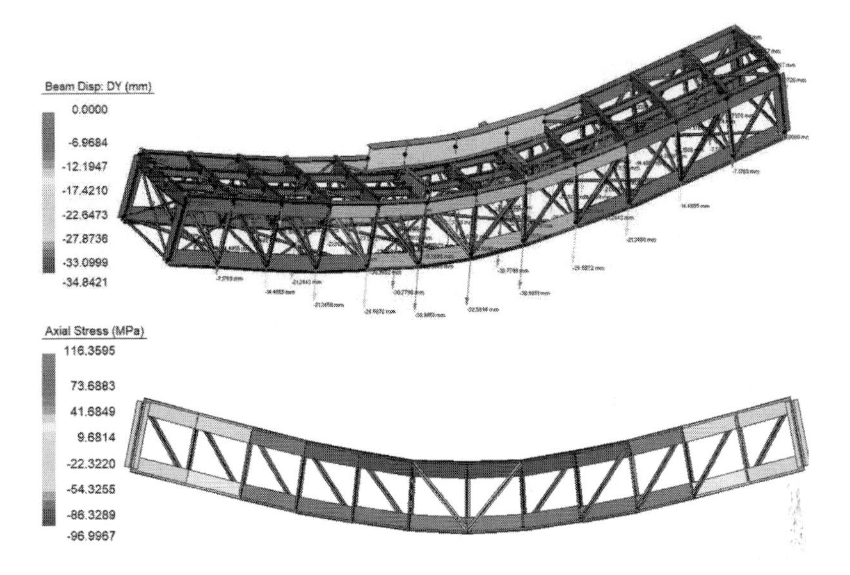

FIGURE 3.12 (Top) Vertical displacements and (bottom) axial stresses under a load of 1250 kN.

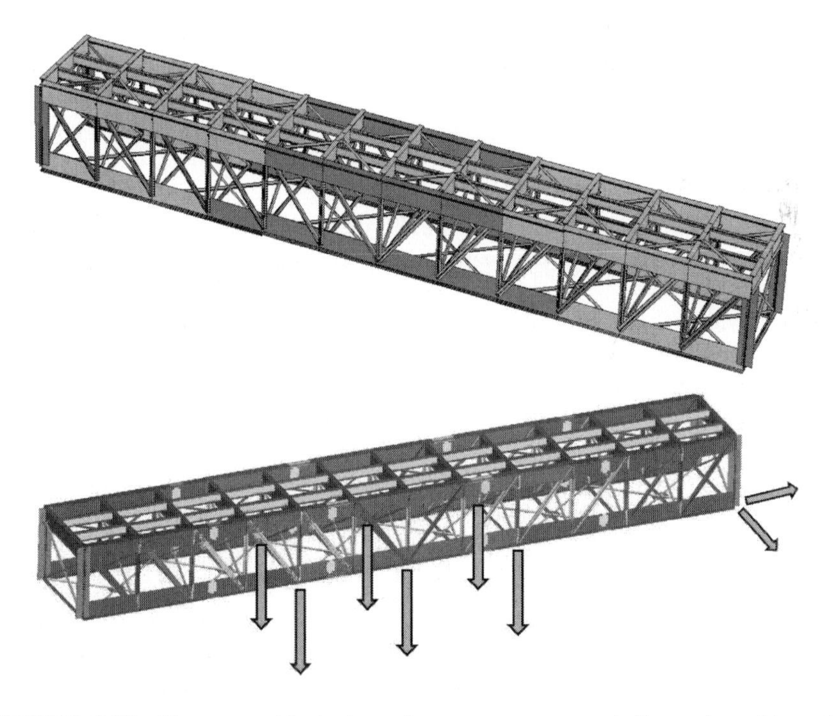

FIGURE 3.13 Elements with (top) strain measurements and (bottom) position of LVDTs. The arrows on the bottom picture represent the direction of the displacement measurements.

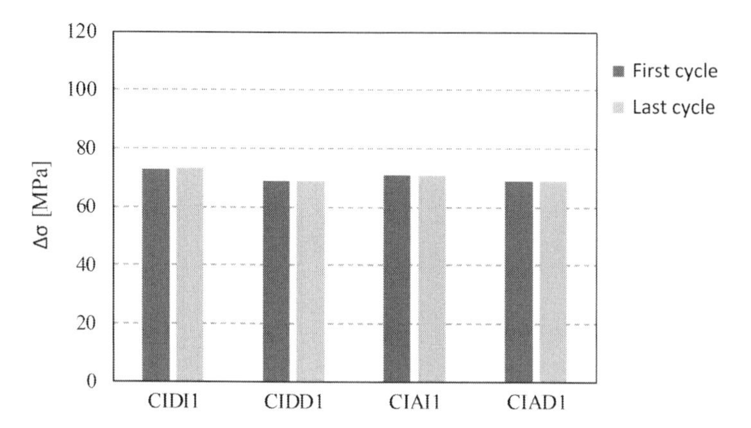

FIGURE 3.14 Example of stress increment measurements in the location of four strain gauges at the first and last cycles.

behavior at both times was practically identical. This was repeated in all the test sensors.

The test clearly showed elastic linear behavior, with no apparent damage and justified by:

- The stresses in the different elements are clearly lower than those of the yielding strength.
- Repetitive deformations and stresses registered from the first to the last cycle. The variations produced were lower than 5% in all cases and lower than 2.5% in 85% of cases.
- There were repeated vertical and horizontal displacements during the entire test; all the variations produced were less than 3%.

The dynamic tests showed a variation in the first structural vibrational mode ranging from 8.40 Hz at the start to 8.05 Hz at the end of the test. As a percentage, this means a reduction of the frequency that implies a slight rise in bridge flexibility of 4.2%.

The values obtained indicate a reduced variation in the overall bridge behavior after the intense fatigue test was carried out, while breaking point was never reached and no signs of weakness were detected in any of the principal bridge elements.

3.5.2.2.3 Remaining Life Based on Experimental Evidence

The tests showed that the most critical elements of the bridge are those that belong to the local distribution of the train loads, which reached fatigue failure. By means of the accumulated damage method, it was estimated that the remaining service life of these elements was around 10.7 years. The remainder of the principal structural elements (trusses) have longer remaining service lives, since

FIGURE 3.15 Computational models (top) for the trigger of a crack and (bottom) for predicting the cracking pattern in comparison with the real failure.

they did not reach failure in the tests that reproduced a damage level equivalent to 27.2 additional years of service.

3.5.2.3 Fracture Mechanics and Failure Patterns

In addition to the tests and simulations described so far, analytical studies and computational simulations were also carried out to evaluate and reinforce the conclusions obtained.

The advanced computational fracture mechanics models carried out on ABAQUS XFEM [17] provided similar conclusions to those obtained experimentally. For example, the numerical prediction established the start of a crack after 7530 cycles (experimentally between 10,000 and 15,000 cycles), and its propagation followed the same pattern as that obtained experimentally (see Figure 3.15).

3.6 PRACTICAL RECOMMENDATIONS FOR SHM

The study was amplified to define SHM recommendations. Up to now, we have described and analyzed the tests and numerical simulations to evaluate the structural response to fatigue stresses, in which only one beam reached failure. However, to allow for any existing uncertainties, there will always be some risk of the failure of an element either due to fatigue or for any other unidentified action. To define SHM recommendations, we opted to amplify the study to other failure cases that could involve structural elements. Since the bridge was available in the ICITECH-UPV test center, it was tested before and after the successive failures of up to two diagonals (see Figure 3.16). The study was amplified computationally to include the failure of other elements (top and bottom chords, vertical elements, braces, beams, etc.; see Figure 3.16). An exhaustive description of these tests can be found in Buitrago et al. [12] and Caredda et al. [13]. In combination with the tests described in the present chapter, these studies were used to establish a series of SHM recommendations.

For existing steel truss-type bridges that need to be controlled or are suffering from some problems (including problems of fatigue life) that could involve the

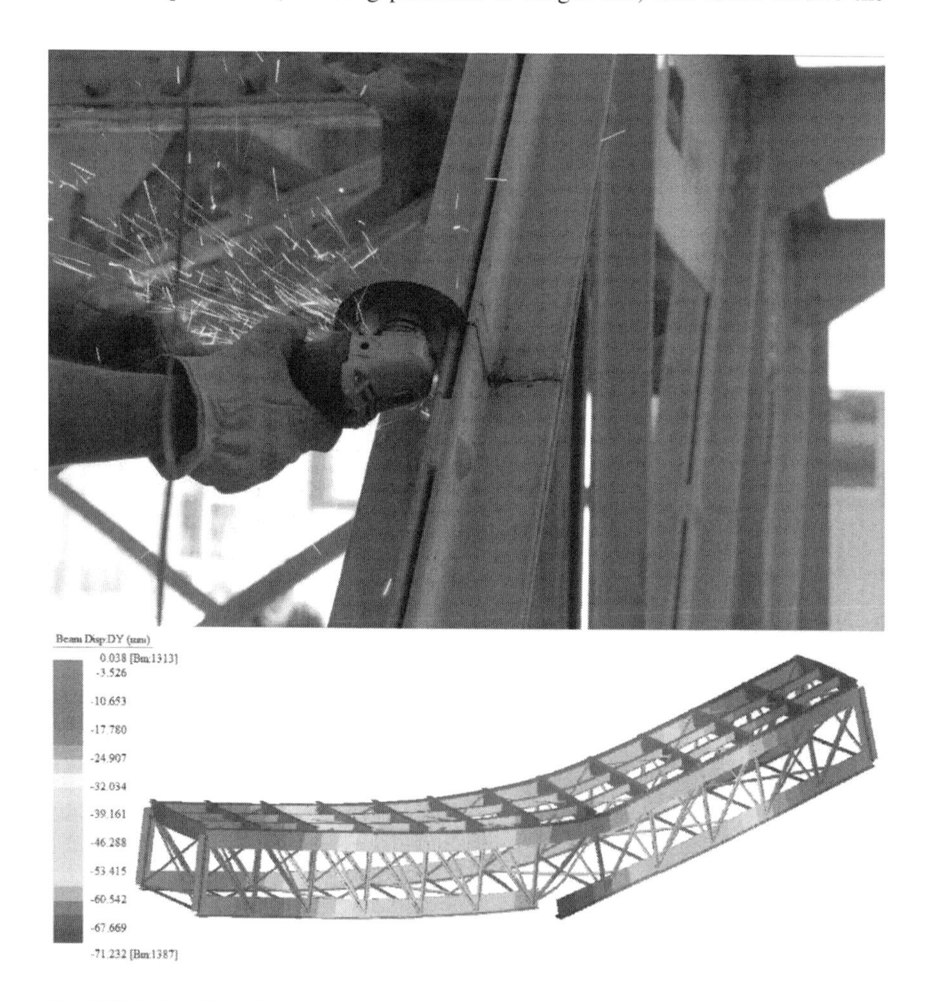

FIGURE 3.16 (Top) Experimental study and (bottom) numerical study of the influence of failed elements in the behavior of the structure.

failure of any structural element, it is recommended to monitor critical elements in real time to anticipate the complete failure of a local element that could initiate a progressive collapse. This last option can be carried out by means of strain, displacement, and acceleration sensors. Based on the results obtained, this section offers a series of recommendations for real-time structural monitoring similar to the bridge type studied here for early failure detection, including aspects such as: (i) parameters to be controlled, (ii) appropriate sensors for different parameters, and (iii) a specific location for each type of sensor.

Firstly, it is recommended to monitor element deformation by strain gauges. As found previously, the breakage or damage process of an element and its neighbors is reflected by its deformation. The strain sensors should be placed at points in the cross section at which strain is higher after or during the failure of an element. As can be seen from all the analyses carried out, the elements generally suffer significantly higher axial and bending stresses when failure occurs. Early detection of the failure of an element will be simpler if the strain sensors are placed far away from the center of gravity of the cross section and near joints to detect the increases in both axial and bending stresses. Table 3.1 gives a summary of all this information with recommendations for monitoring and early failure detection of the different elements.

Secondly, it is recommended to monitor deflections by any of the different methods, for example topography or LVDTs. As in conventional monitoring, it is generally enough to measure deflection at one point at mid-span, although early failure detection may require monitoring deflection at other points. Full monitoring can be carried out by measuring deflection at two points at mid-span and one each at a quarter and three-quarters along the length of the span (four extra points), as in the test described in the present study (see Figure 3.13). This is important to identify the site of the failure.

Finally, as can be extracted from the experimental and numerical results, accelerations can also be measured in a bridge to obtain the principal structural vibrational modes in real time. All types of structural anomalies can be reflected by small changes in the frequencies of the structure's main vibrational modes. It is recommended to install at least two accelerometers for the control of the

TABLE 3.1
Location of Strain Sensors

		Early Failure Detection
Element	General position	Additional details to be measured in case of failure
Chords	Section: Far from the center of gravity	Compression and tension increments in the top and bottom chords, respectively
Diagonals		Tension increments
Vertical Columns	Position: Close to a joint	Compression increments: In general, in the point of the element closer to the center of the bay

first vertical vibrational mode (one accelerometer at mid-span is sufficient). A more complete system would additionally include two additional accelerometers in each span, one in the middle of the span but on the opposite frame to follow the structure's possible torsional mode, with the other at a quarter or three-quarters along the length of the span to follow the second vertical vibrational mode.

3.7 CONCLUSIONS

This chapter summarizes the experimental and computational work carried out on the fatigue limit state of the Quisi Bridge (Alicante) by the CALSENS spin-off company of the Universitat Politècnica de València and the Concrete Science and Technology University Institute (ICITECH-UPV). The study included all the stages recommended in Ref. [14], from the preliminary inspection to an in-depth inspection and analysis. For this, we were able to count on the unique opportunity of having available the Ferrandet Bridge, identical to the bridge under study. This allowed us to carry out an ambitious series of experimental studies, including submitting an entire bridge span to fatigue tests in the lab. Computational tests and simulations were also carried out in which certain representative elements were removed from the structure to analyze the structural response and establish structural health monitoring (SHM) recommendations for the early detection and avoidance of structural failures.

From the tests carried out, the following conclusions can be drawn:

- The Quisi Bridge presents a local load distribution system with high accumulated fatigue damage. However, the overall behavior of the structure presents a bigger safety and structural redundancy margin, which allows attention to be focused mainly on the local load distribution system.
- The SHM recommendations provided here are substantially different from those usually recommended for normal structural control and are based exclusively on early detection of structural failures to avoid structural collapse due to fatigue or any other unidentified action.

The knowledge and recommendations extracted in this study are being applied to three real truss-type bridges in the region of Valencia, Spain, with more than 400 strain, temperature, displacement, and acceleration sensors in a real-time monitoring system with a series of automatically checked alarms. Further results, conclusions, and recommendations are expected to be extracted from this acquired experience and big-data analysis in the coming years.

ACKNOWLEDGMENTS

The authors wish to acknowledge their gratitude to the companies that entrusted this work to the ICITECH-UPV and Calsens: FCC, CHM, and CONVENSA. Special thanks to the FGV and Juan Antonio García Cerezo for all the facilities they gave us to successfully carry out the tests.

REFERENCES

1. Deng L, Wang W, Yu Y. State-of-the-Art Review on the Causes and Mechanisms of Bridge Collapse. J Perform Constr Facil 2016;30:04015005. https://doi.org/10.1061/(ASCE)CF.1943-5509.0000731.
2. Biezma MV, Schanack F. Collapse of Steel Bridges. J Perform Constr Facil 2007;21:398–405. https://doi.org/10.1061/(asce)0887-3828(2007)21:5(398).
3. National Transportation Safety Board. Collapse of the I-5 Skagit River Bridge following a strike by an oversize combination vehicle in mount Vernon, Washington May 23, 2013. Washington D.C.: 2014.
4. Birajdar HS, Maiti PR, Singh PK. Failure of Chauras Bridge. Eng Fail Anal 2014;45:339–346. https://doi.org/10.1016/j.engfailanal.2014.06.015.
5. Wardhana K, Hadipriono FC. Analysis of Recent Bridge Failures in the United States. J Perform Constr Facil 2003;17:144–150. https://doi.org/10.1061/(asce)0887-3828(2003)17:3(151).
6. Ghali A, Tadros G. Bridge Progressive Collapse Vulnerability. J Struct Eng 1997;123:227–231. https://doi.org/10.1061/(ASCE)0733-9445(1997)123:2(227).
7. Cha EJ, Ellingwood BR. Risk-Averse Decision-Making for Civil Infrastructure exposed to Low-Probability, High-Consequence Events. Reliab Eng Syst Saf 2012;104:27–35. https://doi.org/10.1016/J.RESS.2012.04.002.
8. Colajanni P, Recupero A, Ricciardi G, Spinella N. Failure by Corrosion in PC Bridges: A Case History of a Viaduct in Italy. Int J Struct Integr 2016;7:181–193. https://doi.org/10.1108/IJSI-09-2014-0046.
9. Zhuang M, Miao C. Fatigue Reliability Assessment for Hangers of a Special-Shaped CFST Arch Bridge. Structures 2020;28:235–250. https://doi.org/10.1016/j.istruc.2020.08.067.
10. Pipinato A, Pellegrino C, Bursi OS, Modena C. High-Cycle Fatigue Behavior of Riveted Connections for Railway Metal Bridges. J Constr Steel Res 2009;65:2167–2175. https://doi.org/10.1016/j.jcsr.2009.06.019.
11. Bertolesi E, Buitrago M, Adam JM, Calderón PA. Fatigue Assessment of Steel Riveted Railway Bridges: Full-Scale Tests and Analytical Approach. J Constr Steel Res 2021;182:106664. https://doi.org/10.1016/j.jcsr.2021.106664.
12. Buitrago M, Bertolesi E, Calderón PA, Adam JM. Robustness of Steel Truss Bridges: Laboratory Testing of a Full-Scale 21-Metre Bridge Span. Structures 2021;29:691–700. https://doi.org/10.1016/j.istruc.2020.12.005.
13. Caredda G, Porcu MC, Buitrago M, Bertolesi E, Adam JM. Analysing Local Failure Scenarios to Assess the Robustness of Steel Truss-Type Bridges. Eng Struct 2022;262:114341. https://doi.org/10.1016/J.ENGSTRUCT.2022.114341.
14. Kühn B, Lukic M, Nussbaumer A, Günther HP, Helmerich R, Herion S, et al. Assessment of Existing Steel Structures: Recommendations for Estimation of the Remaining Fatigue Life. vol. EUR 23252. Luxembourg: 2008. https://doi.org/10.1016/j.proeng.2013.12.057.
15. Ministerio de Fomento. Instrucción sobre las inspecciones técnicas en los puentes de ferrocarril (ITPF-05) 2005:8.
16. EN 1993-1-9. Eurocode 3: Design of steel structures. Part 1-9: Fatigue 2009.
17. ABAQUS. Abaqus, Theory manual 2019.

4 Sensor-Based Structural Assessment of Aging Bridges

Haris Alexakis, Sam Cocking, Nikolaos I. Tziavos, F. Din-Houn Lau, Jennifer Schooling, and Matthew DeJong

4.1 OVERVIEW

Transport infrastructure managers need to ensure longevity of their networks to meet pressing sustainability demands. Extending the operational life of complex structural systems, such as aging bridges, requires a comprehensive life-expectancy assessment. Given that these structures are suffering from local failures that may not necessarily alter their global response, engineers need to increase their confidence in detecting and characterizing such damage, while assessing deterioration rates in localized regions.

This chapter presents data analysis results from the structural health monitoring (SHM) of three aging bridges: two masonry arch rail bridges, and a half-joint concrete motorway bridge. The aim, in all cases, is to improve deterioration assessment through enhanced sensing of the distributed response across the structures. A core sensing technology used in the three schemes is the development of fiber Bragg grating (FBG) networks, allowing the study of small dynamic strain variations at both the local and global response levels. New ways of installing FBGs are explored for multi-aspect condition monitoring, while their sensitivity in damage detection is enhanced with data analytics and acoustic emission (AE) sensors. The chapter discusses that complementing information from dynamic strain and AE-sensing networks may enable a finer deterioration monitoring of aging structures driven by data.

4.2 INTRODUCTION

Infrastructure managers need to improve the performance and longevity of existing structures to reduce carbon emissions and meet increased sustainability demands. Bridges are vital transport assets where a serviceability concern could lead to prolonged travel and economic disruptions. While ensuring their resilience is essential for the public's safety and prosperity, there are significant uncertainties regarding their structural performance, especially for old bridges that were designed using poorly detailed standards and practices.

 DOI: 10.1201/9781003306924-4

In particular, national design standards for bridges were first introduced as late as the 1960s (BSI, 1962), while wider cross-national standardization efforts started in 1975 by the European Commission led to the publication of the first Eurocodes in 1984. The transition period of the 1960s–1980s marked an era of wide standardization of building codes, which improved the structural quality and performance of bridges while overlooking life-cycle management and sustainability aspects. This has raised concerns for the future of existing bridges, which deteriorate exponentially as they approach their design life expectancy, typically limited to 100–120 years.

Bridges constructed during the 1960s–1980s transition period, occasionally with poor practices, have already reached the second half of their expected operational life as defined in modern codes. Furthermore, a significant number of bridges still in use were constructed much earlier, without design life specifications, especially for railways. For instance, the number of masonry arch bridges and culverts in Europe was estimated to exceed 200,000, corresponding to more than 50% of the total railway bridge stock (Orbán, 2004). Most of these structures have been in operation for more than a century and experience significantly increased operational loads today.

Aging bridges designed with poor standards are of particular concern, as their true operational life expectancy is typically unknown. The vast number of aging structures in transport networks, their unknown structural details, the complexity of structural systems and deterioration mechanisms, and the relatively recent adoption of sustainability goals (UN, 2017) are just some of the reasons explaining the lack of standardization in infrastructure maintenance guidelines, as opposed to design standards. Currently, there are only a few examples of such international standards, which have been developed mostly by individual transport managers (see, e.g., UIC 778-4, 2009).

Moreover, there is an urgent need to deepen our knowledge of the performance of deteriorating structures and reduce the uncertainties regarding their behavior. This uncertainty and the absence of standardization have led to data protectiveness and lack of transparency (in most cases) between stakeholders, as well as poor availability of tools and regulations to support the development of collaborative frameworks that could prevent catastrophic structural failures, such as the Morandi Bridge collapse in 2018 (Calvi et al., 2019). Additionally, this has led to further barriers to the inclusion of new technologies in standards; such barriers further impede investment in novel digital maintenance technologies and delay infrastructure digitalization.

In this uncertain environment, infrastructure managers are searching for comprehensive structural assessment tools to optimize maintenance and operational decisions in their transport networks. Various techniques have been proposed to assess the structural condition of civil infrastructure. Noteworthy developments in the field of SHM include modal-based damage detection, which offers a promising alternative for remote asset management of flexible structures such as long-span bridges (Kaloop and Hu, 2015; Santos et al., 2016). Separately, a wide range of digital technologies for surveying ("from distance") and monitoring can now detect geometric anomalies and cracks, offering tools for rapid assessment (Acikgoz et al., 2017; Acikgoz et al., 2018a; Chaiyasarn et al., 2018). Nevertheless, most steel, concrete, and masonry infrastructure are stiff structures suffering from local deterioration that initially does not affect global performance or response. Masonry bridges suffer from localized failures (e.g., arch-ring separation/delamination, and spandrel–arch-barrel separation) that may induce, at a later stage, partial or global collapse mechanisms. Furthermore, internal reinforcement corrosion is the dominant deterioration mode of concrete bridges. For instance, 66% of deteriorating bridges in Japan are found to suffer mostly from chloride ingress, and 5% from carbonation/alkalinity decrease (Mutsuyoshi, 2001). What is common in all these damage modes is that (i) early detection is crucial to ensure bridge serviceability, public safety, and reduced repair cost; and (ii) such local/underlying damage does not necessarily manifest itself in the global measures of performance or at the material surface (De Santis and Tomor, 2013; Behnia et al., 2014; Alexakis et al., 2021) until the deterioration is extensive and often beyond repair. Hence, a much finer level of deterioration monitoring is necessary to assess the detailed performance state and deterioration rate of aging bridges and to intervene in a timely fashion.

This work presents data analysis results from three aging-bridge SHM projects: two masonry arch rail bridges and one concrete motorway bridge. In all three cases, the central idea is the development of high-sensitivity sensing networks able to provide information on multiple aspects of the structural response, distributed across large critical elements, such as masonry arch barrels or concrete half-joint supports. A core technology used in all bridges for the creation of these networks is FBG sensing, in which the fibers are externally attached via a system of aluminum clamps. New ways of installing FBGs are explored in the three projects to capture different structural response aspects under train or traffic loading, such as local crack behavior, three-dimensional (3D) arch-barrel deformation and identification of collapse mechanisms, out-of-plane arch "pumping," and dynamic deformation of half-joint supports. Identification of mechanical damage is enhanced by introducing new statistical modeling and machine learning for civil SHM to detrend the effect of small train loading and ambient temperature variations. Information regarding local material degradation is enhanced with AE sensing, which is shown to be a complementary sensing technique to strain-based monitoring, as it may provide critical information on the deterioration rate and the likelihood of imminent brittle failures. This chapter discusses how complementary sensing networks could enable data-driven damage assessment of complex aging infrastructure systems, reducing the uncertainties for infrastructure managers regarding the state and deterioration rate of their assets.

4.3 THE MARSH LANE VIADUCT

4.3.1 Sensing Network and Installation Novelties

The Marsh Lane Viaduct is located at the eastern entrance of the central railway station in Leeds, UK, carrying two electrified rail tracks. It was constructed between 1865 and 1869. Figure 4.1 shows the northern view of the investigated

FIGURE 4.1 Northern view of the Marsh Lane Viaduct in Leeds, UK.

section of the viaduct, comprising (from right to left) Arches 37, 38, and 39, while Figure 4.2 is the plan view highlighting the main visible cracks of the arch barrels.

The main damage is concentrated above the relieving arches of the piers, especially between Arches 37 and 38, due to a sagging mechanism that forces the relieving-arch keystone to descend and the pier walls to bow outward (Acikgoz et al., 2018b). This has caused bending cracks that developed along the longitudinal direction from the relieving arches up to the arch-vault keystones (Figure 4.2). Another longitudinal crack has developed in Arch 37 below the north rail track.

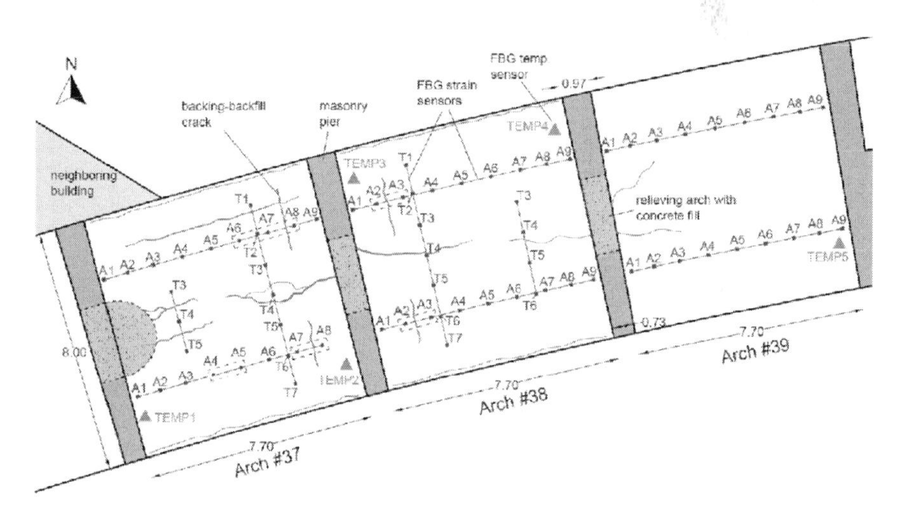

FIGURE 4.2 Plan view of Arches 37–39, showing the main damage and the fiber Bragg grating (FBG) sensor locations.

In addition, cracks in the transverse direction are observed at four symmetric locations around the pier between Arches 37 and 38 at the height where the rigid internal backing meets the backfill (Alexakis et al., 2019a). In an effort to arrest further degradation in these locations, Arches 37–39 were repaired in 2015 by the bridge owner, Network Rail, who filled in the relieving arches with concrete and installed steel ties in the transverse direction to confine the piers and spandrel walls.

Figure 4.2 indicates the installation of the FBG network. The small squares represent the aluminum clamps used to externally attach the fiber-optic cables, also shown in Figure 4.3 for Arch 37. The FBG strain sensors are located between the clamps at a spacing of 1 m, monitoring dynamic strain at a 1 kHz sampling rate (further sensors and system specs are provided in Alexakis et al., 2019b). The FBG arrays were installed in both the longitudinal and transverse directions to intersect perpendicularly with the main transverse and longitudinal cracks, respectively, and thus monitor their behavior. Steel cords with attached FBG sensors were installed between the arch springings, below the longitudinal FBG arrays, to monitor the pier-to-pier arch-span opening response.

Having two FBG arrays in each direction has allowed the study of the 3D dynamic deformation of the vaults during train loading, uncovering the main response mechanisms reported in Acikgoz et al. (2018b). Furthermore, installing FBG arrays in the longitudinal direction, below the two rail tracks, has been particularly useful for train-loading identification and classification (Alexakis et al., 2021).

FIGURE 4.3 FBG strain sensors in Arch 37.

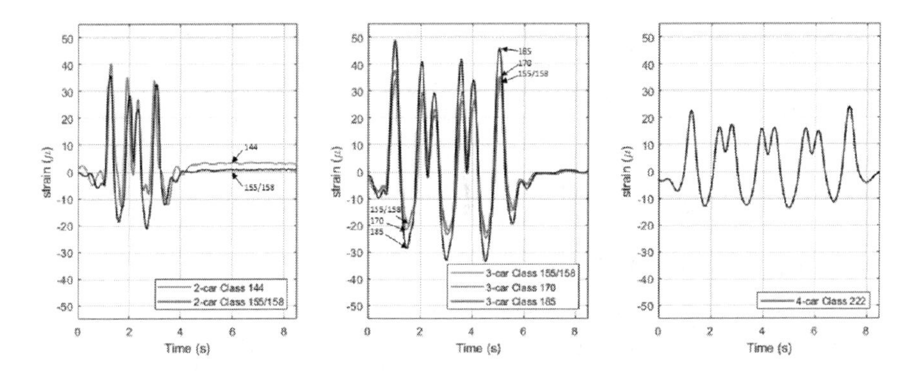

FIGURE 4.4 Response of FBG sensor 37NA5A6, located in the north FBG array of Arch 37, between clamps A5 and A6, for the most common passenger trains with two cars (left), three cars (center), and four cars (right).

For instance, Figure 4.4 presents typical responses of sensor 37NA5A6, which is located in the north FBG array of Arch 37, between clamps A5 and A6 (see Figures 4.2 and 4.3), under the most common passenger trains. In Figure 4.4, the number of positive peaks corresponds to the number of axles, which permits train classification based on the number of cars by means of peak detection analysis. Furthermore, having sensors at symmetric longitudinal locations in subsequent arch spans allows for cross-correlation of signals from different arches to identify time lags, and hence the train speed and direction (Alexakis et al., 2021).

This preliminary train classification allowed the study of statistical variations in the dynamic deformation of the bridge along the FBG network based on the same train loading, in particular the three-car passenger trains that represent around 50% of all data. The analysis indicated five locations where the strain response has been amplified over a period of two years, between 2016 and 2018. These locations are the four transverse cracks and south keystone area of the most damaged arch, Arch 37, indicated in dashed-line boxes in Figure 4.2. In all other locations, relatively uniform strain variations were observed, following seasonal temperature fluctuation (Alexakis et al., 2019c).

This analysis so far has not considered small train-loading variations within the same type of passenger train. For instance, Figure 4.4 (center) shows three different three-car passenger train classes that result in similar responses, which were hard to distinguish in practice. To do so requires a finer signal classification analysis, which would be able to detrend small response changes caused by train-loading and environmental effect variations, and would lead to enhanced mechanical damage detection.

4.3.2 Statistical Shape Analysis and Results

Our analysis proceeds with a dataset that consists of $M = 1151$ three-carriage train passage events heading east that occurred between July 2016 and March 2019. All events have dynamic strain and temperature data from the FBG network, apart from the first 31 events in July 2016 (preliminary dataset) where FBG temperature sensors were not available; for those events, the temperature was taken from the Leeds Weather Archive of the National Centre for Atmospheric Science database (Alexakis et al., 2021).

In order to analyze the bridges' behavior at a sensor location, a method is required to compare train passage events. In our original paper (Alexakis et al., 2021), a modified procedure of ordinary Procrustes analysis (OPA), a key transformation in statistical shape analysis (Dryden and Mardia, 2016), is used to transform one *shape* of dynamic strain values into another. A shape is described by a finite number of points called landmarks (see Figure 4.5) for an example train passage event with landmarks. The modified OPA approach transforms one set of landmarks into another. More precisely, the transformation from landmarks X (a matrix) into Y (another matrix) is given by minimizing

$$D^2(X,Y) = \left\| Y - \beta \circ X - 1_{15}\gamma \right\|^2 \tag{4.1}$$

over $\beta = (\beta_1, \beta_2)$ and γ, where $\|X\| = \sqrt{trace(X^T X)}$, and $\circ$ is the Hadamard product. Minimizing Eq. (4.1) is straightforward and, more importantly, leads to interpretable results.

To assess changes in the bridges' response at a particular sensor location, all $M = 1151$ events are transformed into the first chronological event using Eq. (4.1). This transformation gives the set of estimates $\left\{ \hat{\beta}_1(j), \hat{\beta}_2(j), \hat{\gamma}(j); j = 1, \ldots, M \right\}$. The estimates have the following interpretation: $\hat{\beta}_1(j)$ is the amount of time-scaling required to map the first chronological event onto the jth train passage event. Similarly, $\hat{\beta}_2(j)$ is the amount of strain-scaling required to map the first

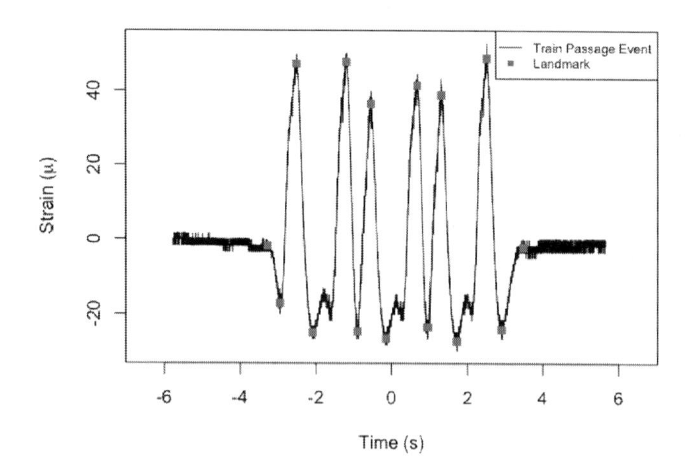

FIGURE 4.5 Dynamic strain readings from a train passage event from sensor location 37NA6A7, along with landmarks.

chronological event onto the jth event. (The parameter γ and its estimate are not of interest in this application.) Thus, the estimated values allow us to compare the shape in FBG signals relative to the first train passage event record at a given sensor location. Comparing dynamic strain values from train events using this modified OPA approach is novel. Typically, events are summarized by a single extreme value such as the maximum strain value, as in previous studies (Alexakis et al., 2019a, 2019b, 2019c). The modified OPA approach uses the strain values from the landmark locations, giving a broader view of the shape of the dynamic strain response. Moreover, the parameters from the modified OPA approach have meaningful interpretations of direct import to SHM. In particular, $\hat{\beta}_1$ represents variation in train speed, and $\hat{\beta}_2$ variation in strain amplitude.

A plot of the estimate strain-scaling values, $\hat{\beta}_2(j)$, against temperature (see Figure 4.6, left), reveals two patterns: as temperature increases, the $\hat{\beta}_2$ value generally decreases; and the magnitude of the $\hat{\beta}_2$ value depends on some latent groups in the data that are suggested by partitioning lines in Figure 4.6 (left). Note that the preliminary 31 events in 2016 have estimated temperature records, and that $\hat{\beta}_1(j)$, relating to train speeds, did not show any pattern over time with temperature or with $\hat{\beta}_2$.

The patterns, presented in Figure 4.6 (left), are captured and predicted using a statistical model, which can be used for damage detection. In order to develop a representative model, the latent groups first need to be labeled. This is achieved using a supervised classification method called a support vector machine (SVM) (Cortes and Vapnik, 1995). The latent groups correspond to a large extent with the different train classes.

First, a SVM is used to classify the two classes of train, (i) Class 185 (heavier, more frequent trains) and (ii) Class 155/158 or 170 (lighter, less frequent trains), using the $\hat{\beta}_2(j)$ and temperature value of each train passage event. The preliminary dataset was excluded since no temperature records exist. The SVM is trained with a limited number of datapoints from the two train classes based on site

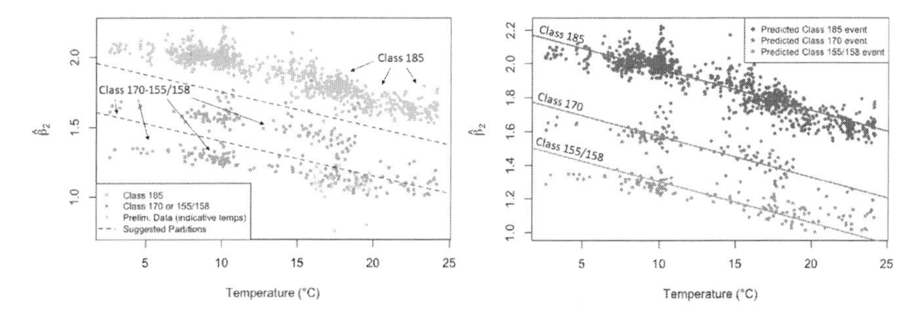

FIGURE 4.6 (Left) Strain-scaling values, $\hat{\beta}_2(j)$; $j = 1, \ldots, M$, against temperature. Suggested groups of data are indicated by the separating partition dashed lines. (Right) Predicted train class labels given by a support vector machine (SVM; dots), along with expected strain-scaling values (solid lines) given by a linear model.

spotting (Alexakis et al., 2021). The trained SVM then predicts the class labels of the remaining datapoints. This approach predicts the class of train with high accuracy—with only 12 incorrectly predicted labels out of 1116 test labels. The SVM approach offers an improved, data-driven approach to classifying trains without the need for "manual" labeling. Next, the SVM procedure is performed using three classes, as the pattern presented in Figure 4.6 (left) suggests three latent groups. The SVM predicted classes are illustrated in Figure 4.6 (right). The three groups are suspected to correspond to the three types of train class—(i) Class 185, (ii) Class 170, and (iii) Class 155/158—which is in accordance with site spotting.

The statistical relationship between $\hat{\beta}_2$ with the temperature and train class is modeled using a linear model. The linear model uses the temperature records and the class of train (one of the three classes predicted by the SVM) as inputs. The preliminary data is not used to fit this model. The expected $\hat{\beta}_2$ value from the fitted linear model is illustrated in Figure 4.6 (right). This model explains the data very well (e.g., its R^2, a goodness-of-fit measure, is 95%).

The fitted linear model can be used to monitor deterioration at a sensor location using the following procedure:

1. Divide the full dataset into two separate sets: a training and test dataset.
2. Fit a linear model to the $\hat{\beta}_2$ values in the training dataset using the temperature and train class as input.
3. Compare the $\hat{\beta}_2$ predictions from the fitted linear model to the $\hat{\beta}_2$ values in the test dataset.

As an example, in Step 1 take the training set as the main set of data from the permanent FBG installation and the test set as the preliminary dataset. This selection was based on the preliminary observations by Alexakis et al. (2019a, 2019b, 2019c) that in five bridge locations, including 37NA6A7, the peak-to-peak amplitude of the dynamic strain response (e.g., Figure 4.5) seemed to have been permanently increased between July 2016 (preliminary dataset) and the permanent sensing installation starting from November 2017 through today.

In Step 2, the train classes are given as one of three classes predicted by SVM. Here, we select the most frequent three-carriage passenger train, Class 185.

In Step 3, the differences between the predicted and the test $\hat{\beta}_2$ values indicate changes in the dynamic deformation over time, by factoring out any seasonal effect and, in this case, the ambient temperature.

The results of this procedure implemented separately at the two symmetric quarter-span locations 37NA6A7 (backfill-backing crack) and 37NA3A4 (no visible deterioration) are presented in Figure 4.7. At sensor 37NA6A7 (Figure 4.7, left), the clear separation between the predicted and the preliminary $\hat{\beta}_2$ values suggests a change in structural behavior. This change corresponds to a peak-to-peak dynamic strain amplification of 32 $\mu\varepsilon$. Note that the resolution of the y-axis is high enough to detect much smaller strain variations (e.g., below 5 $\mu\varepsilon$),

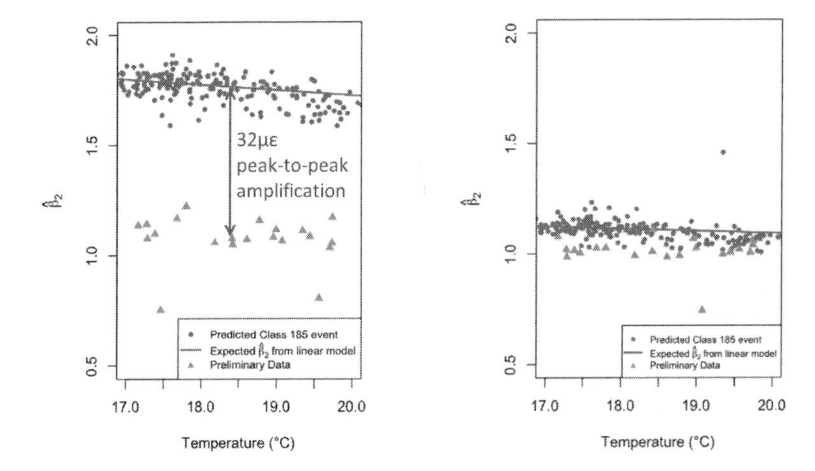

FIGURE 4.7 Comparison between the $\hat{\beta}_2$ estimates for the 2016 preliminary events and the predicted Class 185 curve (solid line given by the linear model) at sensor locations (left) 37NA6A7 and (right) 37NA3A4.

which correspond to the dynamic strain variation due to a typical daily temperature fluctuation of only ~10 °C (Alexakis et al., 2019c). Conversely, at 37NA3A4 the deviation is small, indicating little evidence of any structural change, which is the case for all sensors, apart from the five locations indicated in Figure 4.2 with dashed boxes. Note that some variation of the datapoints around the fitted curves is expected due to the stochastic nature of the data due to, for example, a varying number of passengers or unrecorded seasonal variations.

The entire procedure involves: comparing events using a modified OPA approach, classifying train types using SVM, and monitoring deterioration by investigating changes in $\hat{\beta}_2$ values (strain amplification). This procedure provides a novel data-driven pipeline from sensor network data to SHM information.

4.4 THE CFM-5 SKEWED ARCH BRIDGE

4.4.1 Sensing Network and Installation Novelties

The CFM-5 bridge is a single-span, skewed masonry arch railway bridge in North Yorkshire, UK. As shown in Figure 4.8a, it carries two lines of rail traffic over a highway. CFM-5 was built in 1868, primarily using stone masonry blockwork, although its arch barrel is brickwork, laid helicoidally. The main damage is highlighted in Figure 4.8b and consists of separation cracks between the arch and both spandrel walls, as well as a longitudinal crack on the east side of the arch approximately aligned with the centerline of the south rail track. This damage was addressed in a recent repair in 2016, in which the cracks were stitched at regular intervals and 10 tie bars were installed through

FIGURE 4.8 (a) Elevation view of the CFM-5 bridge and (b) overview of the main cracking damage alongside indications of the recent repair work.

the bridge in its transverse direction. A key goal was to halt growth of the separation cracks and restore connectivity between the spandrels and arch, to reduce the magnitudes of dynamic arch movements in response to trains passing over the bridge. While this has succeeded, the duration over which the repairs will remain effective is unclear. This, in turn, has motivated SHM, although the project has also been used as an opportunity to evaluate a range of sensing technologies for broader application in masonry arch bridge monitoring (Cocking et al., 2019a).

Alongside other technologies (Cocking et al., 2019a), distributed arrays of FBGs were installed on the arch intrados of the CFM-5 bridge using aluminum clamps. Two novel methods were devised to implement these FBGs: (i) rosettes measuring dynamic principal strains, and hence the flow of force, through the arch barrel; and (ii) "FBG pairs" capturing multidimensional movements across the separation cracks. The full FBG installation is reported in Cocking (2021), which also describes complementary analyses leveraging videogrammetry and advanced laser scan analysis, alongside the FBG data.

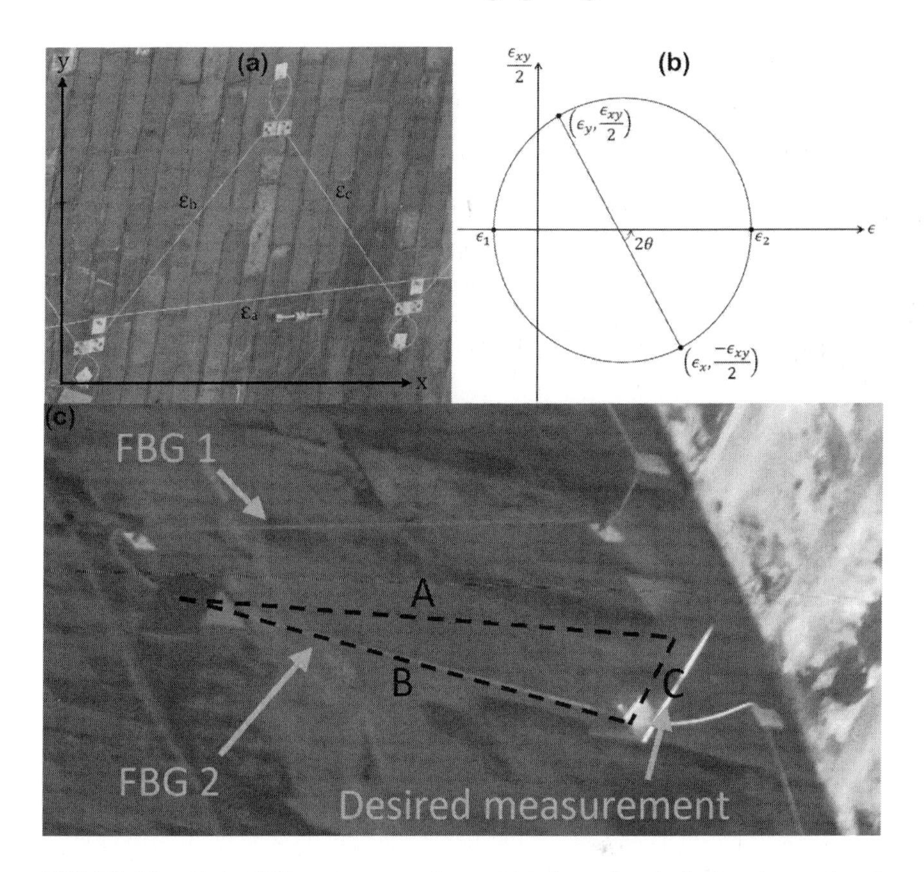

FIGURE 4.9 (a) An FBG strain rosette, from which dynamic principal strains are found using (b) the local Mohr's circle of strain, and (c) an "FBG pair" measuring both in-plane (crack opening) and out-of-plane (shear) movements across a crack.

As in Figure 4.9, FBG rosettes measure a local strain state $\epsilon_{x,y}$, which can be converted, using the Mohr's circle of strain, to find the magnitudes and orientations of principal strains $\epsilon_{1,2}$ (details in Cocking et al., 2019b; Cocking, 2021). *FBG pairs* consist of one FBG directly measuring in-plane crack opening, alongside a second that captures both in-plane and out-of-plane (shear) movements. Considering the geometry of the A-B-C triangle in Figure 4.9c, at each instance of time, dynamic shear deflections can be found (Cocking et al., 2021).

4.4.2 Analysis and Results

The FBG pairs have revealed an asymmetric distribution of dynamic crack-opening displacements that is consistent for common groups of train loading. The high sensitivity of FBGs enables precise characterization of distributed responses, such as these crack displacements and the in-plane arch strains, as well as the quantification of response sensitivity to external variables such as train speed,

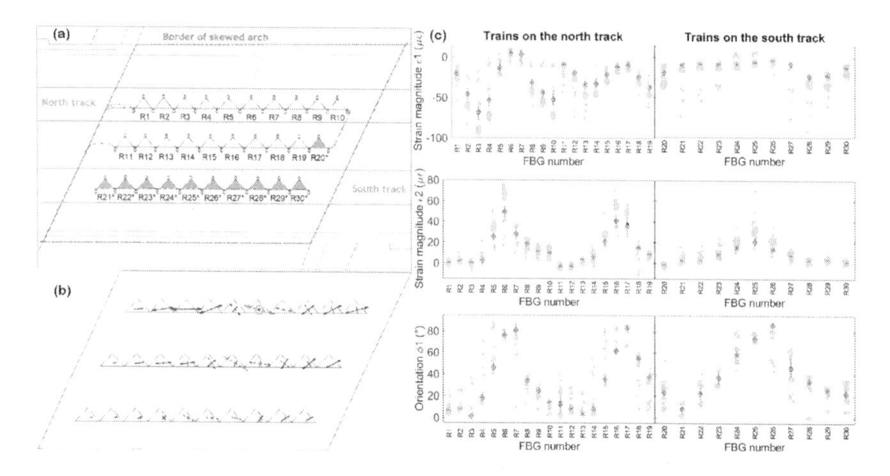

FIGURE 4.10 (a) FBG rosette installation plan for the CFM-5 bridge; (b) median principal strain distribution measured when the critical axles of TransPennine Express trains reach the arch crown (small circle symbol), with tensile and compressive principal strains shown in dashed lines and solid lines with arrow symbols , respectively; and (c) boxplots showing the statistical distribution of principal strains ϵ_1 and ϵ_2 and orientation ϕ_1, for this loading instance.

precise applied loading, and ambient temperature (Cocking, 2021; Cocking et al., 2021; Cocking and DeJong, 2022).

Figure 4.10a shows the installation locations of the FBG rosettes R1 to R19, which are aligned with the northern rail track and longitudinal bridge centerline. Rosette R20 was damaged during installation and not ultimately used. Rotational symmetry in the arch response (demonstrated in Cocking, 2021) allows for data from rosettes R1 to R19, measured for trains traveling on the south track, to be used to infer the response at "imaginary" rosettes R20* to R30* when trains pass on the north track. In this way, distributions of the principal strains measured by FBG rosettes can be mapped across the extent of the arch barrel, to visualize the dynamic flow of force during train-loading events.

Figure 4.10b gives an example that presents the median distribution of instantaneous principal strains that are measured when the critical axles of TransPennine Express trains reach the FBG just to the east of the arch crown (marked with a small circle symbol in Figure 4.10b). A band of tensile (dashed lines) principal strains is observed to occur around this loading location, suggesting that the thrust line in this region of the arch is shifted toward the extrados. Compressive (solid arrow symbols) tensile strains are predominant outside of this region; this is consistent with the anticipated thrust line moving toward the intrados in these locations, particularly on the opposite side of the arch to the loading location. In general, the principal strains are approximately aligned with the skewed span direction, although in some places such as the northeastern (top-right) quadrant of the bridge they are oriented toward the acute corner of the skewed arch.

Figure 4.10c uses boxplots to show the statistical distribution of principal strains for the above loading instance. Data from 1636 TransPennine Express trains has been included in this analysis, with 50% of these traveling on each of the northern and southern tracks. Plotted in dark blue, the dots in circles indicate median responses, thick bars represent the interquartile ranges (e.g., see ϵ_2 for rosette R17), and thin "whiskers" show the data ranges excluding outliers. Outliers are plotted in cyan asterisks. The interquartile ranges are often too small to be visible, as the principal strain distributions are highly consistent for the majority of data. This indicates a stable response under repeated, common applied loading.

It is hypothesized that damage, giving rise to local loss of stiffness, will manifest itself in the principal strain data as a permanent change in the force flow distribution. As already noted, FBGs are sensitive enough to detect damage at an early stage. Therefore, this FBG rosette system offers the means to track changes in the damage state of this bridge over time, as incipient or progressive damage occurs.

To the same end, Alexakis et al. (2020) conducted a parallel experimental study to the field monitoring of the two aging railway bridges described in Sections 4.3 and 4.4, to describe the AE behavior of masonry under cyclic train loading. The study shows that AE sensors can capture increasing AE-energy data trends prior to local brittle failures, such as brick/mortar crushing, joint sliding, and diagonal shear failures, which may not be reflected in the strain responses. This suggests that AE and FBG (dynamic strain) are complementary sensing techniques, with the potential, if combined, to provide a more comprehensive structural assessment at both the local and global response levels. This idea has been the main motivation for the sensing deployment described in the following section.

4.5 REINFORCED CONCRETE HALF-JOINT MOTORWAY BRIDGE

Half-joints are designed with a reduced section depth at the ends and were particularly favored for car parks and bridges owing to a simplified design process and their suitability for precast construction. Despite the advantages, it has been found that this type of joint often deteriorates due to water seepage, resulting in water stagnation in the nibs and ultimately corrosion of the reinforcement (Desnerck et al., 2016). In addition, in many cases they are particularly difficult to inspect as the inner part is not designed to be accessible. In the UK, the National Highways has implemented a dedicated management and inspection regime for half-joints to ensure risk mitigation and sufficient maintenance practices for deteriorated bridge decks with half-joints (English Highway Structures & Bridges Inspection & Assessment, 2020). Within this context, a reinforced concrete half-joint has been selected for monitoring purposes to evaluate the effectiveness of AE testing on damage detection and condition assessment during short- and long-term field campaigns, which is complemented by environmental data and FBG-based dynamic strain sensing.

4.5.1 SENSING NETWORK AND INSTALLATION NOVELTIES

The selected half-joint is within the Strategic Road Network (SRN) in England and is located on a Class A motorway carrying traffic in four lanes, with a width of 15 m in each direction and a total width of approximately 40 m. Pictures of

FIGURE 4.11 (a) Side elevation of the bridge, and (b) traffic on the northbound direction.

the bridge incorporating the half-joints are shown in Figure 4.11, taken during an inspection in December 2019. Access to the bearing shelf is limited on the monitored half-joint, which is under an annual visual inspection regime by the asset operator. Its current condition is considered to be satisfactory (personal communication, asset operator).

The half-joint has been instrumented with a monitoring system comprising AE sensors, FBG strain and temperature sensors, and a weather station. The sensors were installed across the whole width of the half-joint covering both north- and southbound directions. A total of 24 piezoelectric AE sensors of the type PK6i were installed along the nib and the sides of the half-joint. The sensors have a resonant frequency of 55 kHz and an integral preamplifier of 26 dB. Behnia et al. (2014) suggested that this type of sensor has been shown to perform well on concrete bridges. The AE sensors form a linear array with a distance of 2 m between them, whereas the last two sensors were installed on the east side of the joint. A two-part rapid-setting resin was used for acoustic coupling, and a bolted clamp was used to ensure the sensors remain on the concrete surface. The array configuration was found to be appropriate to monitor the entire width of the bridge, taking into account the acoustic signal attenuation within the concrete body following calibration studies. Data is acquired continually with a sampling rate of 1 MHz using the commercially available Sensor Highway III (SH-III) system.

A Vaisala WXT530 series 6 weather station was installed and connected to SH-III to monitor weather parameters such as air temperature, rainfall accumulation, and humidity. Four FBG-cable arrays of 20 sensors each were installed along the half-joint. Three arrays were attached on the concrete surface using aluminum clamps to monitor strain in the longitudinal, transverse, and vertical directions (corresponding to all three spatial directions) in a 2 m spacing along the half-joint. A fourth array was split in two with a 20 m extension and was attached on the east and west sides of the joint. All the arrays were pre-tensioned to a minimum of 500 $\mu\varepsilon$. A sampling frequency of 50 Hz was found appropriate for monitoring. Each group of FBG sensors in three directions was installed adjacent to an AE sensor to ensure that AE data and strain can be correlated. A plan view of the locations of the AE and FBG sensors is shown in Figure 4.12a, along with the locations of AE sensors on the east side (Figure 4.12b).

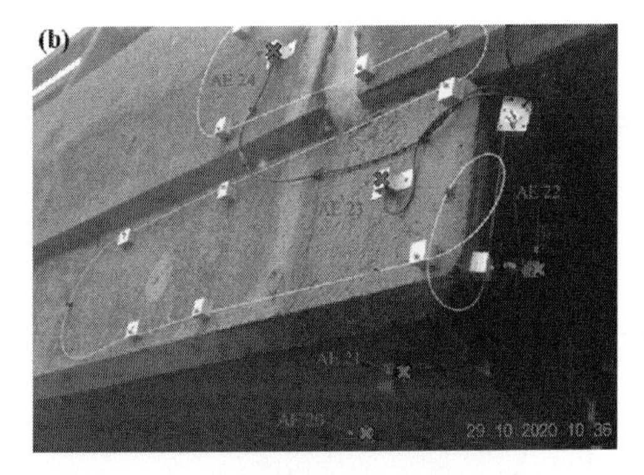

FIGURE 4.12 (a) Locations of AE and FBG sensors, and (b) AE and FBG sensors on the half-joint.

Following successful installation of the AE sensors, a comprehensive calibration protocol was followed to ensure that the acoustic coupling between the sensors and the concrete surface is sufficient and also to determine the background noise, in order for an appropriate AE detection threshold to be set. The calibration was carried out during normal operating conditions using a Hsu–Nielsen source as defined in EN 1330-9 (2017).

To verify the sensor sensitivity, pencil lead breaks were performed next to each individual sensor. Subsequently, pencil lead breaks were carried out at different locations and directions between sensors to obtain the attenuation of the elastic waves within the material. A series of calibration tests was performed involving pencil breaks between consecutive AE sensors. Attenuation was investigated in the longitudinal and transverse directions for sensor AE 2 at fixed intervals of 50, 250, 500, 1000, and 2000 mm (Figure 4.13a). The attenuation across the joint (vertical direction) was studied by performing center punches at the drop in span. Lead breaks are shown in Figure 4.13b, where absolute energy is plotted against the recorded peak amplitudes. Such lead breaks are events of particularly high energy, imitating concrete cracking. Figure 4.13c shows the amplitude decay in the longitudinal directions against distance from sensor AE 2. The indicative pencil breaks between sensors AE 1, 2, 19, and 20 are also shown (Figure 4.13d and 4.13e).

4.5.2 Analysis and Results

The overall strain distribution from the vertical FBG sensors is shown in Figure 4.14. Each line in the boxplot corresponds to the median value of the peak strain recorded per 5 min interval for September 2020 at each sensor location, whereas the box limits define the 25th and 75th percentiles. The corresponding sensor locations are given in Figure 4.11a. It is shown that the strains are distributed symmetrically. Lane one (Figure 4.11a, highlighted in light blue) carries the heaviest traffic, with similar average peak strains recorded for both traffic directions.

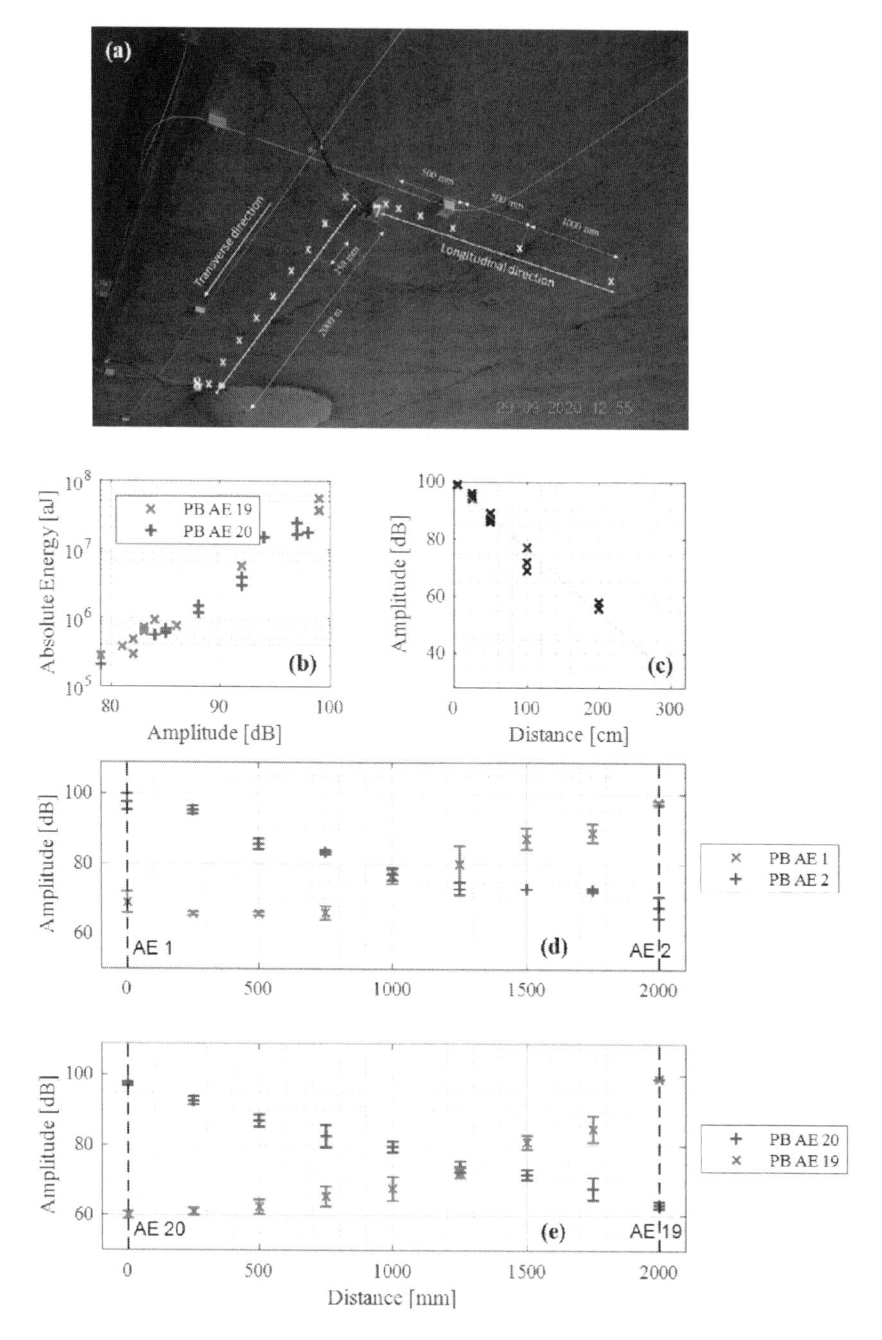

FIGURE 4.13 Calibration of acoustic emission (AE) sensors. (a) Locations of transverse and longitudinal pencil breaks, (b) indicative correlation between absolute energy and amplitude, (c) longitudinal attenuation from sensor AE 2, (d–e) and indicative pencil breaks between sensors AE 1, 2, 19, and 20.

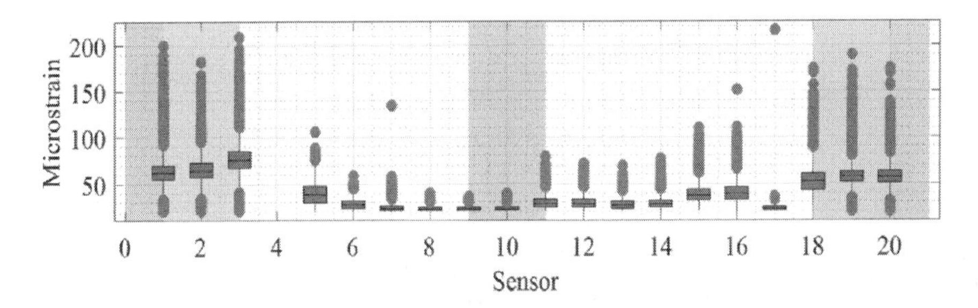

FIGURE 4.14 Boxplots of strain data for each FBG sensor from array 3 during September 2020. Blue regions indicate the lanes with heavier traffic.

For in-service bridges, recorded AE hits are emitted due to traffic, noise, and/or deterioration processes, given that the bridge is continually under load, which is the fundamental principle for acoustic waves being emitted. Hence, one needs to distinguish between different sources, although preliminary findings can be extracted by looking at the total number of AE hits recorded from each sensor. Taking into account the calibration tests, the AE data was pre-processed by removing AE hits with a peak amplitude lower than 59 dB. Furthermore, only hits with counts larger than five and less than 10,000 were considered. This further increase of the threshold was justified from the attenuation data, where the minimum amplitude recorded at the maximum distance between two sensors was 60 dB. In addition, all hits with zero energy were also removed, as this commonly represents spurious data rather than useful data (Tziavos et al., 2020). It was found that the majority of hits with zero energy were lower than 60 dB.

The AE activity on the half-joint expressed by the total number of recorded hits is given in Figure 4.15. The presented data, collected during September 2020 (Figure 4.15a), confirms that AE activity is dependent on the stresses applied on the bridge, as expected. It is also apparent that the AE sensors installed underneath the first lanes in each direction recorded a higher number of hits, indicating that larger stressors on the half-joint were responsible for the higher activity. Although AE hits are consistently higher in the same locations for both northbound and southbound directions, for the latter it is shown that a significantly higher number of hits were recorded, particularly for AE 19.

In Figure 4.15b, the number of hits recorded by each sensor for the first four months of acquisition is illustrated. A total of 17 days of continuous monitoring for each month are included. The findings from September are further confirmed for the subsequent months, indicating that the AE activity is significantly higher on the first two lanes of the southbound direction. This is despite the fact that the recorded peak strains from the FBG sensors are consistent between the two lanes. A further deduction from these two figures is that AE can be also used for short-term monitoring campaigns in order to provide structural engineers with a preliminary analysis of the activity on the structure under in-service conditions, as areas of higher AE activity are commonly associated with deterioration mechanisms as the environmental and traffic influence on both lanes is found to be similar.

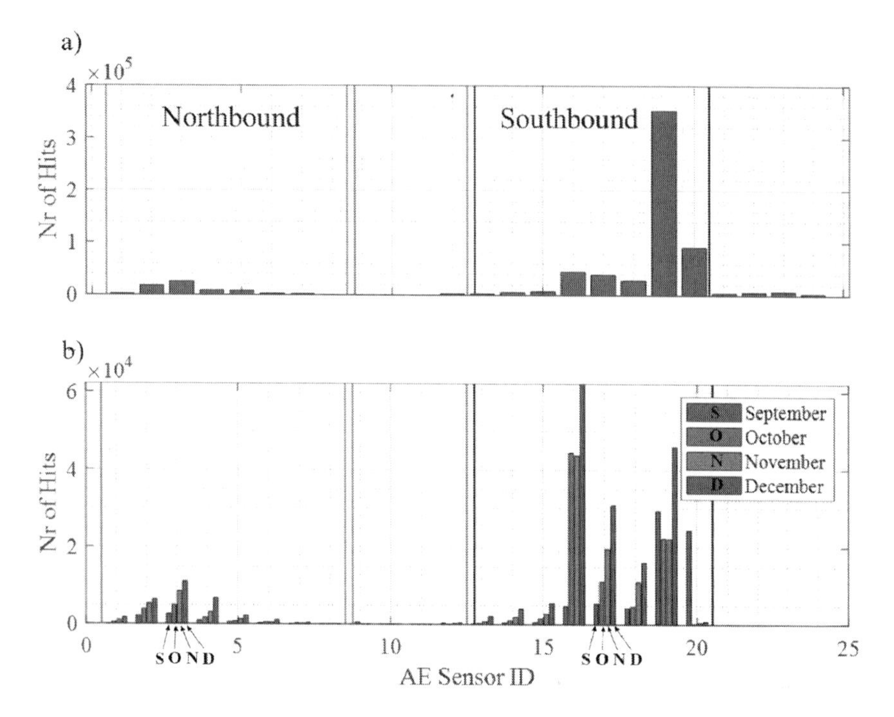

FIGURE 4.15 Number of hits per channel along the AE array for northbound and southbound directions for (a) September and (b) September–December.

4.6 CONCLUSIONS

This chapter summarizes results from three aging-bridge monitoring projects. In all cases, the sensing networks installed cover large areas with high densities of sensing points, monitoring multiple aspects of the structural response at both the local and global levels.

FBG networks are installed in different configurations with the use of aluminum clamps. This technique has been proven to be robust during long monitoring periods (six years so far) and allows versatility in installation. In particular, the technique allows: (i) flexibility in choosing the in-plane sensing direction (e.g., to adjust to a particular cracking pattern or imminent failure mechanism), (ii) the formation of consecutive strain rosettes (e.g., to study the force flow within arch barrels), (iii) out-of-plane displacement monitoring (e.g., to study arch "pumping"), and (iv) altering the sensing physical parameter per FBG location (i.e., to choose between dynamic strain and ambient temperature).

Furthermore, FBG data from selective locations can be used for loading classification (e.g., identification of train speed, class, and direction). When statistical shape analysis was enhanced with a support vector machine, small changes in response due to slight loading and temperature variations could be distinguished, showing potential to detect mechanical deterioration that alters locally the dynamic strain response below any thermal effect.

Analysis of the AE field data of the concrete bridge has identified locations of higher AE activity compared to other symmetric locations of similar loading and dynamic strain response, as shown by the FBG data. This indicates that even when FBG data may not detect a difference in strain, complementary AE sensing may help in detecting underlying deterioration, though this requires further research for validation. Experimental studies on the AE behavior of masonry under cyclic loading, which were conducted in parallel with the field installations, confirm the detection of data trends that are not necessarily shown in strain data, especially for brittle failure modes. On the other hand, while AE sensing alone may enhance the information locally, without FBG sensing it is difficult to interpret. FBG and AE sensing seem to be highly complementary techniques, and creating unified systems that integrate statistical modeling may enhance further the damage assessment of complex deteriorating systems.

ACKNOWLEDGMENTS

The Marsh Lane bridge-monitoring project has been funded by the Lloyd's Register Foundation, the Engineering and Physical Sciences Research Council (EPSRC), and Innovate UK through the Data-Centric Engineering programme of the Alan Turing Institute and through the Cambridge Centre for Smart Infrastructure and Construction (CSIC). Additional funding for IT equipment was provided to the first author by the School of Engineering & Applied Science and the Aston Institute of Urban Technology and the Environment (ASTUTE), Aston University (budget codes 10068 and 10630). The CFM-5 bridge-monitoring project has been funded by Network Rail (NR) and EPSRC via the Doctoral Training Partnership (grant no. EP/M506485/1). The half-joint bridge-monitoring project has been funded by Highways England (HE).

The monitoring installation in all three bridges was performed with assistance from the CSIC, supported by EPSRC, Innovate UK, and industry funding (including grant nos. EP/I019308/1, EP/K000314/1, EP/L010917/1, and EP/N021614/1).

The authors are grateful to NR, HE, AECOM, and Kier Group for supporting the sensing deployment and would like to acknowledge the special contributions of Ian Billington (HE), Robert Dean, Christopher Heap, Sam De'Ath (NR), David Kite (AECOM), Cedric Kechavarzi (CSIC), and their supporting teams, without which this study wouldn't have been possible.

REFERENCES

Acikgoz, S., DeJong, M. J. and Soga, K. 2018a. Sensing dynamic displacements in masonry rail bridges using 2D digital image correlation. *Struct Control Hlth*, 25, e2187.

Acikgoz, S., DeJong, M. J., Kechavarzi, C. and Soga, K. 2018b. Dynamic response of a damaged masonry rail viaduct: Measurement and interpretation. *Eng Struct*, 168, 544–558.

Acikgoz, S., Soga, K. and Woodhams, J. 2017. Evaluation of the response of a vaulted masonry structure to differential settlements. *Constr Build Mater*, 150, 916–931.

Alexakis, H., Franza, A., Acikgoz, S. and DeJong, M. J. 2019a. Structural health monitoring of a masonry viaduct with Fibre Bragg Grating sensors, IABSE SYMPOSIUM GUIMARÃES 2019, Towards a Resilient Built Environment Risk and Asset Management, 1560–1567.

Alexakis, H., Franza, A., Acikgoz, S. and DeJong, M. J. 2019b. A multi-sensing monitoring system to study deterioration of a railway bridge, *9th International Conference on Structural Health Monitoring of Intelligent Infrastructure (SHMII-9), St. Louis, Missouri, USA.*

Alexakis, H., Franza, A., Acikgoz, S. and DeJong, M. J. 2019c. Monitoring bridge degradation using dynamic strain, acoustic emission and environmental data, *International Conference on Smart Infrastructure and Construction, University of Cambridge.*

Alexakis, H., Lau, F. D. H. and DeJong, M. J. 2021. Fibre optic sensing of ageing railway infrastructure enhanced with statistical shape analysis. *J Civil Struct Health Monit*, 11, 49–67.

Alexakis, H., Liu, H. and DeJong, M. J. 2020. Damage identification of brick masonry under cyclic loading based on acoustic emissions. *Eng Struct*, 221, 110945.

Behnia, A., Chai, H. K. and Shiotani, T. 2014. Advanced structural health monitoring of concrete structures with the aid of acoustic emission. *Constr Build Mat*, 65, 282–302.

British Standards Institution—BSI. 1962. Amendment No. 4 (PD 4639) to BS 153-3B4:1958 Specification for steel girder bridges, London: BSI.

Calvi, G. M., Moratti, M., O'Reilly, G. J., Scattarreggia, N., Monteiro, R., Malomo, D., Calvi, P.M. and Pinho, R. 2019. Once upon a time in Italy: The tale of the Morandi Bridge. *Structural Engineering International*, 29, 198–217.

Chaiyasarn, K. et al. 2018. Crack detection in masonry structures using convolutional neural networks and support vector machines, *35th ISARC, Berlin, Germany*, 118–125.

Cocking, S. 2021. Dynamic distributed monitoring of masonry railway bridges (PhD thesis), University of Cambridge, UK.

Cocking, S., Alexakis, H. and DeJong, M. J. 2021. Distributed dynamic fibre-optic strain monitoring of the behaviour of a skewed masonry arch railway bridge. *J Civil Struct Health Monit*, 11, 989–1012.

Cocking, S. and DeJong, M. J. 2022. Long-Term Structural Monitoring of a Skewed Masonry Arch Railway Bridge Using Fibre Bragg Gratings, *EWSHM 2022, the 10th European Workshop on Structural Health Monitoring.* https://link.springer.com/chapter/10.1007/978-3-031-07322-9_69

Cocking, S., Thompson, D. and DeJong, M. J. 2019a. Comparative evaluation of monitoring technologies for a historic skewed masonry Arch Railway Bridge, *ARCH 2019, the 9th International Conference on Arch Bridges, Porto, Portugal*, 439–446.

Cocking, S. H., Ye, C. and DeJong, M. J. 2019b. Damage assessment of a railway bridge using fibre optic sensing and LiDAR data, *ICSIC 2019, the International Conference on Smart Infrastructure and Construction, Cambridge, UK.*

Cortes, C. and Vapnik, V. 1995. Support-vector networks. *Machine Learning*, 20, 273–297.

De Santis, S. and Tomor, A. K. 2013. Laboratory and field studies on the use of acoustic emission for masonry bridges. *NDT&E Int*, 55, 64–74.

Desnerck, P., Lees, J. M. and Morley, C. T. 2016. Impact of the reinforcement layout on the load capacity of reinforced concrete half-joints. *Eng Struct*, 127, 227–239.

Dryden, I. and Mardia, K. 2016. *Statistical Shape Analysis: With Applications in R. Wiley Series in Probability and Statistics*, Wiley.

EN 1330-9. 2017. Non-destructive testing - *Terminology - Part 9: Terms Used in Acoustic Emission Testing, Comité Européen de Normalisation*, 1–13.

English Highway Structures & Bridges Inspection & Assessment. 2020. *CS 466. Risk management and structural assessment of concrete half-joint deck structures* (formerly IAN 53/04 and BA 39/93 (plus part of BD 44/15).

Kaloop, M. R. and Hu, J. W. 2015. Stayed-cable bridge damage detection and localization based on accelerometer health monitoring measurements. *Shock and Vibration*, 102682. https://www.hindawi.com/journals/sv/2015/102680/

Mutsuyoshi, H. 2001. *Present situation of durability of post-tensioned PC bridges in Japan*, Bulletin-Fib.

Orbán, Z. 2004. Assessment, reliability and maintenance of masonry arch railway bridges in Europe. *ARCH'04, Barcelona.*

Santos, J. P., Cremona, C., Calado, L., Silveira, P. and Orcesi, A. D. 2016. On-line unsupervised detection of early damage. *Structural Control and Health Monitoring*, 23, 1047–1069.

Tziavos, N. I., Hemida, H., Dirar, S., Papaelias, M., Metje, N. and Baniotopoulos, C. 2020. Structural health monitoring of grouted connections for offshore wind turbines by means of acoustic emission: An experimental study. *Renewable Energy*, 147, 130–140.

UIC 778-4. 2009. Defects in railway bridges and procedures for maintenance. UIC Code 778-4 R, 2nd Ed., International Union of Railways, Paris, 32.

United Nations. 2017. *Resolution adopted by the General Assembly on 6 July 2017*, A/RES/71/313

5 Pile Integrity Assessment through a Staged Data Interpretation Framework

Qianchen Sun and
Mohammed Z. E. B. Elshafie

5.1 OVERVIEW

Given the inherent nature of how deep cast-in-situ concrete foundations (piles and diaphragm walls) are constructed, evaluating their integrity is difficult. Several well-established methods for testing integrity have been established, but each has its own advantages and disadvantages. Recently, a new integrity test called thermal integrity profiling (TIP) has been put into use in deep-foundation construction. The primary characteristic utilized in this test is the early-age concrete release of heat during curing; abnormalities such as voids, necking, bulging, and/or soil intrusion inside the concrete body lead to local temperature fluctuations. During concrete curing, temperature sensors installed on the reinforcing cage collect precise temperature data along the entire pile, allowing empirical identification of these temperature variations. This chapter proposes a staged data interpretation framework for pile integrity assessment, with the thermal integrity test serving as the initial step. The framework, which is adaptable to different concrete mixtures and pile designs, utilizes the heat of hydration and the theory of heat transmission, as well as numerical modeling with the finite element (FE) method. It also adopts a staged procedure to assess the as-built quality; for a particular pile, more details are revealed about any anomalies being investigated (including location, size, and shape) at each subsequent stage. The primary advantage of this staged process is that it enables practitioners to follow a risk-based approach and decide whether or not to pursue subsequent stages of construction depending on the results they get at the end of each stage. This provides practicing engineers with vital information about the quality of the pile immediately after pile building, thus permitting immediate and less expensive repair and remedial work if required.

DOI: 10.1201/9781003306924-5

5.2 INTRODUCTION

Pile foundations are commonly employed in construction owing to their capacity to overcome poor soil conditions by transferring loads deep into stronger and stiffer soils and so avoiding the weaker soils closer to the surface. In developing pile systems, designers must evaluate how as-built construction works are compared to performance specifications. The integrity and quality of these cast-in-situ foundation piles present significant challenges to the engineering profession. In recent years, pile repair and the associated maintenance accounted for a significant part of the construction cost. In the United Kingdom, infrastructure repair and maintenance costs account approximately to £15 billion each year—a fifth of the total construction costs (Infrastructure UK, 2010). It is therefore crucial to ensure the pile quality at the construction stage. However, the intrinsic nature of these underground structures (limited accessibility, low visibility, large depth, etc.) adds considerable difficulties to structural quality inspection (Matsumoto et al., 2004; Kister et al., 2007). Anomalies present in deep foundations—for example, voids, soil intrusions, material loss, or shaft collapse—could result in structure instability and/or severe durability issues. It is reported that among 10,000 tested bored piles in the United States and Germany, more than 15% of pile test results showed signs of minor defects and 5% of tested piles were confirmed with major defects (DiMaggio & Hussein, 2004; Brown & Schindler, 2007). These defects can have significant financial implications and cause construction program delays, as the evaluation of the pile's bearing capacity, serviceability, and structural safety is dependent on the knowledge of any existing pile anomalies. It is imperative that any construction defects are identified at an early stage, particularly when piles are heavily loaded. However, without adequate monitoring techniques and appropriate identification methods, these defects could very easily go undetected. Thus, testing techniques for the integrity of bored piles are of great value to foundation construction. Traditional integrity testing methods—including sonic pulse echo (SE) testing, crosshole sonic logging (CSL), and gamma-gamma logging (GGL)—are widely used to assess the risk of these geotechnical structures. These integrity tests are expected to detect potential anomalies, such as voids, soil intrusions, and cross-sectional variations. Each technique of testing for integrity has its own advantages and disadvantages. Some are more successful at identifying particular sorts of flaws but incapable of detecting others. In the SE testing method, the maximum depth is limited by the presence of stiff soil or rock as well as a pile length/diameter (L/D) ratio of 20, and it can be difficult to distinguish the soil response from the pile response. In the CSL and GGL testing methods, the connection of long access tubes is a risky activity, and even then only a limited volume of concrete between and around the access tubes can be assessed.

In the last decade, a new integrity test called TIP has been put into use in foundation construction. It monitors the temperature changes and thermal profiles of early-age concrete throughout the casting and curing processes. The heat production and dissipation within the pile's concrete body are influenced by the concrete mix, the ground conditions, and the shaft geometry. If defects are present inside the pile, they will cause local temperature fluctuations relative to the predicted heat produced during the curing process. The observed temperature information is utilized to deduce the as-built shaft shape, the position of the reinforcement cage, and eventually the presence of defects. This novel temperature-based integrity method also has its limitations. The current practice of data interpretation in the piling industry is mostly based on empirical experience. Anomaly detection through direct analysis of temperature profiles is currently indicative or suggestive, and, short of extracting the pile, it is difficult to verify whether the interpretation is valid or not. Temperature signatures are usually similar, and the potential numerous causes are not easily isolated.

This chapter presents a staged-based framework to interpret the temperature data obtained from the field thermal integrity test. The proposed data interpretation framework employs FE numerical modeling to systematically study the pile's as-built quality through one-dimensional (1D) and two-dimensional (2D) analyses. Following this staged investigation procedure, the integrity information, such as overall quality, defect location, and defect size and severity, along the pile is identified and evaluated. The detailed staged data interpretation method is explained in detail in this chapter and demonstrated through a field case study.

5.3 STAGED DATA INTERPRETATION FRAMEWORK

The thermal integrity method utilizes the exothermic characteristic of the cement hydration process. The generated hydration heat increases the concrete temperature. Defects within piles, such as necking, inclusions, or segregation, produce less heat and therefore lower the concrete temperature; in contrast, bulges in the shaft lead to higher concrete temperature.

In order to numerically evaluate pile integrity using monitoring temperature data, a staged data interpretation framework is proposed as illustrated in Figure 5.1. The first stage is to conduct the field thermal integrity test—this comprises collecting continuous temperature data from sensors installed before casting along the entire length of the pile. Following that, a heat-of-hydration test should be conducted in order to calibrate the cement hydration model, which will be integrated into the FE model in the later stages. Upon obtaining temperature data and ground information, the data interpretation process is then performed as follows: Stage 1, the *vertical scanning stage*, employs a 1D FE model to perform a quick evaluation of all temperature data along the pile length. Through this scanning process, the maximum temperatures at the respective sensor locations are converted into the effective pile radii, which are then used to generate a 3D pile shape for more intuitive identification of potentially defective regions. A more detailed explanation of the 1D FE process and the pile effective radii are presented in Sun et al. (2022).

Assuming that potentially defective regions are identified in Stage 1, the investigation moves to Stage 2, the *slicing stage*. Temperature development data is extracted from cross sections within the defective regions, and the defect

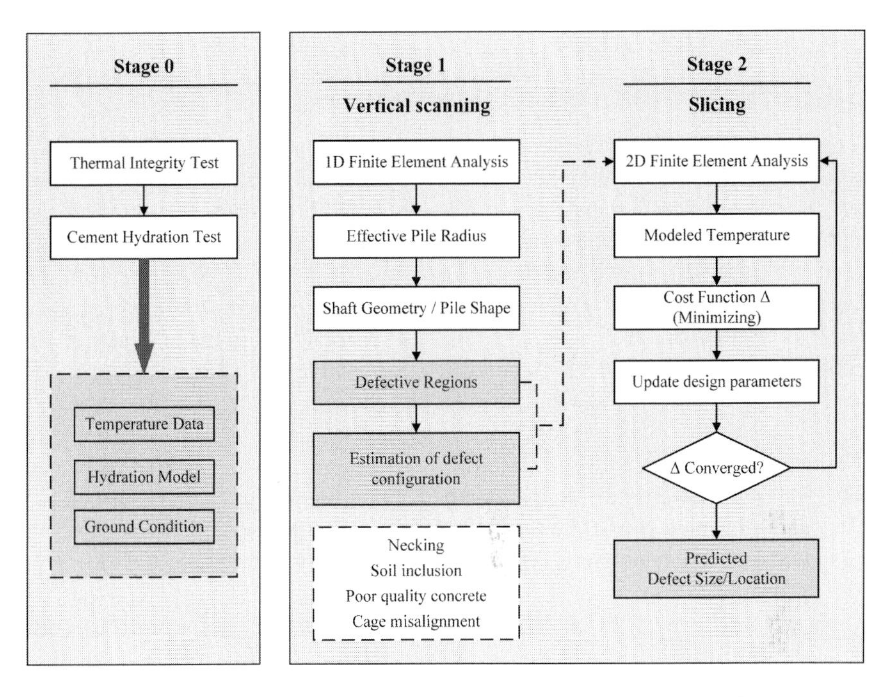

FIGURE 5.1 Staged data interpretation framework.

investigation is performed using a 2D FE model. The information retrieved from Stage 1 is used for setting up (i) an initial defect configuration (assumed size and location) and (ii) suitable search zones within the cross sections. A series of 2D FE simulations for the heat of hydration are then conducted, taking into account the appropriate boundary conditions, with the temperature development data from the model then compared with the actual field data using an appropriate cost function (Δ). It is worth mentioning that more sensors (measuring points) within the cross section provide better results. At the end of each 2D simulation, based on the comparison with the field data, the assumed defect configuration (size and location) is optimized within the search zones through algorithms (such as differential evolution and/or particle swarm optimization) before a new 2D simulation is conducted. This is continued until the cost function is minimized to an appropriate value. It should be pointed out that while the defects are assumed to be circular in shape, in practice, the defects have a more complicated form and could comprise poor-quality concrete. This does not limit the modeling or the framework proposed here. Different shapes (with different sizes) could be assumed in the analyses, and more shapes (with different sizes) could be added systematically within the same cross section to minimize the cost function even further and get the best match possible. More detailed explanation for the 2D FE analysis can be found in Sun et al. (2020). The detailed procedure of the staged method will be demonstrated using a field case study in the following sections.

5.4 FIELD CASE STUDY

5.4.1 Pile Construction and Instrumentation

This field case study investigated the monitoring of a continuous flight auger (CFA) test pile designed to evaluate the capability of thermal integrity testing. The test was performed on a building site in London in 2015. As seen in Figure 5.2, the test pile had an exterior diameter of 900 mm and a length of 20 m. Before casting concrete into the pile, temperature sensors were installed inside the reinforcing cage, which had a 750 mm diameter. A total of three temperature-sensing cables were used; these are designated as TIP-1, TIP-2, and TIP-3 in the pile instrumentation plan view in Figure 5.2. According to the contractor's concreting log, a total of 15.28 m^3 of concrete was cast to construct the pile. The temperature monitoring started about 1 h before concrete casting and lasted for 40 h. As illustrated in Figure 5.2, the testing area's soil stratigraphy comprises Made Ground and River Terrace Deposits overlain by the Lambeth Group and Thanet Sand Formation. The depth of each soil stratum is shown in Table 5.1. A comprehensive set of temperature readings were taken at 300 mm intervals throughout the length of the pile every 15 min.

Figure 5.3 depicts the installation of three designed inclusions to the cage at three distinct positions in order to evaluate the integrity test performance. Figure 5.3a shows inclusion 1 (D1) as a set of 70-mm-thick sandbags filled with Thanet Sand and attached externally to the reinforcing cage. This is used to

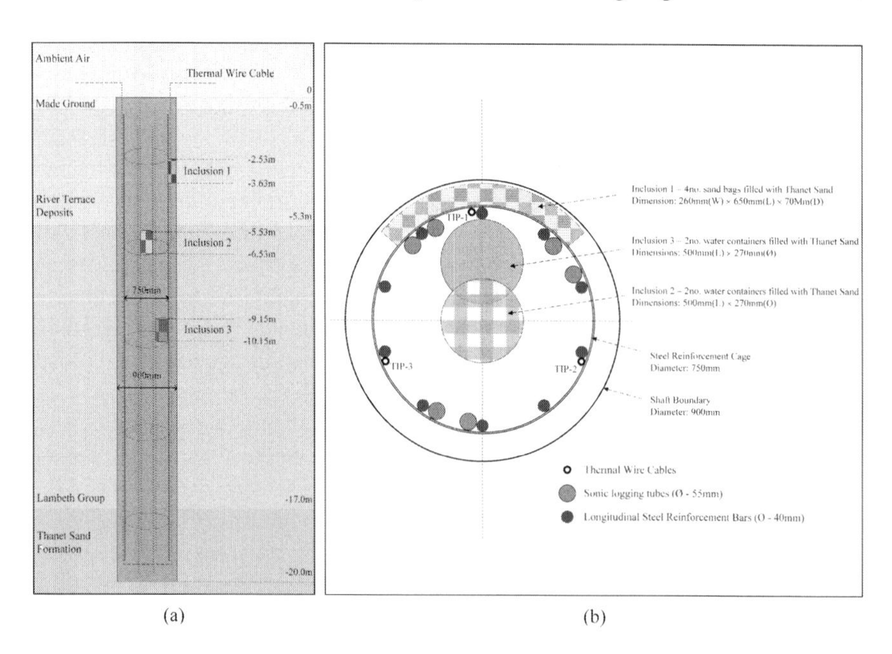

(a) (b)

FIGURE 5.2 (a) Soil stratigraphy and inclusion level and (b) pile instrumentation and reinforcement design.

TABLE 5.1
Soil Stratigraphy

Top of Strata (m)	Strata		Soil Description
+0.00	Made Ground (MG)		Clay and sand
−0.50	River Terrace Deposits (RT)		Sandy gravel and gravelly sand
−5.25	Woolich Formation	Lambeth	Clay overlying sand
−11.0	Reading Formation	Group	Sandy clay
−13.5	Upnor Formation	(LG)	Sand and gravel
−17.0	Thanet Sand (TS)		Sand

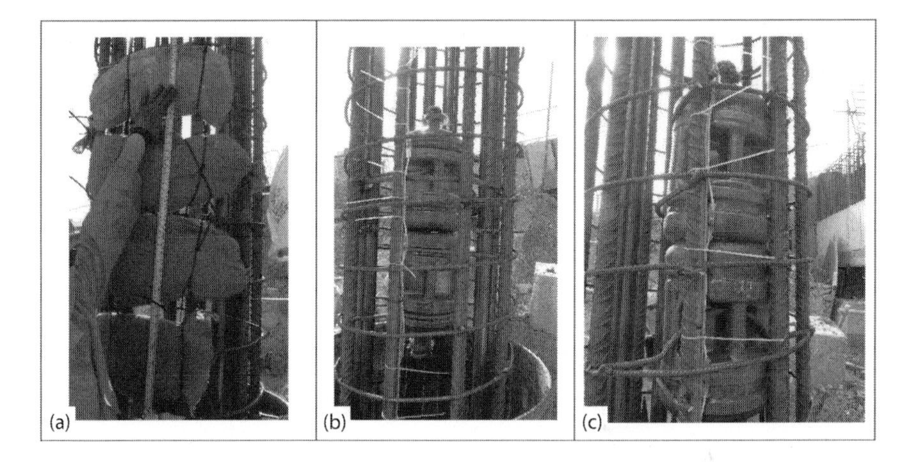

FIGURE 5.3 Installation of inclusions positioned: (a) external to the cage, (b) centrally within the cage, and (c) at the internal circumference of the cage.

replicate soil inclusions outside of the reinforcing cage, one of the most frequent types of pile flaws induced by restricted shaft wall collapse. The second and third inclusions (D2 and D3, respectively) are water containers filled with Thanet Sand. They were approximately 270 mm in diameter each and were positioned at separate locations along the pile (Figure 5.3b and 5.3c). These two defects represent soil inclusions inside the reinforcing cage.

5.4.2 Measured Pile Temperatures

Figure 5.4 illustrates the longitudinal temperature fluctuation profiles recorded over the full length of the test pile at four distinct time periods. The predicted temperature evolution (assuming a perfect pile) derived from a 2D numerical study of a longitudinal cross section of the pile (over its whole depth) has been superimposed on the figure for comparison. Two hours after the commencement of observations, the temperature decreased by about 1 to 2 °C along all

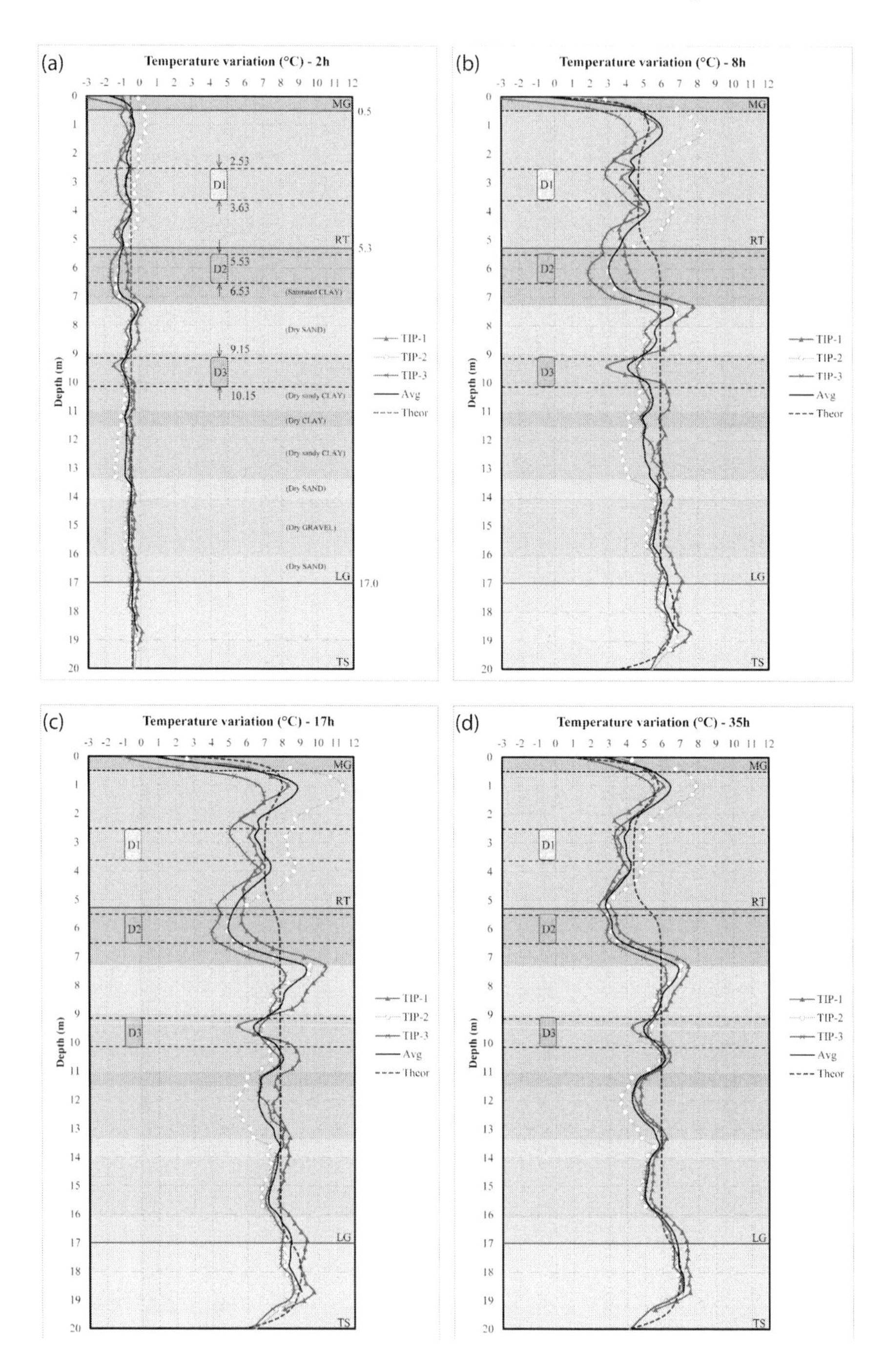

FIGURE 5.4 Longitudinal temperature variation profiles at (a) 2 h, (b) 8 h, (c) 17 h, and (d) 35 h.

three profiles (TIP-1, TIP-2, and TIP-3). The initial curing of the concrete was relatively slow, so it took a few hours for the temperature to achieve equilibrium before enough heat was produced by hydration to compensate for that lost to the cooler ground. The whole pile started to heat up between 4 and 8 h after casting, with the exception of the top half meter, which was exposed to ambient air overnight. At 17 h, the temperature profile of the whole pile reached its maximum value of between 8 and 11 °C. The temperature lowered at a fairly moderate pace after that; at 35 h, the average temperature of all three profiles was around 5 °C. Figure 5.4 also depicts the location of the artificial inclusions (D1, D2, and D3) relative to the soil layers. A theoretical temperature profile (in dashed gray) is displayed in each of the phases to help assess anomalies. This theoretical profile reflects the theoretical temperature changes of an intact pile (no defects) determined numerically assuming thermal conductivities of the different soil layers.

Figure 5.5 demonstrates the change in temperature over the 40 h monitoring period for the three temperature cables on two cross sections at 4.9 m and 15.1 m depth. The observed temperature profiles show a small drop in temperature in the first 2 h immediately after concrete casting, followed by a steep increase for the subsequent 12 h. The maximum temperature changes at the depths shown were about 5–7 °C and 6–8 °C, respectively. As shown, the temperature profiles between different temperature cables were not uniform. The difference ranged between 1 and 2 °C.

The temperature profiles above are the result of a combination of factors, including the hydration heat produced by the concrete and the heat transfer rate between the pile and the surrounding soil. If the measured temperature profile versus depth is consistent and each individual cable is also consistent, the pile is considered to be uniform in shape along the pile length, as shown in Figure 5.6. If not, it is assumed that some kind of defect exists in the pile. In order to assess these defects, a series of FE analyses were then performed to back-analyze the effects of these defects.

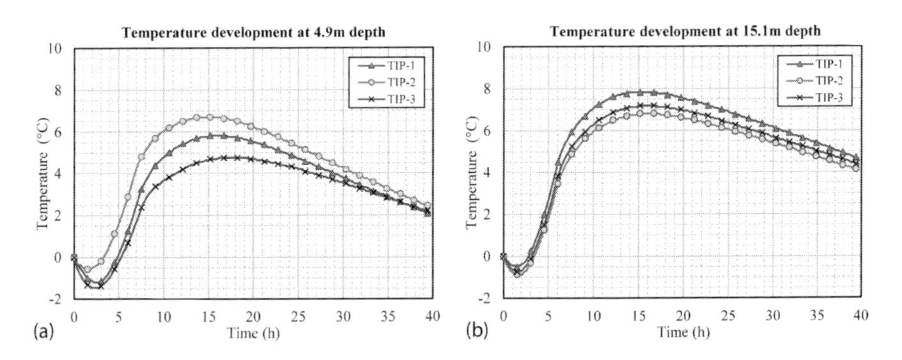

FIGURE 5.5 Thermal wire cable temperature development over time at two depths: (a) 4.9 m, and (b) 15.1 m.

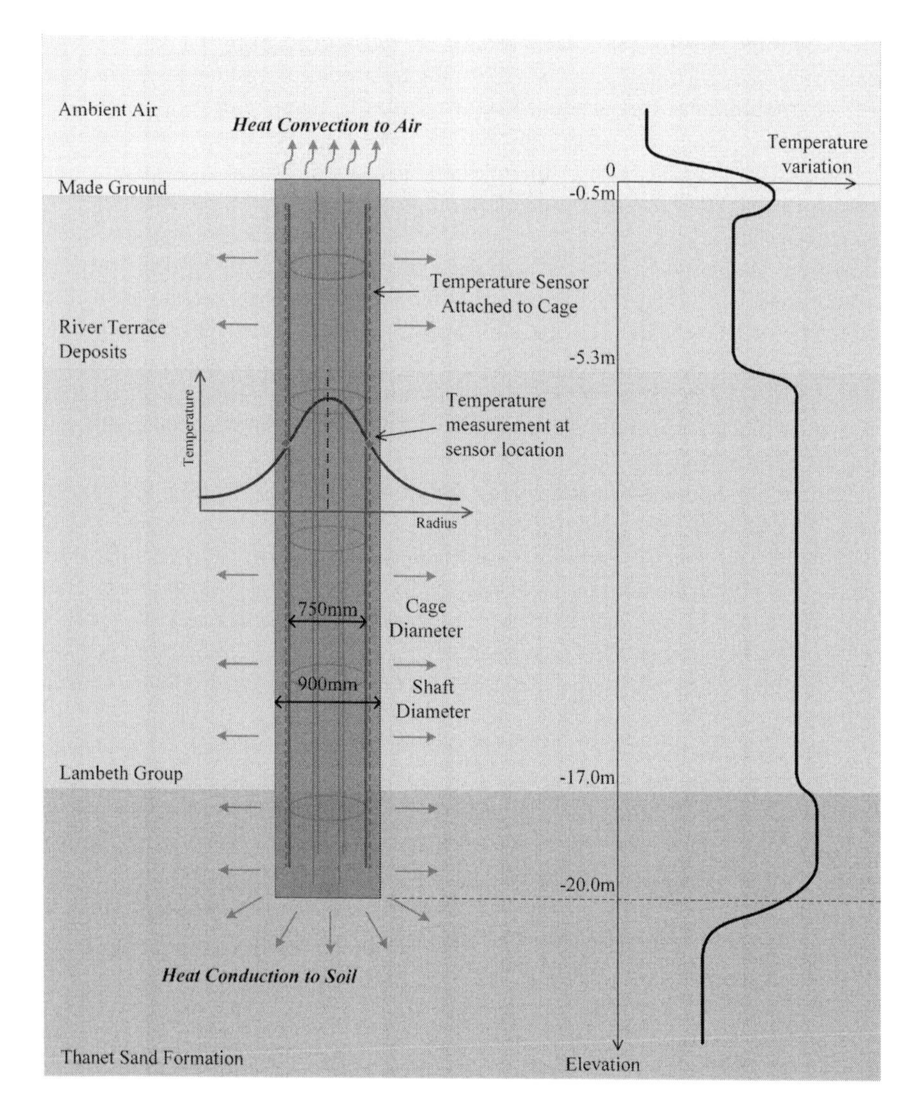

FIGURE 5.6 Temperature profile for a perfect cylindrical pile without defect.

5.5 FIELD TRIAL PILE DATA INTERPRETATION

5.5.1 Pile General Temperature Profiles

The temperature rise of a cast-in-place pile relies on a number of variables, such as the concrete mixture, cement type, shaft diameter, and boundary conditions. The cement composition and concrete mixture govern the hydration model, hence defining the rate of heat production in young concrete. The heat transmission route and rate are determined by the shape and boundaries of the shaft. Figure 5.6 depicts graphically the heat dissipation and temperature distribution for an idealized cast-in-situ pile. The cross-sectional lateral temperature distribution of a cylindrical pile without faults is bell-shaped, with the peak near the shaft center. Soil with larger thermal conductivity permits a quicker heat transfer from the concrete body to the surrounding soil, resulting in a cooler temperature at the center. A bigger shaft size increases the quantity of heat and the route of heat dissipation, and therefore a higher temperature at the center would be expected.

If the thermal conductivity of the soil varies between soil layers, the variable heat dissipation rates will cause a variance in the vertical temperature profile. In general, a large "roll-off" in temperature appears at the top and bottom of the vertical profile, as seen by the observed temperature profiles in Figure 5.4. The bottom of the shaft enables both radial and longitudinal heat dissipation via the end. Since the top of the shaft is exposed to the outside environment (air), internal heat is removed significantly faster than at other points. Consequently, the temperature is often several degrees Celsius lower than the average profile temperature in the top and lower one-diameter lengths of the pile.

5.5.2 Hydration Model Calibration

The hydration of concrete is a thermally induced exothermic reaction. The quantity of heat released during the first few days after casting concrete has a significant impact on the temperature growth of early-age concrete. The precision of the hydration model will have an immediate effect on the FE temperature forecast and the capacity to detect anomalies, so this needs to be considered carefully. Over the years, various formulas have been created to quantify hydration heat production. A complex hydration model based on concrete maturity was created by Schindler (2004). Concrete equivalent age and Arrhenius rate theory were utilized in this model to represent the rate of heat generation. It was then optimized by explicitly expressing some material properties based on data regression from many cement tests (Schindler & Folliard, 2005; Riding et al., 2011). A different model that examined the hydration process at the molecular level was later developed by Tomosawa (1997) and Tomosawa et al. (1997). It focuses more on defining chemical reaction coefficients: the rates of development and degradation of an initial impermeable layer, the chemical reaction process that is triggered by activation, and the diffusion-controlled process.

A classic hydration model developed by De Schutter & Taerwe (1995, 1996) was employed in the work presented in this chapter. This model is derived from a series of isothermal cement calorimetry experiments and concrete adiabatic tests. It contains explicit and relatively straightforward mathematical expressions. The rate of heat generation in concrete $\dot{Q}$ (J/gh) is a function of the actual temperature and the degree of hydration:

$$\dot{Q}(t) = q_{max,20}.c.\left[\sin(\alpha_t \pi)\right]^a .e^{-b\alpha_t} .e^{-\left[\frac{E}{R}\left(\frac{1}{T_c} - \frac{1}{T_r}\right)\right]} \tag{5.1}$$

where a, b, and c are the material constants controlling the distribution of hydration heat production; α_t is the degree of hydration, defined as the fraction of the heat of hydration that has been released ($Q_{released}/Q_{total}$); E is the apparent activation energy; R is the universal gas constant; $q_{max,20}$ is the maximum heat production rate at 20°C; T_c is the temperature of concrete (K); and T_r is the reference temperature (293 K).

In practice, the use of the hydration model encounters several challenges, including improperly recorded concrete mixtures, a lack of knowledge regarding the thermal characteristics of the ground, and unclear boundary conditions for piles. These uncertainties make it difficult to optimize the hydration parameters, which have a direct impact on the method's ability to detect pile anomalies. In order to evaluate the parameters of the concrete hydration model, an inverse technique based on evolutionary algorithms is proposed. The concrete hydration heat is defined by a total of six parameters, grouped as a vector $\{a, b, c, E, q_{max20}, Q_{total}\}$, the magnitudes of which can be determined by treating the model calibration as an optimization problem, that is, to obtain a set of hydration parameters (θ^*) that minimize the difference between the hydration test data and the model simulation:

$$\theta^* = \arg\min_{\theta} \sum_{t} \left\| T(\theta) - T_b \right\| \tag{5.2}$$

where T is considered as a function of θ in the FE hydration model; and T_b denotes the baseline temperature obtained from the hydration test.

Ideally, a calorimeter hydration test following ASTM C1702 Standard (2017) should be conducted on-site following the concrete pour. The test data can be used to evaluate the hydration parameters. However, this needs costly equipment, complex procedures to be followed, and a fully controlled test environment, which makes it unsuitable for field applications. An alternative approach is proposed here to calibrate the hydration model. Considering Figure 5.4, as the region between 14 and 18 m is designed with no engineered defects, and the temperature profile at this depth was relatively stable, the average temperature data of the three profiles along 14–18 m was selected in this case study as the baseline temperature (T_b) for optimization. For more general piles with no prior information, the overall average temperature along the whole pile can be used as the baseline temperature; the calibration numerical model uses the average pile radius determined from the total volume of poured concrete recorded in the concreting log.

This proposed calibration method applies the heuristic technique known as differential evolution for efficient parameter optimization. The differential evolution algorithm is explained in detail by Storn and Price (1997). Primarily, a population of potential solutions is first produced at random throughout the optimization procedure. The possible solutions for the optimization problem are vectors of the six variables (sometimes known as trial vectors). At each iteration (generation), "mutant vectors" are computed by linear interpolation

TABLE 5.2
Calibrated Hydration Model Parameters

Parameters	a	b	c	E (kJ/mol)	q_{max_20} (J/gh)	Q_{total} (J/g)
Cement	0.787	3.3	3.0	28.0	9.91	161

or extrapolation of randomly selected trial vectors from the population. The crossover mechanism then creates a new generation of trial vectors by combining the components of the mutant vectors with those of the trial vectors from the prior generation. The fitness of trial vectors from the two generations (old and new) is assessed and compared using a cost function that assesses the inconsistencies between the FE model simulation and the temperature profiles derived from the field test results in the 14–18 m depth range. The fitness of a specific solution affects its survivability: the fitter solutions remain in the population, while the weaker ones are eliminated. This technique iterates until all solutions in the population converge on a single, global optimal solution, which in this research is the optimal set of six parameters (θ^*). Using the above technique, the optimized hydration model parameters indicated in Table 5.2 can be produced. With these parameters, it is possible to simulate the temperature at 16 m, as illustrated in Figure 5.7, which shows a strong correlation with the baseline temperature.

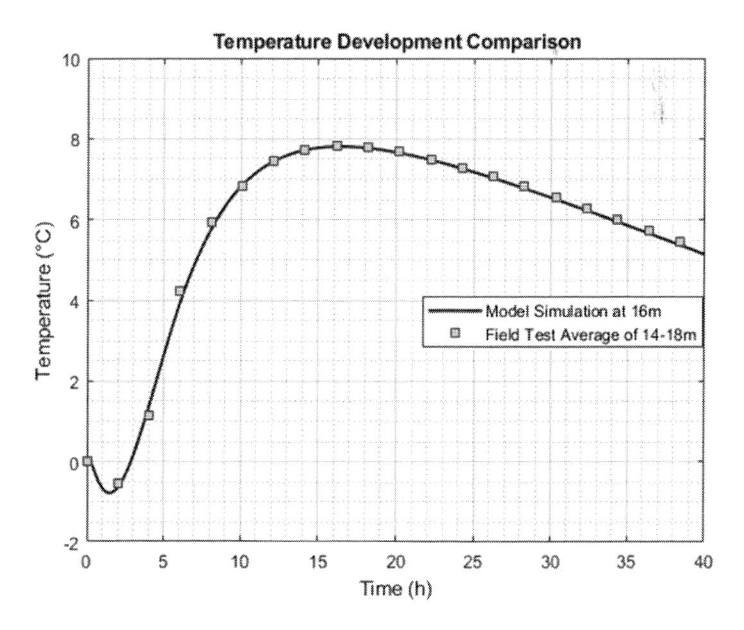

FIGURE 5.7 Model temperature and field test baseline temperature comparison.

5.5.3 The Stage 1 Analysis

The Stage 1 (S1) analysis employs a 1D FE model to perform a quick evaluation of all temperature data along the shaft length. It aims to assess the overall pile quality and identify the defective zones for subsequent analysis. This stage is also called the *vertical scanning stage*. Multiple parameters, including shaft diameter, boundary conditions, and the occurrence of anomalies, impact the observed temperature at the sensor position. Typically, it is difficult to identify one element from another. In particular, the temperature measurement of a predicted intact concrete pile might be comparable to that of a deficient concrete pile with a wider radius. Figure 5.8 depicts a situation in which the temperature recorded at S1 in the first pile is close to the temperature recorded at S2 in the second pile near the flaw.

In this section, the temperature profile variation in each soil layer will be seen as a change in the *effective pile radius*, which is the radius of an intact pile that generates the same recorded temperature at the sensor position. Consequently, the value of the effective pile radius at position S1 ($R_{eff\text{-}S1}$) should be close to that ($R_{eff\text{-}S2}$) at location S2, as seen by the red dashed line in Figure 5.8. For pile quality assurance, the effective pile radius throughout the whole shaft is an important parameter. Engineers may quickly determine whether or not extensive integrity studies are necessary for the pile under consideration by using the effective pile radius. Given that the temperature sensors are mounted to the reinforcing cage with a fixed radius, the effective pile radius (R_{eff}) may be linked to the effective pile cover thickness (D_{cover}) using this equation:

$$R_{eff} = r_{cage} + D_{cover} \tag{5.3}$$

where r_{cage} is the radius of the reinforcement cage.

All variations in temperature profiles are assumed to result from the variation in the effective pile radius. The temperature profiles are obtained using 1D FE studies of heat transport (in three radial directions corresponding to the three

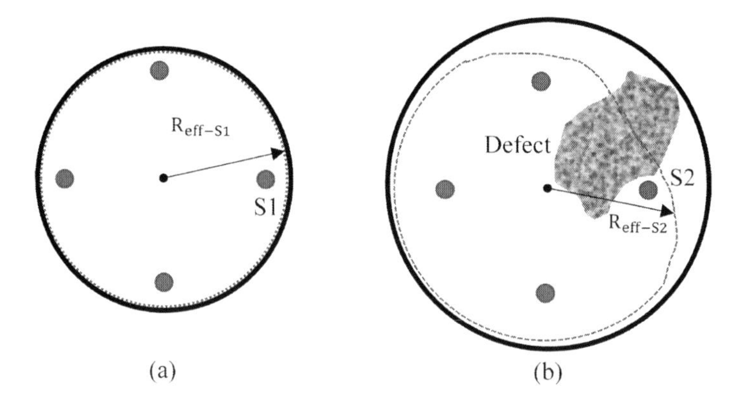

FIGURE 5.8 (a) A smaller pile without defect, and (b) a larger pile with defect.

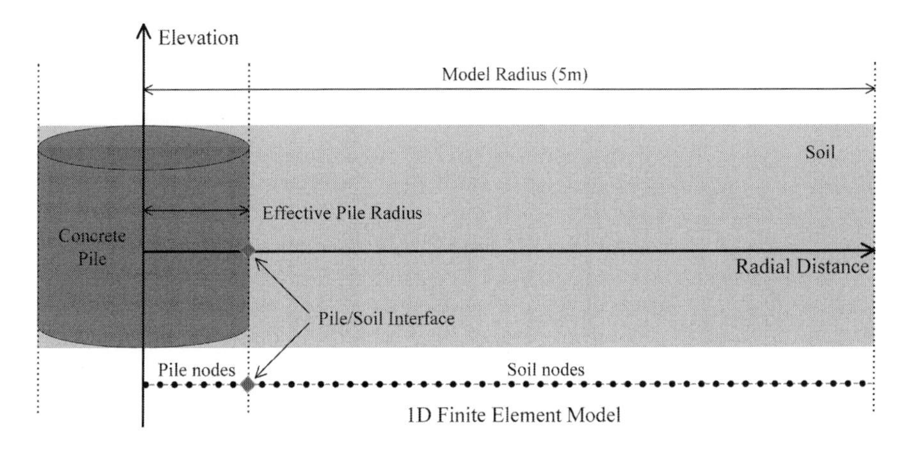

FIGURE 5.9 1D axisymmetric heat transfer finite element model.

sensor sites inside the pile) with effective pile radii assumed along the pile depth. At various positions along the pile depth, the effective pile radius is modified to provide a satisfactory match between the observed temperature profile and that estimated by the FE analysis. This approach allows for the prediction of the effective radius along the depth of the pile in three sensor radial directions corresponding to the three sensor arrays. After analyzing all cross sections, the pile's geometry can be calculated. As illustrated in Figure 5.9, due to axial symmetry, the FE model can be reduced to a 1D model. The FE model consists of pile and soil elements, and the hydration heat source is applied to the pile elements to simulate the generation of heat hydration. In the FE analysis, the pile–soil interface position (or concrete cover thickness) is modified to match the temperature predicted by the FE model to the actual monitoring data.

Typically, the temperature sensor cables are mounted to the reinforcement cage's outer layer in a pile. The temperatures at these locations are used to infer the concrete quality near the measurement points. Therefore, it is necessary to explore the relation between the temperature measurement and the effective shaft diameter. Figure 5.10 illustrates a theoretical relationship between cage radius, shaft radius, and temperature at 16 h after the concrete pour. This relationship, obtained through numerical simulation using the hydration model suggested by De Schutter and Taerwe (1995, 1996) with the parameter data, is presented in Table 5.2. The time of analysis is selected as 16 h, when the whole shaft reached the maximum temperature.

The dashed lines depict the cage position where temperature sensors are connected, which is 0.375 m in this field trial. The solid lines show various shaft radii, with the red line being the designed shaft radius of 0.45 m. The intersection of the solid and dotted lines is the theoretical temperature measurement. With a fixed cage radius of 0.375 m, the theoretical temperature measurement changes linearly as the shaft radius moves around the designed value of 0.45 m. Normally, pile concrete cover thickness would vary between 25 mm and 200 mm with or

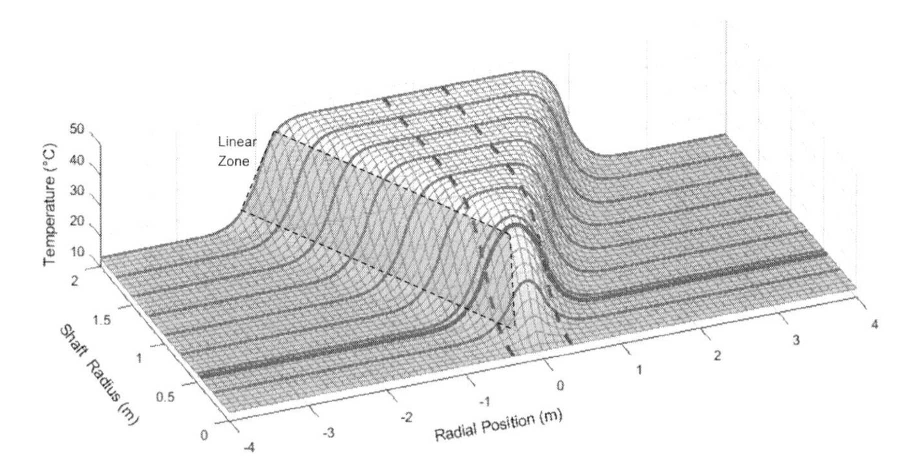

FIGURE 5.10 Relationship between cage position, shaft size, and temperature measurement at 16 h.

without defects; thus, the measured temperature should fall within the linear zone shown in Figure 5.10. The temperature measurement at the cage position is particularly sensitive to the concrete cover thickness, which may serve as an excellent signal for evaluating the as-built pile shape.

The work presented here uses the highest temperature at each measurement point across the full curing period instead of using the temperature distribution at a particular moment (e.g., 16 h in Figure 5.10). The maximum temperature at each sensor point is also highly correlated to pile size, concrete quality, and boundary conditions. A series of 1D FE models were established in four layers of soil: Made Ground, River Terrace Deposits, Lambeth Group, and Thanet Sand Formation. In each soil layer, a number of FE simulations with different concrete cover thicknesses were performed. The thickness varied between 1 cm and 20 cm at intervals of 0.5 cm. All other criteria, including the position of reinforcing cages and the thermal characteristics of the concrete and the soil, must adhere to the pile design and construction standards.

Figure 5.11 presents the highest measured temperature at the sensor position for different concrete cover thicknesses at four soil layers. It reveals a strongly linear relationship that can be expressed in the following equation:

$$D_{\text{cover}} = \alpha_k * T_{\max} - \beta_k \tag{5.4}$$

where D_{cover} represents the concrete cover thickness (in cm); T_{max} denotes the maximum temperature (°C) at the measurement point; and α_k (cm/°C) and β_k (cm) are parameters dependent on factors including sensor location and soil material. The values of these parameters were obtained through linear regression of the FE analysis data and are listed in Table 5.3. The obtained relationship allows the back-calculation of the cover thickness at three different sensor locations on

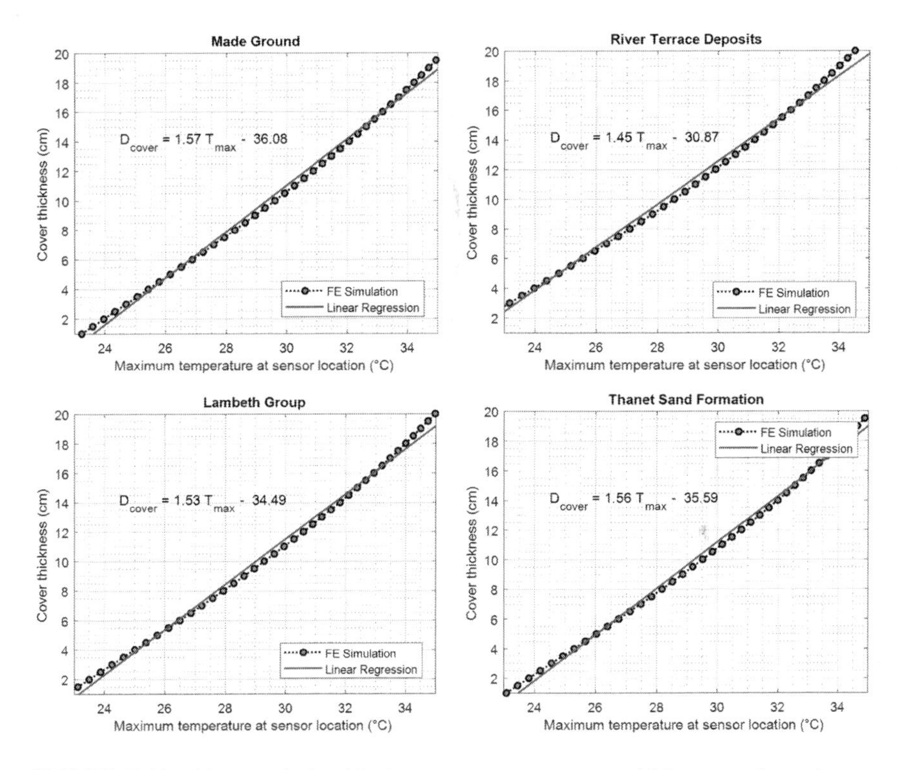

FIGURE 5.11 Linear relationship between concrete cover thickness and maximum temperature.

the cross sections along the pile depth. The effective pile radius can be subsequently obtained using Eq. (5.2).

Figure 5.12 demonstrates the 3D pile shape according to the calculated effective radius method. The circles indicates an expanded pile radius, and solid lines represents a contracted pile radius smaller than the average value; these are marked as defective zones. Two minor defective zones can be seen in Figure 5.12 between 2 and 4 m and 9 and 12 m below the ground level; and one severe defective zone, shown as a dark blue area between 6 to 8 m, appears to be a significant

TABLE 5.3
Values of Thermal Parameters

Parameters	α_k (cm / °C)	β_k (cm)
Made Ground	1.57	36.08
River Terrace Deposits	1.45	30.87
Lambeth Group	1.53	34.49
Thanet Sand Formation	1.56	35.59

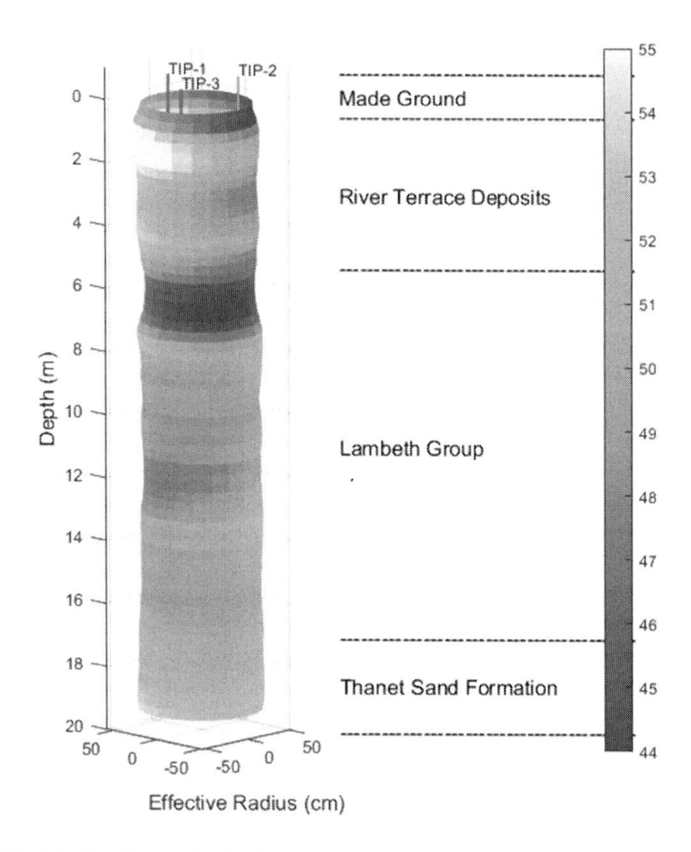

FIGURE 5.12 Predicted 3D pile shape.

necking problem. The pile radius varies significantly along the upper half of the pile, where it ranges from 43 to 57 cm. The shape of the lower half of the pile is relatively consistent at around 51 cm in radius.

5.6 THE STAGE 2 ANALYSIS

The 2D FE analysis in this stage aims to find the detailed defect location and size within each defective zone identified in the first stage. In this stage, known as the *slicing stage*, the defective zones are cut into multiple cross sections to evaluate the defects at the 2D level. Normally, if the as-built pile geometry is uniform and no defects are present, the measured temperature profile versus depth at each individual cable should be consistent, and the temperature development at three sensors within the same cross section should be relatively similar. The temperature data at each cross section can be a good indicator for pile integrity at the corresponding level. As shown in Figure 5.13, if a defect is located near sensor 1, the measured temperature development should be lower than the expected values, owing to a loss of heat source. For those sensors away from defects, the reduction of temperature

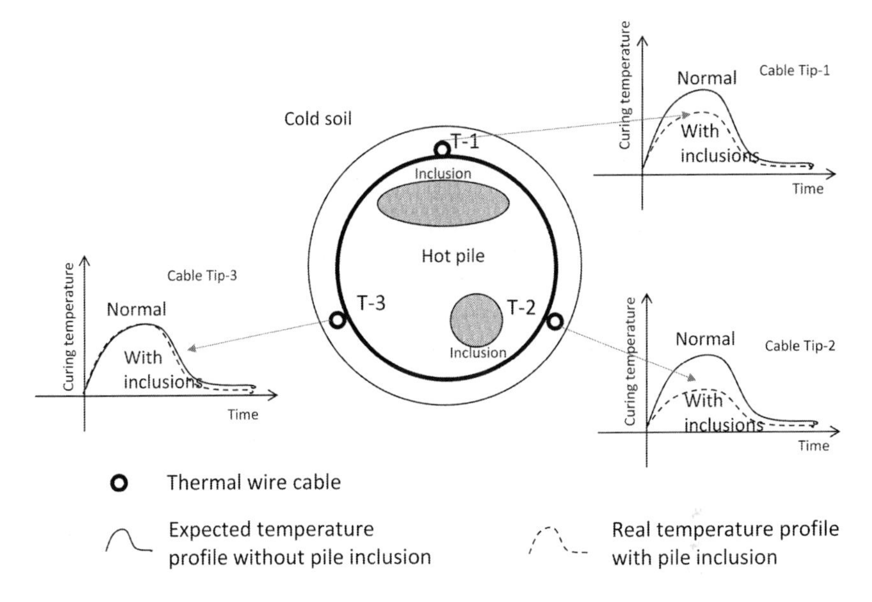

FIGURE 5.13 Conceptual relationship between temperature development and pile integrity.

measurement should be insignificant, such as sensor 3 in the figure. Based on that, a search grid zone is then set up around the expected defect location, and a series of FE simulations are systematically conducted where a circular anomaly of changing radius is centered at each of the search grid zone points.

A comprehensive assessment of anomalies is detailed here:

a. Taking a potential anomaly location into consideration, a numerical model is set up for a cross section corresponding to each location within the pile depth within the defective zone. From the Stage 1 analysis, these depths correspond to 2~4 m, 6~8 m, and 9~12 m (Figure 5.12). Therefore, multiple 2D numerical models are set up spaced at 300 mm for these three defective zones.
b. For each cross section, the actual temperature readings from the three cables in the cross section can give a rough indication of the location of an assumed circular anomaly within the cross section. Cross sections at depths of 2.8 m, 6.1 m, and 9.4 m are taken as examples where the most significant temperature change can be found along the temperature profiles. As observed at 6.1 m depth, the three cables all reported a reduction in temperature at the same time; this suggests that the anomaly is centrally located within the section. If one cable shows a larger decrease compared to the other two, such as at 9.4 m depth (TIP-1 shows a larger decrease in temperature compared to the other two cables), this suggests the anomaly is internal (inside the reinforcement cage), is closer to that

cable, and so on. The inference of defect location in this step is based on the assumption of a uniform shaft boundary. Necking and/or poor-quality concrete (among other reasons) could cause similar temperature variations. It should be noted that while the temperature data could be fitted with a large number of equivalent interpretation models, the interpretation method proposed in this chapter focuses on the most common defects based on practical experience.

c. Based on this, for the sections at 2.8 m and 6.1 m depth, an assumed starting *search origin* (0,0) is located at the center of the cross sections, while for the section at 9.4 m the starting search origin is located at (0,19 cm), as shown in Figure 5.14. The above selection of the origin depends on the estimated defect locations discussed in (b). Note that it is also possible that the anomaly is external to the reinforcement cage; however, in this scenario (which will be investigated later for 2.8 m), the anomaly is expected to have a significant effect on the closest cable and little to no effect on the other two. This is the reason why the search origin at 2.8 m is not located close to the reinforcement cage, even though the engineered Inclusion 1 is known to be in the concrete cover region.
d. For each cross section, a search grid zone is then set up around the search origin and a series of 2D FE simulations are systematically conducted where a circular anomaly of changing radius is centered at each of the search grid zone points. For the cross section at 6.1 m depth, the search grid (6 cm by 6 cm) had a total number of 49 search points. Numerical models were used to simulate circular anomalies centered at each of the 49 search points with the radius changing in size from 12 cm to 18 cm (adopting a step of 0.5 cm for each simulation)—a total of 637 2D FE simulations for the cross section. The total number of simulations for the three cross sections is 2469 (1377 FE simulations at 2.8 m depth and 455 FE simulations at 9.4 m depth).
e. After each simulation has been conducted (using the appropriate boundary conditions for the cross section under consideration), the predicted

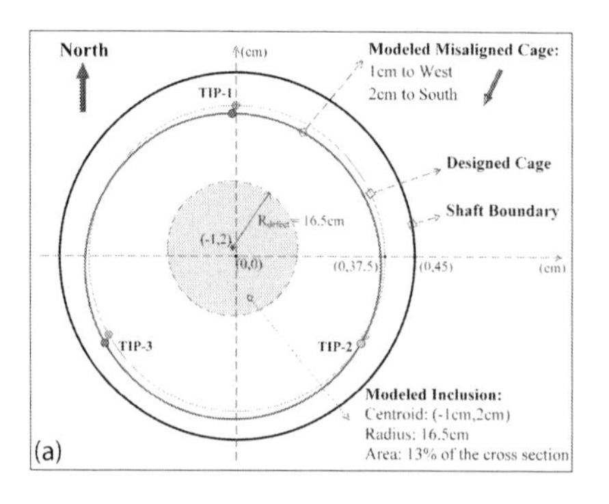

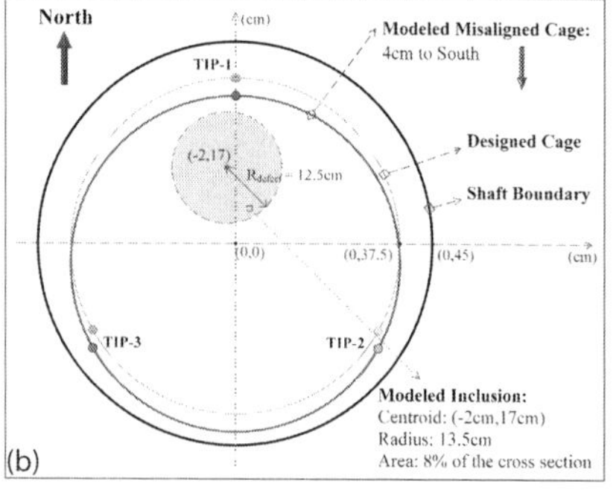

FIGURE 5.14 Predicted anomaly configurations at (a) 6.1 m and (b) 9.4 m.

temperatures at the location corresponding to TIP-1, TIP-2, and TIP-3 are compared to the actual field temperatures, and their cumulative temperature difference is set as the cost function that minimizes the difference/error.

f. The reinforcement cage misalignment needs to be taken into account. As such, a search zone for the locations of TIP-1, TIP-2, and TIP-3 is then established within the numerical model in order to search for the minimum value of the cost function—this is to take the cage reinforcement offset into account at the end of each of the 2469 simulations.
g. At the end of FE simulations (taking into account any cage misalignment), the anomaly configuration corresponding to the minimum cost function at each cross section is regarded as the detected anomaly.

Figure 5.14 demonstrates the results of the 2D FE defect predictions at 6.1 m and 9.4 m depth. At the cross-section of 6.1 m, an assumed circular anomaly was found centered at (−1 cm, 2 cm) with a radius of 16.5 cm accounting for 13% of the cross-sectional area and a cage eccentricity of 1 cm to the west and 2 cm to the south. At the cross section of 9.4 m, the FE simulation results predicted an inclusion center at (−2 cm, 17 cm) with a size of 8% of the cross section and a cage eccentricity of 4 cm to the south. The Stage 2 FE analysis gave an accurate prediction for Inclusions 2 and 3 in terms of size and location. Compared with the actual size of the engineered inclusions (9% cross-sectional area), the area for Inclusion 2 is slightly underestimated by 1%, while the FE analysis overpredicted Inclusion 3 by 4%. The FE studies did not detect Inclusion 1—four sandbags were utilized to simulate an externally attached concrete cover defect on the reinforcement cage. According to the on-site engineer, these sandbags were damaged during the reinforcing cage installation process. At this depth, sand spreads within the concrete body. As the volume of sand was negligible, totaling only 3% of the cross-sectional area of the pile, the concrete quality did not appear to have been considerably harmed.

In this chapter, the anomalies are supposed to be circular in shape; however, in practice, the flaws have a more intricate shape and may consist of low-quality concrete. This does not restrict the modeling or methodology given above. Different shapes (of varying sizes) may be assumed in the studies, and more shapes (of varying sizes) could be gradually added inside the same cross section to reduce the cost function and get the best possible match. In addition, a new technique of topology optimization has been devised to discover faults in both 2D and 3D space. It can evaluate the quality of concrete in a precise manner and predict defects in any shape. The details can be found in the authors' future publications.

5.7 CONCLUSIONS

Thermal integrity testing, which employs temperature change induced by hydration heat as a measurement of integrity, is a promising relatively new technique. This chapter proposed a staged data interpretation framework for the test and presented a case study of this test on a trial pile using this new approach.

The Stage 1 analysis uses a 1D FE analysis to interpret the data while accounting for the parameters of the concrete mix and soil. The FE model establishes a linear relationship between maximum temperature and effective pile radius. The effective pile radii throughout the whole shaft could be determined using the linear relationship, and the expected as-built pile's 3D shape could then be recreated, making it possible to more effectively identify problematic areas. After that, the Stage 2 analysis, which used 2D FE analysis, was able to accurately estimate the sizes and positions of the anomalies with less than 10% of the cross-sectional areas of the sections under consideration.

More information regarding the defects, such as their position, size, and shape, can be disclosed in each stage. This staged procedure enables practitioners to take a risk-based approach and choose whether or not to go on to the next stage based on the data they get at the conclusion of each stage. This innovative method may provide practicing engineers with vital test data about the quality of the pile immediately after pile building, thus permitting immediate and less expensive repair and remedial work.

REFERENCES

Brown, D., & Schindler, A. (2007). High performance concrete and drilled shaft construction. *Contemporary Issues in Deep Foundations*. ASCE. https://doi.org/10.1061/40902(221)31

De Schutter, G., & Taerwe, L. (1995). General hydration model for Portland cement and blast furnace slag cement. *Cement and Concrete Research*, *25*(3), 593–604.

De Schutter, G., & Taerwe, L. (1996). Degree of hydration-based description of mechanical properties of early age concrete. *Materials and structures*, *29*(6), 335–344.

DiMaggio, J. A., & Hussein, M. H. (2004, July). *Current Practices and Future Trends in Deep Foundations*. American Society of Civil Engineers.

Infrastructure UK, H. T. and I. and P. A. Infrastructure cost review: Technical report. (2010).

Kister, G., et al. (2007). "Methodology and integrity monitoring of foundation concrete piles using Bragg grating optical fibre sensors." *Engineering Structures*, *29*(9), 20482055.

Matsumoto, T., et al. (2004). "Monitoring of load distribution of the piles of a bridge during and after construction." *Soils and foundations*, *44*(4), 109–117.

Riding, K. A., Poole, J. L., Folliard, K. J., Juenger, M. C., & Schindler, A. K. (2011). New model for estimating apparent activation energy of cementitious systems. *ACI Materials Journal*, *108*(5), 550–560.

Schindler, A. K. (2004). Effect of temperature on hydration of cementitious materials. *Materials Journal*, *101*(1), 72–81.

Schindler, A. K., & Folliard, K. J. (2005). Heat of hydration models for cementitious materials. *ACI Materials Journal*, *102*(1), 24.

Storn, R., & Price, K. (1997). Differential evolution—A simple and efficient heuristic for global optimization over continuous spaces. *Journal of Global Optimization*, *11*, 341–359.

Sun, Q., Elshafie, M., Barker, C., Fisher, A., Schooling, J., & Rui, Y. (2020). Thermal integrity testing of cast in situ piles: An alternative interpretation approach. Structural Health Monitoring, https://doi.org/10.1177/1475921720960042.

Sun, Q., Elshafie, M. Z., Barker, C., Fisher, A., Schooling, J., & Rui, Y. (2022). Integrity monitoring of cast in-situ piles using thermal approach: A field case study. *Engineering Structures*, *272*, 114586.

Tomosawa, F. (1997). Development of a kinetic model for hydration of cement. *Proc. of the 10th Int. Cong. on the Chem. of Cem., Gothenburg, Sweden, 1997.*

Tomosawa, F., Noguchi, T., & Hyeon, C. (1997). Simulation model for temperature rise and evolution of thermal stress in concrete based on kinetic hydration model of cement. Proceedings of Tenth International Congress Chemistry of Cement (Vol. 4, pp. 72–75).

6 Data-Centric Monitoring of Wind Farms

Combining Sources of Information

Lawrence A. Bull,[1] *Imad Abdallah,*[1] *Charilaos Mylonas, Luis David Avendaño-Valencia, Konstantinos Tatsis, Paul Gardner, Timothy J. Rogers, Daniel S. Brennan, Elizabeth J. Cross, Keith Worden, Andrew B. Duncan, Nikolaos Dervilis, Mark Girolami, and Eleni Chatzi*

[1]These authors contributed equally.

6.1 OVERVIEW

This chapter summarizes recent developments in data-centric monitoring of wind farms. We present methodologies which share information from multiple sources. Problems include inter-turbine modeling of wind speed and wake effects; methods for combining multiple emulators; spatially distributed virtual sensing, for the estimation of dynamic loads; hierarchical Bayesian modeling of power relationships at the systems level. We demonstrate how an increasingly holistic approach to wind farm monitoring brings a series of advantages.

6.2 INTRODUCTION AND BACKGROUND

It is clear that increasingly efficient wind energy infrastructure is required to continue to advance renewable resources. In the UK alone, the government has outlined a national plan to extend the offshore wind sector (OWS) to generate enough electricity to power every UK home by 2030 [1]. The Climate Change Act 2008 sets a demanding 2050 target for an 80% cut in UK carbon emissions [2]. It is, therefore, critical that advances in operations and maintenance evolve to meet the demands of OWS growth while reducing or maintaining costs. As wind farms are built in increasingly remote offshore locations, deeper water will require new approaches to life-cycle management, decision making, and remote operations and maintenance.

DOI: 10.1201/9781003306924-6

There will be an inevitable shift toward strategies that automate asset management, while incorporating domain and engineering expertise as far as possible.

In view of incoming demands on the wind energy sector, this chapter presents recent developments in data-centric monitoring—focusing on methods to combine information from multiple sources and assimilate knowledge between:

- Collected turbines in a wind farm and their interactions
- Multiple simulations of wind energy systems
- Physics-based modeling augmented with machine learning methods

This data-centric and *systems* view of turbine monitoring should allow us to reimagine many practical applications of design, control, and maintenance for wind farms. We prioritize interpretability and uncertainty quantification via statistical modeling or machine learning procedures. Importantly, the digital representations should only predict outcomes that reflect what is possible in reality [3]. Each section covers examples of virtual sensing, load prediction, wake modeling, simulation, and longitudinal analysis of turbine population data. The chapter layout is summarized below.

6.2.1 Layout

Section 6.3 initiates from the problem of data-driven assessment and diagnostics at the individual wind turbine level. It utilizes a highly instrumented turbine to predict the damage equivalent load (DEL) on neighboring, downwind turbines in a simulated case study. The method exploits Gaussian process regression (GPR) with Bayesian estimates of the model hyperparameters, which allow for interpretable sensitivity analysis of turbine interactions.

Section 6.4 harnesses another data-driven approach, implemented at the wind farm (population) level for the purpose of generative modeling using graph representations. A graph neural network (GNN) is utilized to learn relational structures from wind farm measurements and to infer the distribution of operational characteristics, conditioned on the wind farm layout. The suggested generative approach allows alternative farm arrangements (geometries) to be simulated—demonstrating major implications for design, monitoring, and control.

Section 6.5 utilizes simulators (e.g., physics-based models) for estimation and prediction tasks. An ensemble method is proposed to combine the outputs from multiple stochastic simulators, to diversify the hypothesis space of predictions and quantify uncertainty. In an illustrative example, the ensemble collects 10 finite element models (FEMs), predicting the dynamic equivalent load for varying wind speeds. A key ingredient of this scheme is a probabilistic clustering to associate the outputs of simulator subgroups with different areas of the predictive space.

Section 6.6 presents a hybrid approach for augmenting physics-based models with data-centric techniques—in the context of virtual sensing. The investigated virtual-sensing problem considers an unknown (distributed) latent force, acting on a turbine blade, while also learning from sparse measurements. The latent force is estimated with GPR, while a Kalman filter approach estimates the blade response in unmeasured locations for a simulated example.

Section 6.7 considers the nested structure of datasets recorded from operational wind farms, relating to operating condition or turbine subgroups. By considering the wind farm data as a whole, the value of measurements is extended. Two methods are presented: (i) mixture models, which share information via (task-wise) data pooling only; and (ii) multilevel models, which share information using *shared* parameters and partial pooling over the predictive tasks. By modeling the inter-turbine correlations, uncertainty quantification of interpretable parameters is improved, and the covariance structure of the model provides insights into wind farm similarities.

6.3 DATA-DRIVEN SCHEMES FOR ASSESSMENT AND DIAGNOSTICS OF INTERACTIONS

This section considers previous work, originally presented in Ref. [4]. We initiate an overview of data-driven tools for the assessment at the (individual) wind turbine level, where interactions with neighboring turbines are accounted for. We exploit a GPR scheme with Bayesian hyperparameter calibration to obtain a surrogate mapping from input variables to output quantities of interest—considering the blade and tower components of adjacent upwind or wake-affected turbines.

6.3.1 Method

The approach [4] aims to infer short-term fatigue DELs on wake-affected turbines on the basis of supervisory control and data acquisition (SCADA) data (wind-field inflow) and structural-monitoring measurements from sensors deployed on upwind (neighboring) turbines, as suggested in Figure 6.1.

The inference is achieved using a GPR framework, which includes Bayesian hyperparameter identification. This is achieved by adopting a Metropolis–Hastings (MH) scheme for the quantification of modeling uncertainty. In the methodological section (Section 6.2.1), we describe how the GPR hyperparameters can be used to evaluate the sensitivity of the regressed outputs to individual inputs, and how a formal comparison between the extracted GPR models can be achieved using the Kullback–Leibler (KL) divergence.

In essence, this approach delivers a virtual-sensing method for inferring DELs on wake-affected turbines, on the basis of input from upwind, monitored turbines. This framework can also be utilized to optimally design a minimally invasive, yet maximally informative, set of sensors—used to deliver accurate DEL predictions for wake-affected turbines.

6.3.1.1 Aeroelastic Simulations

The simulations rely on a model of two interacting turbines: (i) a primary, upwind turbine; and (ii) a secondary turbine positioned in the wake of the first (wake-affected) turbine. A full description of these aeroelastic simulations is offered in

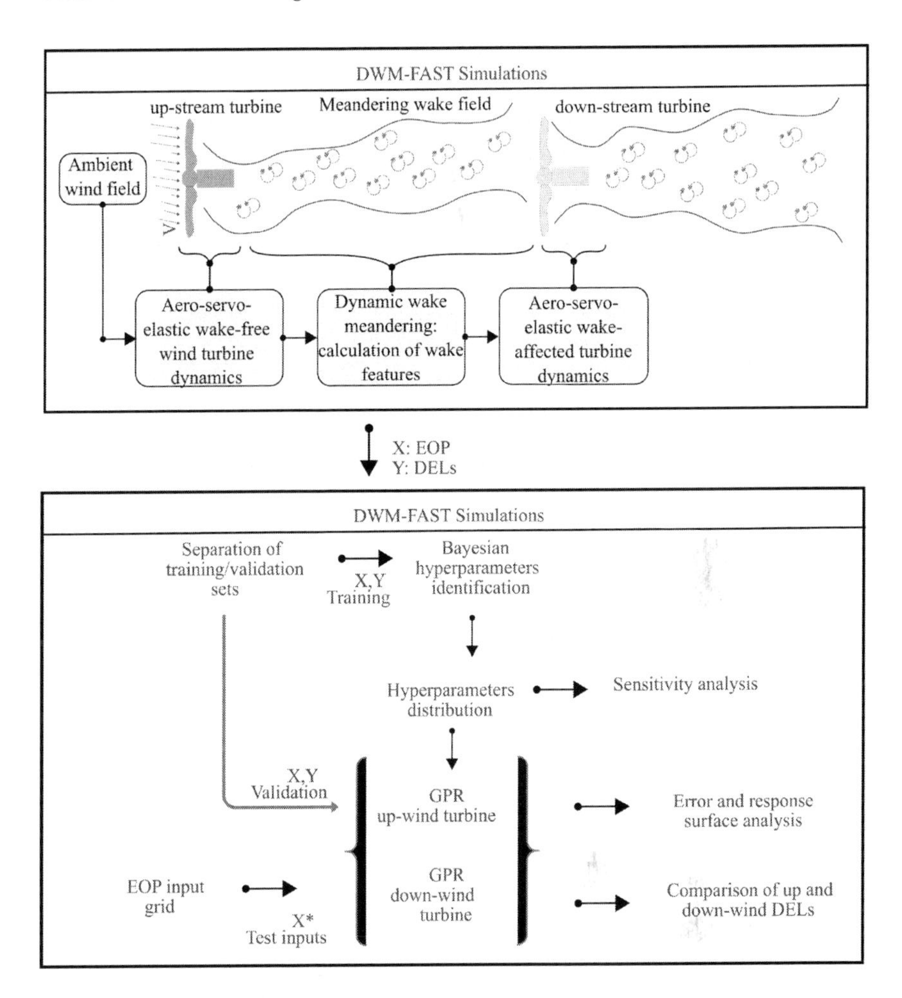

FIGURE 6.1 Wind farm-level virtual sensing: The objective is to estimate the short-term fatigue damage equivalent loads (DELs) at unmeasured locations of the downwind (wake-affected) turbine from direct measurements on the upwind turbine. The details of the Gaussian process regression (GPR)-based approach are elaborated in Section 6.3.1.2. (Adapted from Ref. [4].)

Ref. [4]. In brief, the simulations rely on three classes of random variables (RVs). First are inflow RVs, which influence the structural response of turbines: mean wind speed U, turbulence intensity σ, wind shear α, and horizontal inflow skewness ψ. Second are aerodynamic RVs, which correspond to a stochastic model for the lift and drag coefficients. This defines the distribution of aerodynamic axial and tangential induction over the rotor and, in turn, the transport of the wake. Third are fatigue RVs, which pertain to the short-term fatigue DEL. The DEL offers a scalar quantity of interest (QoI), which condenses information contained in multiple time-series records [5]. The exponent of the S-N curve (Wöhler exponent) is assumed to be an RV, with different values describing the tower and blade materials, on the basis of which the DEL is computed. Finally, the dynamic wake meandering (DWM) and aero-servo-elastic simulations are carried out in DWM-FAST (fatigue, aerodynamics, structures, and turbulence). Two NREL reference three-bladed turbines were considered with a rotor diameter of 126 m and a hub height of 90 m at 5 MW rated power. The retained output from the aeroelastic simulations focused on select QoIs, namely, the tower base's fore–aft (*TwrBsMyt*) and side–side (*TwrBsMxt*) bending moments, alongside the root's flap-wise (*RootMyb*1) and edgewise (*RootMxb*1) bending moments for one of the blades. Complete details of the simulated data are offered in Ref. [4].

6.3.1.2 Gaussian Process Regression

A function $f(\boldsymbol{x}) \in \mathbb{R}$ of an input vector $\boldsymbol{x} \in \mathbb{R}^n$ is referred to as a Gaussian process (GP) if its sampled value $\boldsymbol{f}(\boldsymbol{X}) = \begin{bmatrix} f(\boldsymbol{x}_1) & f(\boldsymbol{x}_2) & \cdots & f(\boldsymbol{x}_N) \end{bmatrix}^T$, over a finite number of inputs $\boldsymbol{X} = \begin{bmatrix} \boldsymbol{x}_1 & \boldsymbol{x}_2 & \cdots & \boldsymbol{x}_N \end{bmatrix}$, follows a multivariate normal distribution $\mathcal{N}\left(\mu(\boldsymbol{X}), \boldsymbol{K}(\boldsymbol{X},\boldsymbol{X})\right)$, with mean and covariance defined as follows [6, Sec. 2.2]:

$$\mu(\boldsymbol{X}) = \begin{bmatrix} \mu(\boldsymbol{x}_1) \\ \mu(\boldsymbol{x}_2) \\ \vdots \\ \mu(\boldsymbol{x}_N) \end{bmatrix} \boldsymbol{K}(\boldsymbol{X},\boldsymbol{X}) = \begin{bmatrix} k(\boldsymbol{x}_1,\boldsymbol{x}_1) & k(\boldsymbol{x}_1,\boldsymbol{x}_2) & \cdots & k(\boldsymbol{x}_1,\boldsymbol{x}_N) \\ k(\boldsymbol{x}_2,\boldsymbol{x}_1) & k(\boldsymbol{x}_2,\boldsymbol{x}_2) & \cdots & k(\boldsymbol{x}_2,\boldsymbol{x}_N) \\ \vdots & \vdots & \ddots & \vdots \\ k(\boldsymbol{x}_N,\boldsymbol{x}_1) & k(\boldsymbol{x}_N,\boldsymbol{x}_2) & \cdots & k(\boldsymbol{x}_N,\boldsymbol{x}_N) \end{bmatrix} \tag{6.1}$$

where $k(\boldsymbol{x}_i,\boldsymbol{x}_j)$ is the symmetric positive definite covariance function between $\boldsymbol{x}_i$ and $\boldsymbol{x}_j$. For the purposes of this scheme, a zero mean function and squared exponential covariance kernel are adopted, defined as [6, pp. 83–84]:

$$k(\boldsymbol{x},\boldsymbol{x}') = \sigma_f^2 \cdot \exp\left(-\frac{1}{2} \sum_{i=1}^{n} \frac{1}{l_i^2} \cdot (x_i - x_i')^2 \right) \tag{6.2}$$

where $\sigma_f^2 := k(\boldsymbol{x},\boldsymbol{x})$ is the function variance; and l_i reflects the length scale for each one of the input dimensions, which defines the *smoothness* of the latent function. In general, extrapolation is not reliable more than l_i units away from the data, especially when a zero mean function is used. A large length scale implies smoothness, or minor variability, in the latent function.

In the context of GPR, the covariance kernel can be used to exploit function values observed on a finite set of (training) points, $\boldsymbol{X} = \begin{bmatrix} \boldsymbol{x}_1 & \boldsymbol{x}_2 & \cdots & \boldsymbol{x}_N \end{bmatrix}$, in order to infer an estimate of the function value $f(\boldsymbol{x}_*)$ at a test point $\boldsymbol{x}_*$. The noisy observations vector $\boldsymbol{y} := \begin{bmatrix} y_1 & y_2 & \cdots & y_N \end{bmatrix}^T$ and the test function value $f(\boldsymbol{x}_*)$ are jointly normally distributed variables. Under exploitation of the properties of the multivariate normal distribution, it can be shown [6, Sec. 2.2] that the distribution of the function on the test input, conditioned on the noisy observation set $\boldsymbol{y}$, is also Gaussian:

$$p(f(\boldsymbol{x}_*)|\boldsymbol{y},\boldsymbol{X}) = \mathcal{N}\left(\bar{f}(\boldsymbol{x}_*),Q(\boldsymbol{x}_*)\right) \tag{6.3}$$

with the conditional mean $\bar{f}(\boldsymbol{x}_*)$ and variance $Q(\boldsymbol{x}_*)$ computed as:

$$\bar{f}(\boldsymbol{x}_*) = \boldsymbol{k}_* \cdot \left(\boldsymbol{K} + \sigma_w^2 \boldsymbol{I}_N\right)^{-1} \boldsymbol{y} \tag{6.4a}$$

$$Q(\boldsymbol{x}_*) = k_* - \boldsymbol{k}_* \cdot \left(\boldsymbol{K} + \sigma_w^2 \boldsymbol{I}_N\right)^{-1} \cdot \boldsymbol{k}_*^T \tag{6.4b}$$

where $k_* := k(\boldsymbol{x}_*,\boldsymbol{x}_*)$, $\boldsymbol{k}_* := \boldsymbol{k}(\boldsymbol{x}_*,\boldsymbol{X})$, and $\boldsymbol{K} := \boldsymbol{K}(\boldsymbol{X},\boldsymbol{X})$.

The GPR performance depends on the choice of the hyperparameters, or kernel parameters, summarized in $\mathcal{P} := \left\{\sigma_w^2, \sigma_f^2, l_1^2, \cdots, l_n^2\right\}$ (with a joint domain of Ω). The *marginal likelihood*, defined below, can be maximized to select these parameters [6, Sec. 2.3]:

$$\ln p(\boldsymbol{y}|\boldsymbol{X},\mathcal{P}) = -\frac{1}{2}\boldsymbol{y}^T\left(\boldsymbol{K} + \sigma_w^2 \boldsymbol{I}_N\right)^{-1} \boldsymbol{y} - \frac{1}{2}\left|\boldsymbol{K} + \sigma_w^2 \boldsymbol{I}_N\right| - \frac{N}{2}\ln 2\pi \tag{6.5}$$

This forms a nonlinear optimization problem requiring appropriate methods, such as gradient-based nonlinear optimization, which delivers deterministic estimates. Alternatively, Bayesian schemes can offer a probabilistic characterization of the hyperparameters, on the basis of available data and certain assumptions regarding the prior hyperparameter distribution $p(\mathcal{P})$ [7, pp. 12–13]. In this case, the posterior hyperparameter distribution $p(\mathcal{P}|\boldsymbol{y},\boldsymbol{X})$ is as follows:

$$p(\mathcal{P}|\boldsymbol{y},\boldsymbol{X}) = p(\boldsymbol{y}|\boldsymbol{X},\mathcal{P}) \cdot p(\mathcal{P}) \cdot p^{-1}(\boldsymbol{y}|\boldsymbol{X}) \tag{6.6}$$

where $p(\boldsymbol{y}|\boldsymbol{X})$ describes the model evidence, obtained as:

$$p(y|X) := \int_{\Omega} p(y|X,\mathcal{P}) \cdot p(\mathcal{P})\,\mathrm{d}\mathcal{P} \tag{6.7}$$

Since this problem is nonlinear, an analytical solution to the posterior distribution is intractable. An approximation can be sought via sampling, for example via Markov chain Monte Carlo (MCMC) methods [7, Ch. 6–7]. Here, we adopt an MH algorithm.

As previously mentioned, the GPR length scale, $1/l_i^2$, determines the kernel smoothness for a particular input x_i. Low positive values of l_i^2, or equivalently large positive values of $1/l_i^2$, indicate highly non-smooth behavior, or low correlation between input points. We exploit this to compute the sensitivity of a GPR-approximated function to each input, shown below.

Because of the exponential form of the kernel function in Eq. (6.2), it can be factorized in a way that allows for decoupling the contribution of each input x_i:

$$k(\boldsymbol{x},\boldsymbol{x}') = \sigma_f^2 \cdot \prod_{i=1}^{n} \exp\left(-\frac{1}{2l_i^2}\cdot(x_i - x_i')^2\right) = \sigma_f^2 \cdot \prod_{i=1}^{n} k_i(x_i, x_i')$$
$$k_i(x_i, x_i') := \exp\left(-\frac{1}{2l_i^2}\cdot(x_i - x_i')^2\right) \tag{6.8}$$

Following our derivation in Ref. [4], let us define $\Delta x_\rho \in \mathbb{R}^+$ as the increment in the input x_i, required to decrease the maximum covariance value by an amount of ρ:

$$k_i(x_i, x_i + \Delta x_\rho) = k_i(x_i, x_i) - \rho \tag{6.9}$$

where $0 < \rho \ll 1$. Using Eq. (6.9), we obtain:

$$\Delta x_\rho = l_i \cdot \sqrt{-2 \cdot \ln(1-\rho)} \tag{6.10}$$

On the basis of its definition, Δx_ρ reflects the amount of shift in the i-th input required to decrease the correlation between $f(x_i)$ and $f(x_i + \Delta x_\rho)$ by an amount of ρ. Values of Δx_ρ that exceed the actual range of the specific input in the dataset are not acceptable. A value lying close to zero indicates insensitivity of the function $f(\boldsymbol{x})$ to the individual input x_i.

Assuming two GPR sets, defined as $\mathcal{M}_a := \{\boldsymbol{y}_a, \boldsymbol{X}_a, \mathcal{P}_a\}$ and $\mathcal{M}_b := \{\boldsymbol{y}_b, \boldsymbol{X}_b, \mathcal{P}_b\}$, corresponding to different training datasets and hyperparameters, we wish to evaluate the success of each trained algorithm when inferring the function value at a given test point $\boldsymbol{x}_*$. Considering that GPR prediction corresponds to a Gaussian distribution, the KL divergence can be adopted as an appropriate metric to determine whether both predictive distributions, corresponding to $\mathcal{M}_a$ and $\mathcal{M}_b$, are equivalent [8, p. 57]. The KL divergence for Gaussian distributions assumes the form:

$$D_{KL}(\boldsymbol{x}_* \mid \mathcal{M}_a, \mathcal{M}_b) = \frac{1}{2}\left(\frac{Q_a(\boldsymbol{x}_*)}{Q_b(\boldsymbol{x}_*)} + \frac{(\bar{f}_a(\boldsymbol{x}_*) - \bar{f}_b(\boldsymbol{x}_*))^2}{Q_b(\boldsymbol{x}_*)} + \ln\frac{Q_b(\boldsymbol{x}_*)}{Q_a(\boldsymbol{x}_*)} - 1\right) \tag{6.11}$$

where $\bar{f}_i(\boldsymbol{x}_*)$ and $Q_i(\boldsymbol{x}_*)$ (with $i = \{a, b\}$) are the respective GPR predictive mean and variance calculated via Eq. (6.4).

Under the simplifying assumption of no cross-correlations in the predictive distribution (i.e., $\mathcal{E}(f(\boldsymbol{x}_{*_1}) - \bar{f}(\boldsymbol{x}_{*_1}))\cdot(f(\boldsymbol{x}_{*_2}) - \bar{f}(\boldsymbol{x}_{*_2})) \mid \boldsymbol{y}, \boldsymbol{X} = 0$ for $x_{*_1} \neq x_{*_2}$), we may approximate the global and marginalized KL divergence as:

$$\text{Global:} \quad D_{KL}(\mathcal{M}_a, \mathcal{M}_b) = \int_{\mathcal{X}} D_{KL}(\boldsymbol{x} \mid \mathcal{M}_a, \mathcal{M}_b)\,\mathrm{d}\boldsymbol{x} \tag{6.12}$$

$$\text{Marginalized:} \qquad D_{KL}(x_i|\mathcal{M}_a,\mathcal{M}_b) = \int_{\mathcal{X}_{\sim i}} D_{KL}(\mathbf{x}|\mathcal{M}_a,\mathcal{M}_b)\,\mathrm{d}\mathbf{x}_{\sim i} \tag{6.13}$$

where $\boldsymbol{x}_{\sim i}$ represents the input vector after eliminating input x_i, and $\mathcal{X}_{\sim i}$ denotes the respective domain. We highlight that the assumption of an uncorrelated predictive distribution does not comply with the GP definition, but it is adopted as a compromise. This simplifies the computation of the KL-divergence approximations, which are eventually numerically computed, via sampling.

6.3.2 Results

In this section, we implement the presented GPR-based methodology on the aeroelastic simulations, summarized in Figure 6.1. These concern a simplified example of two interacting turbines. In the first step, the GPR hyperparameters are inferred via the Bayesian approach, which relies on the use of an MH sampling scheme for approximating the posterior distribution. A sensitivity analysis follows, which is performed on the resultant distribution. Response surfaces (or slices) can be constructed for the prediction of DELs in the upwind and wake-affected turbine. In this brief overview, we do not elaborate on the sensitivity analysis or GPR comparison by means of the KL-divergence metrics—for more information, the interested reader is referred to Ref. [4].

6.3.2.1 Hyperparameter Identification

The proposed approach is implemented for inference of the posterior distribution of the GPR hyperparameters, by exploiting the previously described DWM-FAST simulations. The MH sampling algorithm is described in Ref. [4]. The marginal likelihood is estimated via the generation of 180 random input–output pairs, which feed the MH sampling loop. The remaining samples are used for posterior model validation. An agglomerative hierarchical clustering tree is built, forming 180 clusters on the input set, based on Ward's linkage and the Euclidean distance. This is implemented to ensure that the training samples are evenly distributed by randomly generating samples for each cluster. Table 6.1 presents the assumed configuration of the distribution of hyperparameters. A total of 10,000 samples are simulated with this procedure.

Figure 6.2 summarizes the hyperparameter distributions for the blade DELs in the edgewise and flap-wise directions of the upwind turbine. Despite similar prior assumptions, the posterior distributions converge to different intervals. Narrow distributions are obtained for the noise variance, the kernel variance, and the first three length-scale parameters (ℓ_1^2: wind speed; ℓ_2^2: turbulence intensity; and ℓ_3^2: shear exponent), while ℓ_4^2 (horizontal inflow skewness) presents a broader distribution in both the edgewise and flap-wise DELs. This implies that the horizontal inflow skewness bears a more limited influence on the computed DELs.

TABLE 6.1
Configuration of the Metropolis–Hastings Algorithm for Estimation of the Gaussian Process Regression (GPR) Hyperparameter Posterior

Hyperparameter	Prior	Proposal
Kernel variance σ_f^2	$\ln\sigma_f^2 \sim \mathcal{N}(-1,10^2)$	$\ln\sigma_f^2 \mid \ln(\sigma_f^2)_- \sim \mathcal{N}(\ln(\sigma_f^2)_-,0.4)$
Kernel scaling $\frac{1}{l_i^2}, i=1,\ldots,4$	$\ln\frac{1}{l_i^2} \sim \mathcal{N}(0,10^2)$	$\ln\frac{1}{l_i^2} \mid \ln\left(\frac{1}{l_i^2}\right)_- \sim \mathcal{N}\left(\ln\left(\frac{1}{l_i^2}\right)_-,0.4\right)$
Noise variance σ_w^2	$\ln\sigma_w^2 \sim \mathcal{N}(-1,10^2)$	$\ln\sigma_w^2 \mid \ln(\sigma_w^2)_- \sim \mathcal{N}(\ln(\sigma_w^2)_-,0.4)$

Note: Number of Monte Carlo samples: 10^4. The symbols $(\sigma_f^2)_-$, $\left(\frac{1}{l_i^2}\right)_-$, and $(\sigma_w^2)_-$ indicate the values of the same quantity drawn in the previous iteration of the MH sampling algorithm.

6.3.2.2 Estimated DELs Based on the Derived GPs

Once the hyperparameter posterior distributions are approximated, these can be used to generate DEL samples across the input space. This allows for the construction of one-dimensional (1D) slices, which portray the DEL as a function of single-input variables, assuming the remaining variables to be fixed. Figure 6.3 illustrates such DEL slices for the blade flap-wise direction, extracted using *maximum a posteriori* (MAP) hyperparameter estimates from the GPR of the upwind and downwind turbines. Each subplot reflects a particular hyperparameter that is varied, while the rest are set equal to the sample median. A spacing of 11 rotor diameters is assumed between, while the Wöhler exponent is assumed to be $m = 9$. The GPR predictive mean results are generally similar

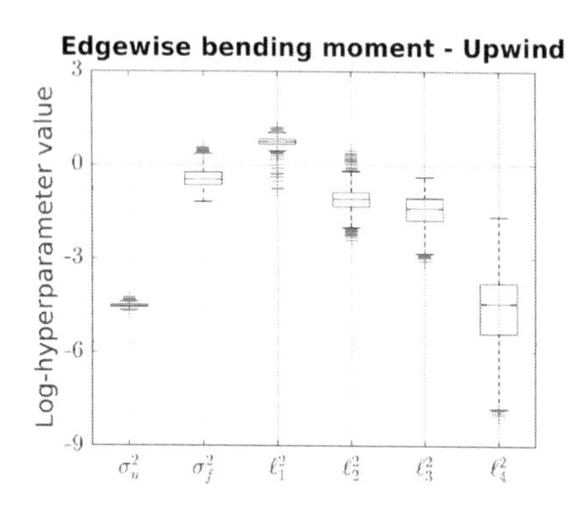

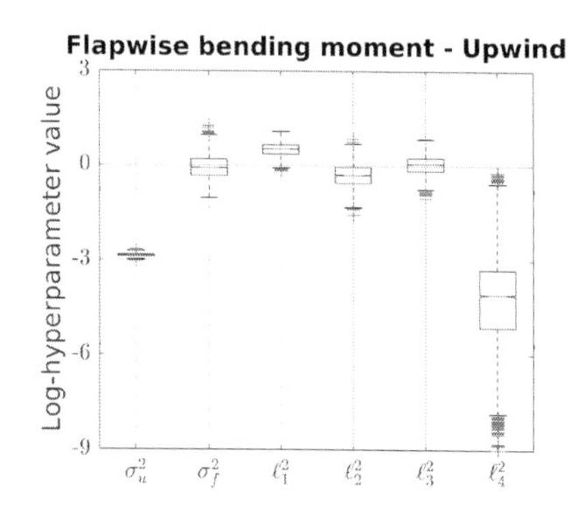

FIGURE 6.2 Boxplots displaying the distribution of the GPR hyperparameters for the DELs in the blade edgewise and flap-wise directions of the upwind (non-wake-affected) turbine. Wöhler exponent: 9. The notation $\frac{1}{l_i^2} = \ell_i^2$ is adapted in the figure axes. (Reproduced from Ref. [4].)

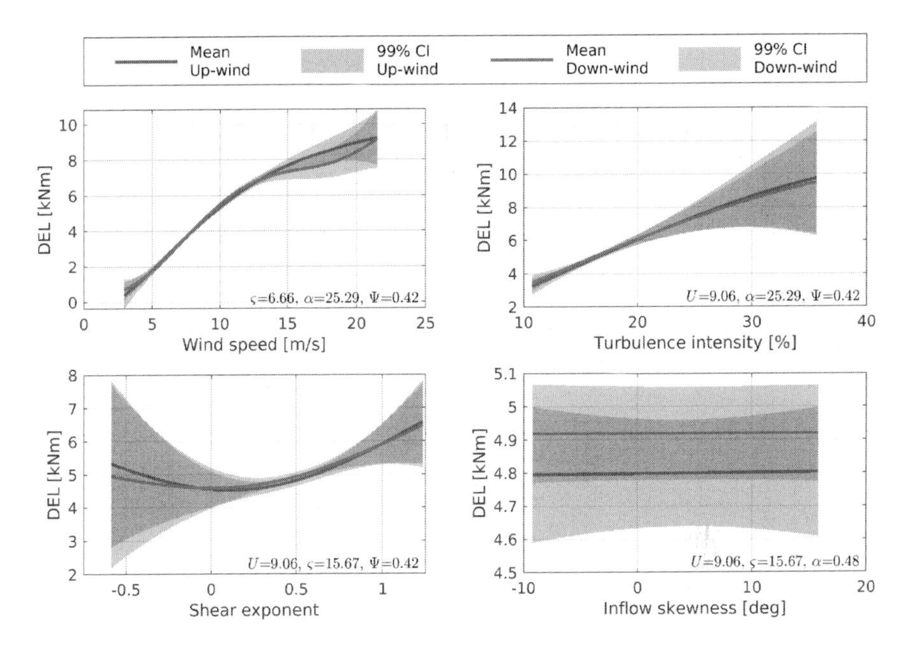

FIGURE 6.3 Slices of the DEL mean, along with a shading of the 99% confidence intervals (CIs), based on a GPR with *maximum a posteriori* (MAP) hyperparameter estimates. The DEL is computed in the flap-wise direction of the blade for the upwind and downwind turbines. The remaining inputs are assumed to be fixed at their training set sample median, with the corresponding values reported in each subplot. Turbine spacing: 11 diameters; Wöhler exponent: 9. (Reproduced from Ref. [4].)

for the upwind and wake-affected turbines. A notable DEL decrease is noted, however, for the wake-affected turbine at high wind speeds. Higher confidence is noted in the midrange of the input parameters, with the confidence intervals becoming broader near the input bounds. This is expected, given the lower number of samples around these bounds.

6.3.2.3 DEL Prediction in the Wake-Affected Turbine from Remote Measurements

In a realistic setting, we are interested in inferring DELs for turbines without sensors. To explore the feasibility of such a task, we assume that the upwind turbine in this example is fully instrumented, and we use these observations in order to estimate the DELs in the wake-affected (downwind) turbine. We assume different input combinations, as listed in Table 6.2. Scenario 1 in the table corresponds to an idealized case, with observations from all wind-field input features of the upwind turbine (including speed U_{up}, turbulence intensity σ_{up}, shear exponent α_{up}, and horizontal inflow skewness ψ_{up}). Scenarios 2 and 3 assume limited availability of wind-field inputs from the upwind turbine (namely, wind speed and turbulence)

TABLE 6.2
Upwind Observed Variables for Prediction of Damage Equivalent Loads (DELs) in the Wake-Affected Turbine for Each Scenario

Scenario	Input Variables (Measured Only in Upwind Wind Turbine)							
	U_{up}	σ_{up}	α_{up}	ψ_{up}	DEL_{BEW}	DEL_{BFW}	DEL_{TFA}	DEL_{TSS}
1	√	√	√	√				
2	√	√			√	√		
3	√				√	√		
4					√	√	√	√

along with the loads on the blade of the same turbine. Finally, scenario 4 corresponds to a case where information exclusively exists on structural-monitoring (load) measurements from both the blade and the tower of the upwind turbine. This information is then used to estimate the loads in the nonmonitored, wake-affected turbine via GPR.

We report the normalized mean squared error (NMSE) for the predicted DEL in the wake-affected turbine, based on local and upwind environmental and operational parameter (EOP) measurements in Figure 6.4. The horizontal axis reflects increasing spacing between turbines, with the Wöhler exponent set to 9 for the blade loads and 3 for the tower loads. Each subplot reflects a DEL computed for the tower and blade in different directions. The blade DELs in Figure 6.4a and 6.4b reveal that the best-performing scenarios are the second and third, which include the wind speed and the blade loads of the upwind turbine, while the remaining scenarios seem to follow closely. For the tower DEL estimates, a similar performance is noted. These results imply that the prediction of loads for a wake-affected turbine is possible, based on measurements obtained from an adjacent, upwind structure.

Note that the local minimum appearing in all plots (at a spacing of 11) is due to the recovery of the velocity deficit, and diffusion related to the wake-meandering phenomenon. In other words, at such a spacing, the aerodynamic loads in the upwind and downwind turbines are equivalent (and trivial to infer). For spacing in the range of 3–5, the prediction error is reduced when fusing information of wind flow (e.g., mean wind speed) with relevant loads data from the upwind turbine. This is justified when considering the mechanisms of the formation and transport of wakes.

6.3.3 Comments and Discussion

This section has presented a data-driven GPR-based approach for predicting DELs on wake-affected wind turbines. The method relies on local or remote wind-field

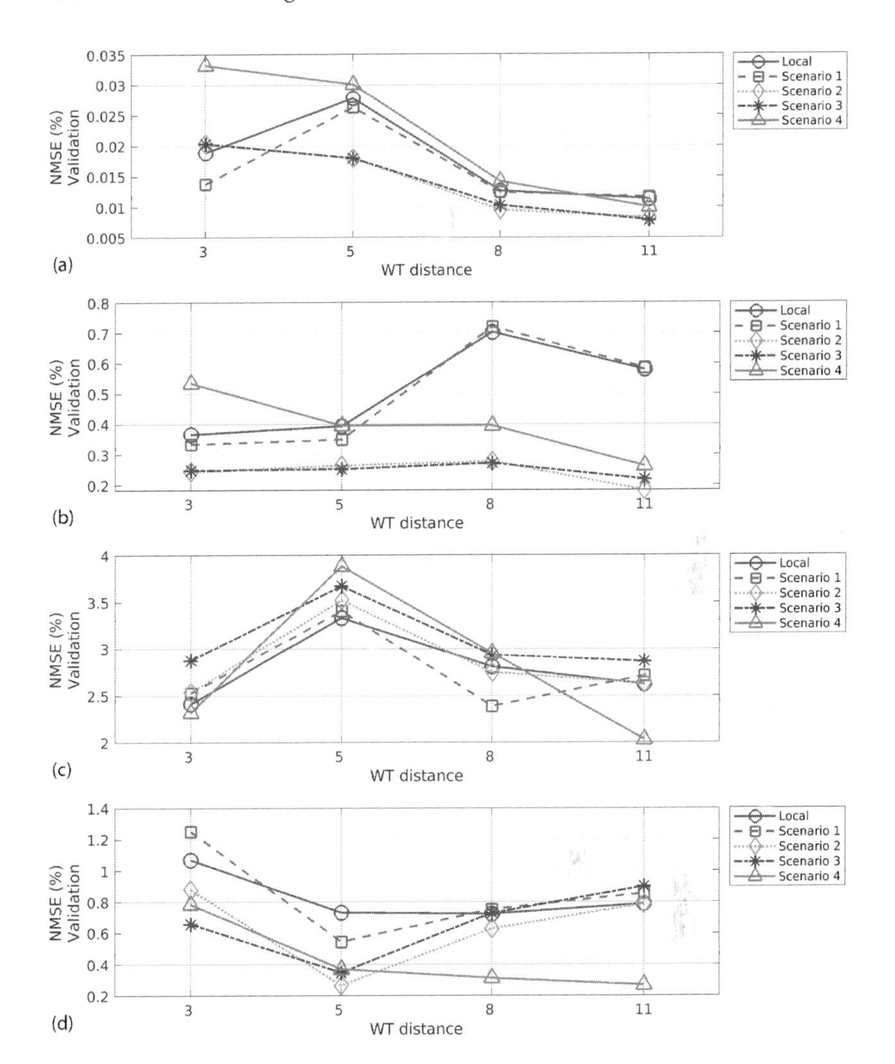

FIGURE 6.4 Comparison of the normalized mean squared error (NMSE) for the predicted DELs for different spacings between turbines. The DEL values are computed for the wake-affected turbine via the use of a GPR on the following inputs: Local environmental and operational parameter (EOP) measurements and remote measurements from an upwind turbine corresponding to scenarios 1–4 in the (a) blade root edgewise direction, (b) blade root flap-wise direction, (c) tower base fore–aft direction, and (d) tower base side-to-side direction. (Reproduced from Ref. [4].)

information and structural-monitoring measurements in adjacent, upwind turbines. A simplified example is used, which reveals low prediction errors under the use of a GPR framework. The initial results motivate an approach to aid the optimal instrumentation of turbines in a farm, while accounting for local interaction effects, such as those induced via wakes.

6.4 GRAPH NEURAL NETWORKS FOR GENERATIVE MODELING AT THE WIND FARM LEVEL

Graph neural networks (GNNs) [9, 10] combine flexible function approximations with relational modeling abilities and prior knowledge. In this section, based on the work of Ref. [11], we present an expansion of the general graph network (GN) model to probabilistic modeling with variational Bayes (VB) for graph data. We concentrate on generative modeling and uncertainty representation for continuous attributes ascribed to directed graph data. In essence, the strategy suggested allows for learning flexible distributions over entity and relation attributes while utilizing the relational structure of the data. There are numerous solutions to this problem in existing literature [12–14]. The aim of this work is to propose a *conditional relational variational auto-encoder* (CRVAE) approach, which capitalizes on the generalization capabilities of GNNs and their potency in modeling conditional distributions of structured data. The prevalence of noisy structured data and systems with stochastic or partially observable interactions among wind turbines in wind farms served as the initial impetus for this research. We demonstrate our model on simulated and real wind farm monitoring data.

6.4.1 METHOD

In the suggested scheme, modeled entities (*nodes*) and relations (*edges*) correspond to partially observed and/or stochastic *states*, which may be inferred from the available data. The partial observability further holds true for global (*graph*) attributes. We iterate the main ingredients of the CRVAE approach—originally presented in Ref. [9]. We denote global attribute augmented graphs as $G = (\mathcal{V}, \mathcal{E}, \mathbf{u})$, where $\mathcal{V}$ denotes the nodes (vertices) of the graph; $\mathcal{E}$ is the set of edges, with edge attributes $\mathbf{e}_k$, and $\{s_k, r_i\}$ representing the head (sender) and tail (receiver) nodes, respectively; while $\mathbf{u}$ stands for the global attribute. The GN (or GraphNets) block involves the following operations: (i) an edge update (represented by ϕ^e), (ii) a node update (ϕ^v), and (iii) a global update block (ϕ^u). The interested reader is referred to Refs. [9, 11] for a more detailed view of these operations. A graphical summary of the proposed CRVAE is shown in Figure 6.5.

In the CRVAE model, nodes, edges, and global attributes reflect a mixture of both deterministic and stochastic variables, which may be fully or partially observable. We refer to those variables that are observed as conditioning variables, hence the term *conditional relational VAE*. The observed node, edge, and global attribute quantities are denoted, respectively, as $\mathbf{v}^h, \mathbf{e}^h, \mathbf{u}^h$, where h denotes an observed (conditioning) variable. The remaining variables are noted as $\mathbf{v}^d, \mathbf{e}^d, \mathbf{u}^d$. The observed attributes can be used to create a *conditioning graph* $G_h = (\mathcal{V}_h, \mathcal{E}_h, \mathbf{u}_h)$ and a *state graph* $G_x = (\mathcal{V}_x, \mathcal{E}_x, \mathbf{u}_x)$. The full graph state is denoted by $G_d = (\mathcal{V}_x \cup \mathcal{V}_h, \mathcal{E}_x \cup \mathcal{E}_h, \mathbf{u}_x \cup \mathbf{u}_h)$, where $\cup$ denotes set union. Since part of the node, edge, and global attributes may be stochastic (or unobservable or unknown), a latent graph representation $G_z = (\mathcal{V}_z, \mathcal{E}_z, \mathbf{u}_z)$ is assumed. The joint

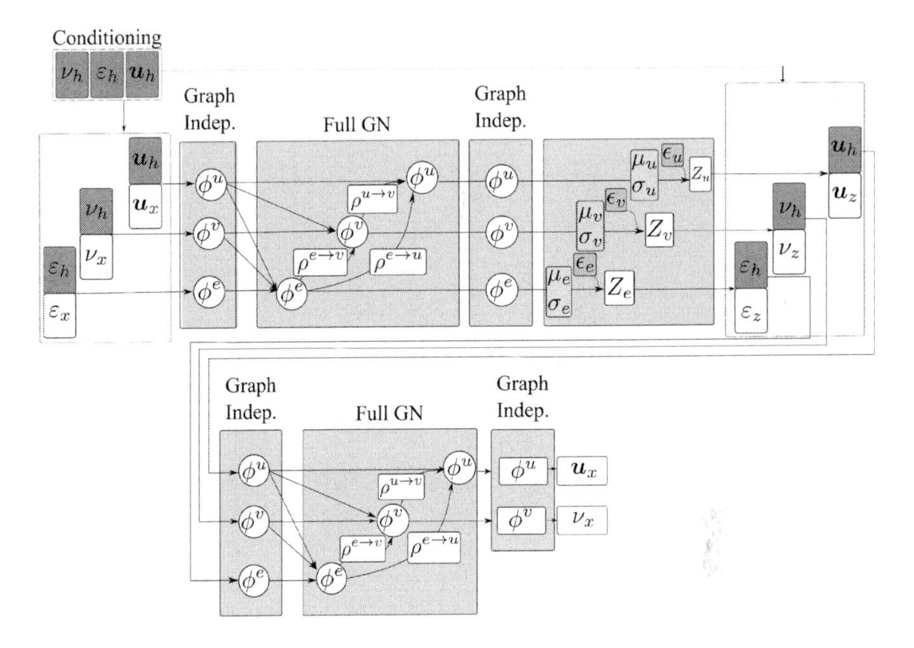

FIGURE 6.5 Graphical summary of the architecture of the proposed *conditional relational variational auto-encoder* (CRVAE) with a single message-passing step in the encoder and decoder. (Reproduced from Ref. [11].)

distribution of the graph-structured observations is derived on the basis of the following model:

$$p(G_x;G_h) = \int p(G_x \mid G_z;G_h)p(G_z;G_h)\,dG_z \tag{6.14}$$

where $p(G_z;G_h) = p(\mathcal{V}_z;\mathcal{V}_h)p(\mathcal{E}_z;\mathcal{E}_h)p(\mathbf{u}_z;\mathbf{u}_h)$ is the distribution of the latent variables given as G_h. In this context, we assume a prior distribution for the latent variable conditioned on G_h, which is further factorized along each node, edge, and global latent variable separately. The interested reader can consult Ref. [11] for details of the factorization of the joint probability density function.

An approximate posterior (i.e., *recognition model*) is assumed for G_z as $q_\phi(G_z \mid G_x;G_h)$ together with a generative model for G_x, $p_\theta(G_x \mid G_z;G_h)$, whose model parameters θ and inference parameters ϕ are learned synchronously following the steps established in a variational autoencoder (VAE) [15]. Assuming independent identically distributed (i.i.d.) graph observations $\left\{G_x^{(1)},\ldots G_x^{(i)}\right\}$, the evidence lower bound (ELBO) for the marginal log-likelihood reads:

$$\begin{aligned}\mathcal{L}(\theta,\phi;G_x^{(i)},G_h^{(i)}) = {} & \mathbb{E}_{q_\theta(G_z \mid G_x^{(i)};G_h^{(i)})}[\log p_\theta(G_x^{(i)} \,|\, G_z;G_h^{(i)})] \\ & - D_{KL}(q_\phi(G_z \,|\, G_x^{(i)};G_h^{(i)}) \,\|\, p_\theta(G_z;G_h^{(i)}))\end{aligned} \tag{6.15}$$

We intend to perform efficient approximate inference over the G_z graph while taking advantage of the existing *relational structure* in the data. A convenient choice for G_z is to assume parametric distributions over edges, nodes, and global attributes. In this work, we adopt Gaussian distributions for the graph-structured observation G_x, the graph-structured conditioning G_h, and the graph-structured latent G_z, as follows:

$$\mathcal{V}^z \sim q_\phi^{(\mathcal{V})}(G_z \mid G_x; G_h) = \mathcal{N}(f_{q\phi}^{\mu(\mathcal{V})}(G_x; G_h), f_{q\phi}^{\sigma^2_{(\mathcal{V})}}(G_x; G_h)) \tag{6.16}$$

$$\mathcal{E}^z \sim q_\phi^{(\mathcal{E})}(G_z \mid G_x; G_h) = \mathcal{N}(f_{q\phi}^{\mu(\mathcal{E})}(G_x; G_h), f_{q\phi}^{\sigma^2_{(\mathcal{E})}}(G_x; G_h)) \tag{6.17}$$

$$\mathbf{u}_z \sim q_\phi^{(\mathbf{u})}(G_z \mid G_x; G_h) = \mathcal{N}(f_{q\phi}^{\mu(\mathbf{u})}(G_x; G_h), f_{q\phi}^{\sigma^2_{(\mathbf{u})}}(G_x; G_h)). \tag{6.18}$$

A GN is proposed for inferring the parameters. The functions $f_{\cdot}^{\mu(\cdot)}$ and $f_{\cdot}^{\sigma^2_{(\cdot)}}$ are implemented as a GN for accounting for relational information while inferring over $\mathcal{V}_z, \mathcal{E}_z$ and $\mathbf{u}_z$. The parameterization for nodes, edges, and global variables comprises the corresponding states of the GN at the final message-passing step, as depicted in Figure 6.5. Furthermore, a GN generator network $g_{p\theta}(\cdot)$ is used for p_θ. Since the prior and posterior are factorized over nodes, edges, and the global variable of each graph data point, the ELBO in Eq. (6.15) is split accordingly as:

$$\begin{aligned}
\mathcal{L}(\boldsymbol{\theta}, \phi; G_x^{(i)}, G_h^{(i)}) = \; & \mathbb{E}_{q_\theta(G_z \mid G_x^{(i)}; G_h^{(i)})}[\log p_\theta(G_x^{(i)} \mid G_z; G_h^{(i)})] \\
& - \beta_{\mathcal{V}} D_{KL}(q_\phi^{(\mathcal{V})}(G_z \mid G_x^{(i)}; G_h^{(i)}) \,\|\, p_\theta^{(\mathcal{V})}(G_z; G_h^{(i)})) \\
& - \beta_{\mathcal{E}} D_{KL}(q_\phi^{(\mathcal{E})}(G_z \mid G_x^{(i)}; G_h^{(i)}) \,\|\, p_\theta^{(\mathcal{E})}(G_z; G_h^{(i)})) \\
& - \beta_{\mathbf{u}} D_{KL}(q_\phi^{(\mathbf{u})}(G_z \mid G_x^{(i)}; G_h^{(i)}) \,\|\, p_\theta^{(\mathbf{u})}(G_z; G_h^{(i)}))
\end{aligned} \tag{6.19}$$

where $\beta_{\mathcal{V}}, \beta_{\mathcal{E}}, \beta_{\mathbf{u}}$ are regularization parameters that can be used for controlling disentanglement [16] or for preventing posterior collapse and aiding training [17, 18]. The parameterization of distributions over $\mathcal{V}, \mathcal{E}$ and $\mathbf{u}$ allows for defining alternative ELBOs for applying the VB approach—variational inference (VI) is outlined in Section 6.7.3.2. Note that the distribution does not need to be factorized along the latent vector components, which allows for flexibility using a suite of distribution options [19–21].

6.4.2 Results

6.4.2.1 Wind Farm Operational Data

We demonstrate the developed approach on an application that inherently benefits from the incorporation of structure into the modeling framework, namely, wind turbine structures within a farm. We treat individual wind turbines as nodes of the GN, which are described by a set of variables, including information on the position, wind speed, and power production. We can also assume that operational SCADA data can be exploited to derive information on the state of the turbine, as commonly described by 10-min statistics information. This information is partial (coarse measurements) and stochastic in nature, as it is contaminated with significant noise. Further to the inherent uncertainties relating to observations, the state of wind turbine structures is a result of interaction effects, such as the wake, which affects the wind field (speed and turbulence), power production, and dynamic response of downstream turbines. The wake effects are defined, albeit in a stochastic manner, on the basis of the following variables: (i) turbulence (global variable), (ii) wind orientation (global variable), (iii) upwind turbine nacelle orientation (node variable), and (iv) the relative position between the two turbines, the rotor diameter, and the distance between the two turbines (edge static variable). The interaction between two wind turbines is one-way directional but can change directionality depending on the wind orientation. Conditional on the turbines' characteristics and farm layout, our objective is to infer the distribution of operational characteristics of wind turbines in a wind farm. An important aspect of learning graph-structured data is to include stochasticity in, for example, the edges and interactions of the considered graph. As noted previously, static graph edges consider features such as the distance and position of turbine pairs. This type of geometrical and spatial parameterization allows us to generalize the model in terms of unseen farm layouts, while learning from monitoring data from different farm geometries.

6.4.2.2 Real Wind Farm SCADA Dataset

Six months' worth of 10-min average SCADA measurements were used as the training data for CRVAE models. During training, 20% of the turbine data was randomly masked. All runs utilized the Adam optimizer [22] with default parameters and a small learning rate of $5 \cdot 10^{-5}$.

The ELBOs for all trained models, featuring different aggregation functions and different numbers of encoder and decoder layers, are shown in Table 6.3, where the proposed CRVAE models are compared against a two-layer multilayer perceptron (MLP)-based conditional VAE (CVAE) [23]. The composite aggregation function corresponds to a concatenation of *mean*, *max*, and *min* aggregator functions. Despite their expected larger network site, the performance of composite aggregators is superior, aligning with recent literature observations [24]. We note that the physics of the problem is the main motivation for using composite aggregators; by using such aggregators, it is easier to discriminate between non-wake-affected and wake-affected turbines within a farm. This is meaningful, as upstream turbines bear a higher potential for power production. We observe that the largest CVAE model was the worst-performing of the evaluated CVAE models. The majority of the CRVAE models largely outperform the CVAE alternatives, due to the incorporation of relational inductive biases. To further support this claim, we plot the results of an imputation study for masked turbines depending on upstream turbines. Qualitative results of the imputation of wind speed data are shown in Figure 6.6.

TABLE 6.3
Test Set ELBO of the Learned Models on the Anholt SCADA Dataset

Model	MLP Units	N^{G_z} Size	Enc.	Dec.	Agg.	No. of Parameters	ELBO
CRVAE	64	32	0	1	Mean	184,717	1.96(0.30)
	64	32	1	1	Mean	341,517	6.99(0.29)
	64	32	2	2	Mean	498,317	7.48(0.61)*
	64	32	2	2	(Comp.)	522,893	**8.11(0.48)***
	64	32	3	3	(Comp.)	679,693	7.70(0.53)*
CVAE	128	64	—	—	—	77,194	2.12(0.10)
	256	64	—	—	—	252,554	1.17(0.16)
	384	96	—	—	—	563,146	1.23(0.09)

Numbers in parentheses are the standard deviations of the ELBO estimates in the test set (higher is better). The same number of node, edge, and global attributes were used in (N^{G_z}). The $\cdot^*$ superscript denotes results that were not derived from early stopping.

Abbreviations: CRVAE: Conditional relational variational autoencoder (VAE), CVAE: Conditional VAE, ELBO: Evidence lower bound, MLP: Multilayer perceptron, SCADA: Supervisory control and data acquisition.

6.4.2.3 Simulated Wind Farm Dataset

We further test how the trained CRVAE generalizes on arbitrary wind farm layouts (i.e., for spatial configurations that have not formed part of the training set). The simulated data was generated using the steady-state wind farm wake simulator FLORIS (FLOw Redirection and Induction in Steady State) [25]. We use a single farm layout for training and a different configuration for testing. Both farms are simulated with random global wind attributes, such as wind direction and average wind speed. We construct a graph, where wind turbines are the nodes and edges are created between the turbines by truncating an all-to-all graph to a

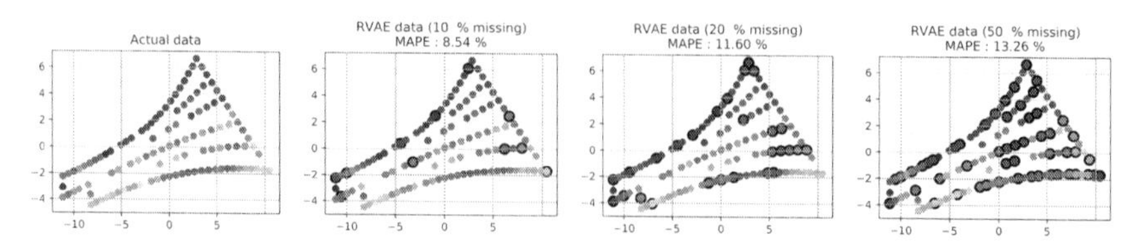

FIGURE 6.6 Performance of the CRVAE for imputation of wind speed prediction on the well-known Anholt supervisory control and data acquisition (SCADA) dataset. The imputed points are marked with dark circles. The mean absolute percentage error (MAPE) is reported, which is computed as $1/N^T \sum_{i=1}^{N^T} \left(\left|\mathbf{v}_i^T - \hat{\mathbf{v}}_i^T\right|\right)/\left|\mathbf{v}_i^T\right|$, where $\mathbf{v}_i^T$. is the actual value of node i, N^T is the number of target turbines, and $\hat{\mathbf{v}}_i$ is the CRVAE prediction. (Reproduced from Ref. [11].)

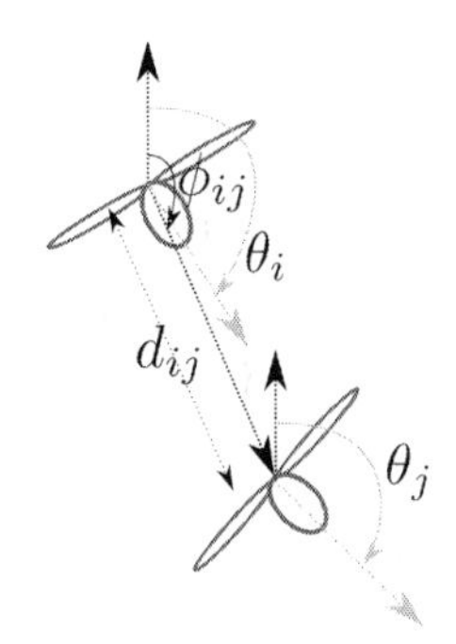

FIGURE 6.7 (Left) The edge attributes $\mathcal{E}_h$ for an edge with $s = i, r = j$ are $(\cos(\phi_{ij}), \sin(\phi_{ij}), d_{ij})$. The nacelle yaw angle θ_i is included as a node attribute. (Middle) Simulated farm layout for the training set. (Right) Simulated farm layout for the test set. (Reproduced from Ref. [11].)

cutoff distance of $100 \cdot d$, where d is the turbine rotor diameter. The 5 MW NREL prototype turbine with a diameter $d = 126$ m [26] was used for the simulations.

To test model generalization, its implicit ability to learn how to use the relative position of the turbines, the nacelle wind orientation, and the power generated by upwind turbines, the model is tested on a different farm layout compared to the training layout, as shown in Figure 6.7. We perturb the yaw directions of the turbines with $\mathcal{N}(0.,5.^{\circ})$ around the global wind orientation to mimic stochasticity corresponding to yaw measurement errors. The same mean wind and wind orientation global conditions (u_h) are employed for the simulations in both the train and test wind farms.

We explore learning of the orientation-dependent wake deficit for individual wind turbines within the wind farm. As indicated in Figure 6.7 (left), the edge features that are employed are the angle ϕ_{ij}, defined by $\cos(\phi_{ij})$, $\sin(\phi_{ij})$, and the distance d_{ij}. Since in a real setting, yaw angles may vary due to controller settings on individual turbines, we further use the yaw angle θ_i with respect to the north as a node feature, along with the mean and standard deviation of hub-height wind speed and power production.

Figure 6.8 reveals that the CRVAE model is successful in generating the orientation-dependent wake deficit for individual turbines in a farm of a different (previously unobserved) spatial layout. To further visualize the wake deficit, when predicting on the basis of a model trained for a single simulated farm configuration, we use an individual turbine as a probe and move it along a regular grid while keeping a further (source) turbine at a fixed central position (0,0). By using the predicted wind speed at the probe turbine as input, we are able to compute and map the wake deficit two-dimensionally (2D) behind the source turbine, using the described approach (see Figure 6.9). The figure further indicates the spatial dependence of the wake deficit; for distances larger than 200 m, the wind deficit is accurately predicted. Close distances between turbines (<200 m) lead to a higher error in the wakes' prediction, which is expected since the CRVAE

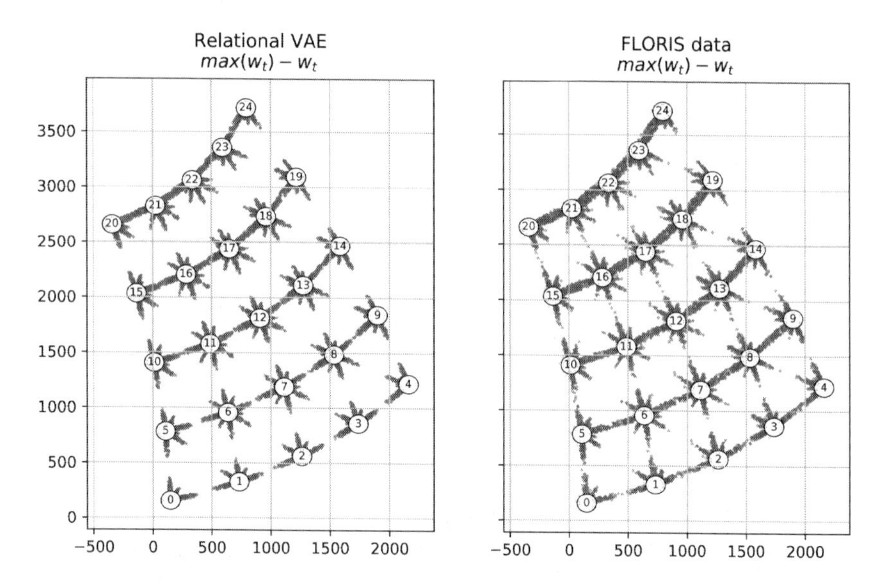

FIGURE 6.8 Comparing the CRVAE prediction and FLOw Redirection and Induction in Steady State (FLORIS) reference values for the wind deficits on the simulated test farm. Each wind turbine is plotted in a 2D polar coordinate system, toward the orientation of the *incoming* wind. The distance from the origin is proportional to the wake deficit, estimated as $max(v) - \hat{v}$, where $\hat{v}$ corresponds to the mean wind. (Reproduced from Ref. [11].)

is asked to extrapolate to previously unseen ranges of spacing. In Figure 6.10, CRVAE predictions and reference data from the simulated test wind farm are shown for conditioning on the global latent variable $\mathbf{u}_h$. The CRVAE seems to accurately reproduce the wind deficits on all wind turbines, despite the fact that this takes place on an unseen (new) configuration.

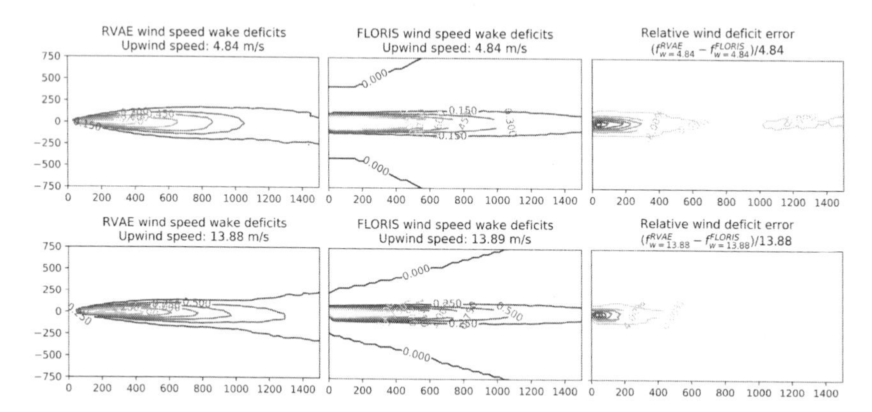

FIGURE 6.9 Comparing the CRVAE learned and FLORIS reference spatial distribution of the wake-related wind deficit values, evaluated as $w_{(0,0)} - w_{(x,y)}$, where $w_{(0,0)}$ reflects the wind speed at the upwind turbine, and (x, y) denotes the wind speed for the *probe* turbine located at $w_{(x,y)}$. (Reproduced from Ref. [11].)

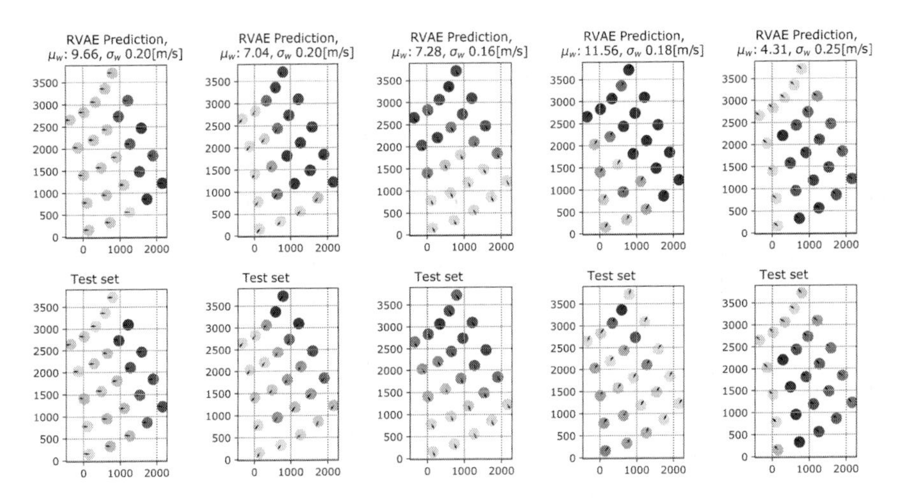

FIGURE 6.10 Examples of mean wind (top) as predicted from the CRVAE for an unseen farm, and (bottom) plot of the corresponding mean wind, as simulated in the test dataset. Lighter colors correspond to higher wind speeds. (Reproduced from Ref. [11].)

6.4.3 Comments and Discussion

In this section, we have presented an attributed graph approach that can probabilistically represent relationships amongst wind turbines in a wind farm, allowing for uncertainties in the modeling. At the same time, the scheme is bestowed with generalization (transfer) capabilities, since the aspect of conditioning allows for parameterizing, among other variables, with respect to the farm's spatial configuration. The proposed method fuses physical intuition (inductive bias), flexible function approximation through GNs, and the VB scheme through re-parameterized gradients. The approach is validated on a real-world dataset, the Anholt wind farm SCADA data, and a simulated wind farm for demonstrating generalization (transfer) potential. Such a scheme can enable computationally efficient and precise modeling of wake effects for improving wind farm siting [27], farm layout optimization [28], wind farm control [29], and power production potential, offering a powerful generative tool for operation and maintenance planning.

6.5 UNSUPERVISED LOCAL CLUSTER-WEIGHTED BOOTSTRAP AGGREGATION OF OUTPUTS FROM MULTIPLE STOCHASTIC SIMULATORS

We review investigations, originally presented in Ref. [30], which explore the reduction of uncertainty when physics-based (engineering) modeling is exploited for estimation purposes. To this end, the output from multiple stochastic forward simulators (in this case, numerical physics-based models) is fused for inferring a quantity of interest, while reducing the model-form uncertainty. In the absence of complementary information from measurements, this is achieved by essentially weighing the different modeling options, which come at different precision, granularity, and even performance for different input ranges. Engineering models of any form (analytical, numerical, or empirical) are valuable as simulators for the prediction of a system's response, but they invariably come with approximations and associated uncertainties. In this sense, it is almost impossible to choose an optimal simulator for any operation setting of a dynamical system, particularly under the availability of limited sampled data. However, engineers often rely on outputs from multiple simulators for corroborating their assessment. Capitalizing on this idea, we have proposed an unsupervised local cluster-weighted bootstrap aggregation (ULC-BAG) method, which employs ensemble learning based on local *Clustering* and bootstrap aggregation (*Bagging*). The framework funnels stochastic predictions of individual simulators in an ensemble estimate rather than trying to select an individual optimal performance, which allows for diversifying the hypothesis space.

6.5.1 Method

The ULC-BAG scheme is described in Algorithm 6.1, which was originally proposed in Ref. [30], and is overviewed in the context of reviewing data-centric monitoring tools, which mix different sources of information, at the wind farm level. The first step consists of clustering by means of a variational Bayesian Gaussian mixture (VBGM). This allows us to categorize the outputs of similarly performing simulators and further derive the probability map (i.e., the weights) of the clusters. Since we are interested in differentiating performance across input ranges, the clustering operation is carried out for stochastic model outputs across individual bins of the input space. This implies that the probability map can be established for each local input bin, which delivers an adaptive solution, able to select alternate simulators according to the current region of the input space. The second step of this framework pertains to a local cluster-weighted bootstrap aggregation, which serves for delivering the weighted estimate of the clustered ensemble of simulator outputs.

Algorithm 6.1: Unsupervised local cluster-weighted bootstrap aggregation (ULC-BAG) of outputs from stochastic simulators

Input: $\{\mathcal{Y}_l\}_{l=1}^{s}$: vector of stochastic observations from each simulator $\{\mathcal{S}_l\}_{l=1}^{s}$, $nMCs >> 1$: Number of repeats, $\boldsymbol{\mathcal{X}} = \{\mathbf{x}^{(i)}\}_{i=1}^{N}$: d-dimensional explanatory input variables i.i.d. draws from the joint distribution, Δx: input bin size, Q: a large prespecified number, M: a large prespecified number

Result: $\hat{g}_{w,bagg}$: unsupervised local cluster-weighted bagged estimator

1 Set $\mathcal{Z} = \left[\mathcal{X}^T \left\{ \mathcal{Y}_l^T \right\}_{l=1}^{s} \right]$;
2 **for** *iter* $= 1,\ldots,nMCs$ **do**
3 Randomly permute the cases in $\mathcal{Z}$, and then select n training samples such that $n = 0.7N$;
4 Assign every $\left\{ \mathbf{x}^{(i)} \right\}_{i=1}^{n}$ to be the center of bins of size Δx;
5 **for each** $\Delta\mathbf{x}$ **do**
6 *//Local Clustering*
7 **for** $q = 1,\ldots, Q$ **do**
8 Randomly draw 90% of local responses of simulators $\left\{ \mathcal{Y}_l^T \right\}_{l=1}^{s}$ without replacement;
9 Identify local clusters $C_q\left(\left\{ \mathcal{Y}_l^T \right\}_{l=1}^{s}, K_q \mid \Delta \mathrm{x} \right)$, using a variational Bayesian Gaussian mixture;
10 Compute local weights $P_q(\mathcal{S}, C \mid \Delta\mathbf{x})$;
11 **end**
12 *//Expected local cluster weights*
13 Cluster stability: Use majority vote $out-of-Q$ to establish the expected local number of clusters $K_{\Delta\mathbf{x}}$;
14 Compute expectation of local weights: $P(\mathcal{S}, C \mid \Delta \mathrm{x}) = \mathbb{E}\left[P_q(\mathcal{S}, C \mid \Delta \mathrm{x}) \right]_Q$;
15 *//Local weighted bagging*
16 **for** $k = 1,\ldots, M$ **do**
17 Construct a weighted bootstrap sample $\left(\mathrm{x}_*^{(1)}, \mathcal{Y}_*^{(1)} \right), \ldots, \left(\mathrm{x}_*^{(n_{\Delta\mathrm{x}})}, \mathcal{Y}_*^{(n_{\Delta\mathbf{x}})} \right)$ by randomly drawing $n_{\Delta\mathbf{x}}$ times with replacement from the local clustered data, the local weights being $P(\mathcal{S}, C \mid \Delta\mathbf{x})$;
18 Compute the bootstrapped estimator $h_n(\cdot)$:

$$\hat{g}_*^k = h_n\left(\left(\mathrm{x}_*^{(1)}, \mathcal{Y}_*^{(1)} \right), \ldots, \left(\mathrm{x}_*^{(n_{\Delta\mathrm{x}})}, \mathcal{Y}_*^{(n_{\Delta\mathrm{x}})} \right) \right)$$

19 **end**
20 Compute the local cluster-weighted bagged estimator:

$$\hat{g}_{w,bagg} = \frac{\sum_{k=1}^{M} \hat{g}_*^k}{M}$$

21 **end**
22 **end**

The clustering step of the ULC-BAG serves for (1) categorizing the individual simulator outputs (naturally, some approaches contain approximations that render them more similar to others) and (2) deriving the weights, which essentially assign a probability map for each output cluster. Different clustering options are available in the existing literature, including self-organizing maps, k-means, Gaussian mixtures, and hierarchical or deep learning clustering schemes. The ULC-BAG specifically exploits VBGM clustering.

Finite Gaussian mixtures can flexibly account for nontypical probabilistic densities sampled from heterogeneous stochastic populations. This is achieved by mixing a finite number K of Gaussian distributions of unknown parameters:

$$p(\mathcal{Y} \mid \pi, \mu, \Sigma) = \sum_{k=1}^{K} \pi_k \mathcal{N}(\mathcal{Y} \mid \mu_k, \Sigma_k) \tag{6.20}$$

where π_k is the mixing coefficient, and each mixture component $\mathcal{N}(\mathcal{Y} \mid \mu_k, \Sigma_k)$ reflects a Gaussian density, defined by its mean μ_k and covariance Σ_k [31]. Maximization of the likelihood function of the Gaussian mixture can be used to estimate the parameters of the mixture components:

$$p(\mathcal{Y} \mid \pi, \mu, \Sigma) = \prod_{i=1}^{N} \left[\sum_{k=1}^{K} \pi_k \mathcal{N}(\mathcal{Y}^{(i)} \mid \mu_k, \Sigma_k) \right] \tag{6.21}$$

Due to the presence of singularities in the event of a collapse of one of the mixture components to a specific data point, this problem is often ill-posed [31]. Further complications relate to knowing which component a sample belongs to during training, typically solved via the expectation–maximization (*EM*) algorithm [32], as well as the lack of a clear rule for the selection of the number of Gaussian mixtures K. Bayesian variational treatment can solve the latter issue by automatically determining K via model selection. Each of these components involves a mixing coefficient $\{\pi_k; k = 1,\ldots,K\}$. The end goal is to evaluate the posterior distribution,

$$P(\pi, \mu, \Sigma | \mathbf{Y}) = \frac{P(\mathbf{Y}, \pi, \mu, \Sigma)}{P(\mathbf{Y})} = P(\pi, \mu, \Sigma) \frac{P(\mathbf{Y} \mid \pi, \mu, \Sigma)}{P(\mathbf{Y})} \tag{6.22}$$

which is generally intractable.

In view of this, variational methods can be applied for determining a tractable lower bound on $P(\mathbf{Y})$, by minimizing the KL divergence $KL\left(q(\Theta) \parallel P(\pi, \mu, \Sigma \mid \mathbf{Y})\right)$, where $q(\Theta)$ generally corresponds to a simple parametric family of the posterior probability density. The family $q(\Theta)$ aims to approximate the true posterior distribution, and Θ is a vector collecting all parameters. For a more detailed discussion of variational inference (VI), the reader can refer to Section 6.6.4.2.

The mixture inference is solved using the algorithm in Ref. [33], allowing us to approximate the posterior distribution of the clusters across the binned input space.

The choice of the approximate posterior distribution is a core problem in VI. Instead, normalizing flows are a potential alternative for specifying flexible, arbitrarily complex, and scalable posterior distributions compared to the known limitations of VI [34, 35]. It should be noted that the cluster stability is intermittently examined for perturbations of the data in a given bin. A simple approach is adopted here, where we repeatedly draw subsamples from the population and apply the clustering algorithm, and subsequently perform a majority vote to select the most commonly occurring cluster's mixture structure.

We offer a simple illustration to demonstrate the derivation of the probability map (weights) of the clustered simulator outputs. Let us assume the availability of three simulators, each one comprising 12 data (output) points, which are clustered as shown in Figure 6.11. Instead of considering the total simulator probabilities $P(\mathcal{S}_i)$, or the conditional probabilities $P(\mathcal{S}_i \mid C_j, \Delta\mathbf{x})$, we here present a framework that pushes the aggregate estimate toward larger density clusters (i.e., areas of consensus). This implies that concurring simulators reinforce each other, or, conversely, neutralize each other in case of a lack of consensus. We can, therefore, calculate the weights, which are computed as the joint probability $P(\mathcal{S}_i, C_j \mid \Delta\mathbf{x})$ of a cluster $\{C_j; j = 1,2\}$ and a simulator $\{\mathcal{S}_i; i = 1,3\}$,

$$P(\mathcal{S}_i, C_j \mid \Delta\mathbf{x}) = P(S_i \mid C_j, \Delta\mathbf{x}) \cdot P(C_j \mid \Delta\mathbf{x}) \tag{6.23}$$

where

$$\begin{aligned} P(C_j \mid \Delta\mathbf{x}) &= \frac{N_{C_j \mid \Delta\mathbf{x}}}{N}, \\ P(S_i \mid C_j, \Delta\mathbf{x}) &= \frac{N_{S_i \mid C_j, \Delta\mathbf{x}}}{N_{C_j \mid \Delta\mathbf{x}}} \end{aligned} \tag{6.24}$$

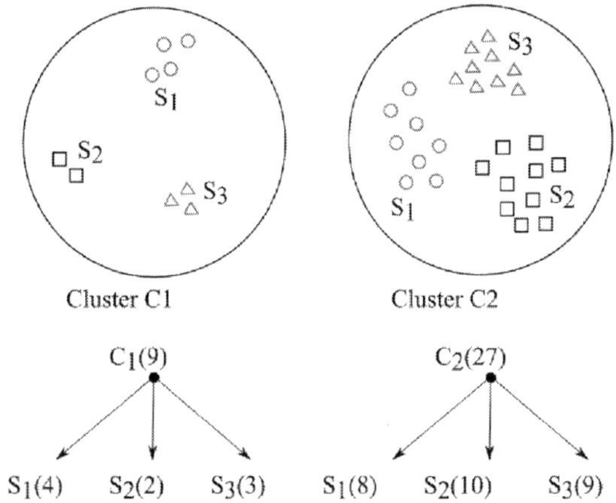

FIGURE 6.11 Illustration of two clusters $\{C_1, C_2\}$ of the stochastic output from three simulators $\{S_1, S_2, S_3\}$. (Reproduced from Ref. [30].)

TABLE 6.4
Assigned Weights, $P(\mathcal{S}_i, C_j \mid \Delta \mathbf{x})$, to the Clustered Simulators Output

	Simulator $\mathcal{S}_1$	Simulator $\mathcal{S}_2$	Simulator $\mathcal{S}_3$
Cluster C_1	0.111	0.056	0.083
Cluster C_2	0.222	0.278	0.250

where $N_{C_j|\Delta \mathbf{x}}$ denotes the number of data points in cluster C_j; N is the total number of available simulator outputs in $\Delta \mathbf{x}$; and $N_{\mathcal{S}_i|C_j,\Delta \mathbf{x}}$ is the number of cluster C_j data points corresponding to simulator $\mathcal{S}_i$. Figure 6.11 illustrates the outcome with: four data points in C_1 corresponding to $\mathcal{S}_1$, two data points corresponding to $\mathcal{S}_2$, and three data points corresponding to $\mathcal{S}_3$. In C_2, eight data points correspond to $\mathcal{S}_1$, 10 data points correspond to $\mathcal{S}_2$, and nine data points correspond to $\mathcal{S}_3$. The resulting joint probabilities of individual clusters and simulators are outlined in Table 6.4.

The final step in this method corresponds to local cluster-weighted bootstrap aggregating (bagging) [36], to fuse the ensemble of outputs from each bin—based on the previously extracted weights—into one single aggregated predictor. The bagging algorithm is composed of the following steps:

1. Construct a weighted bootstrap sample $\left(\mathbf{x}_*^{(1)}, \mathcal{Y}_*^{(1)}\right), \ldots, \left(\mathbf{x}_*^{(n_{\Delta \mathbf{x}})}, \mathcal{Y}_*^{(n_{\Delta \mathbf{x}})}\right)$ by randomly drawing $n_{\Delta \mathbf{x}}$ times with replacement from the local clustered data.
2. Compute the bootstrapped estimator $h_n(\cdot): \hat{g}_* = h_n\left(\left(\mathbf{x}_*^{(1)}, \mathcal{Y}_*^{(1)}\right), \ldots, \left(\mathbf{x}_*^{(n_{\Delta \mathbf{x}})}, \mathcal{Y}_*^{(n_{\Delta \mathbf{x}})}\right)\right)$. $h_n(\cdot)$ defines an estimator as a function of the data.
3. Repeat steps (1) and (2) M times to infer $\left\{\hat{g}_*^k, k = 1, \ldots, M\right\}$. The final local cluster-weighted bagged estimator is obtained as $\hat{g}_{w,bagg} = \frac{\sum_{k=1}^{M} \hat{g}_*^k}{M}$.

The estimator $h_n(\cdot)$ assumes a flexible form, in that it can represent any learning algorithm $\Psi: \gamma \rightarrow \phi$, which produces a predictor $\Phi = \Psi(L) \in \phi$ for given input data $L \in \gamma$. Alternative forms of such an estimator can include the expected value (or any other quantile), a function that fits a probability distribution, a surrogate, and so on. When the complete available dataset is used for training, $h_n(\cdot)$ is designated as the base learner. In Ref. [37], bagging has already been demonstrated to outperform the base learner, particularly in cases where the base learner is unstable with respect to the random training data.

6.5.2 Results

Our proposed method is demonstrated and compared to stacking and various flavors of Bayesian model averaging (BMA) on toy analytical examples in Ref. [30].

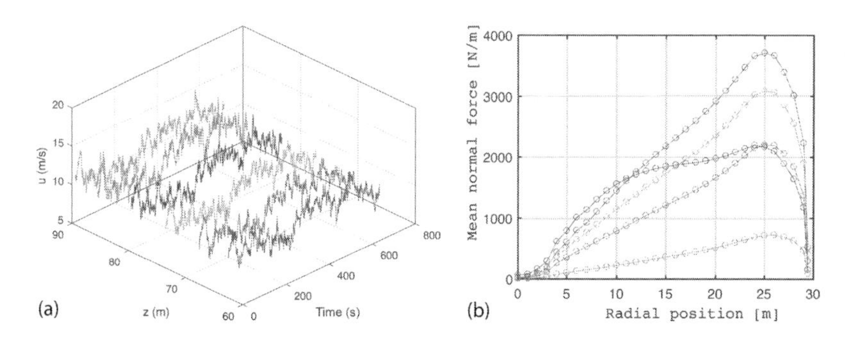

FIGURE 6.12 (a) A sample of the free stream turbulent wind speed time series at various radial positions along the span of the blade. (b) A sample of the mean value of the normal force F_N along the span of the blade for different mean wind speeds. (Reproduced from Ref. [30].)

Here, we demonstrate the method on an engineering application by evaluating the fatigue DEL on a 30-m-long wind turbine blade, using 10 FEM-based simulators.

Wind loads: We generate the turbulent wind field using spatially correlated wind inflow time series along the span of the blade, as shown in Figure 6.12a, based on an exponential coherence model and the Kaimal turbulence auto-spectrum, to account for the spatial correlation structure of the longitudinal velocity component [38]. Thereafter, we compute the time-varying normal F_N and tangential F_T aerodynamic forces along the span of the blade (Figure 6.12b) using a quasi-static blade element momentum (BEM) model.

FEM simulators: For simplicity, we convert the complex structure of the blade into an equivalent tapered clamped-free beam. We developed 10 finite element simulators of the beam to compute the time-varying blade root in-plane bending moment M_X from the applied nodal normal F_N and tangential F_T aerodynamic forces along the span of the blade. The constitutive parameters of the FEM simulators are shown in Table 6.5, with a further illustration of three characteristics of such simulators included in Figure 6.13. We generated 500 times series of normal F_N and tangential F_T aerodynamic forces as inputs, and the forces are assumed to be exerted onto the cross-sectional aerodynamic center. Each time series is 600 sec long, with a time step of 0.01 sec.

Fatigue: The short-term fatigue DELs at the blade clamped root end are calculated on the basis of the M_X output times series. For a given mean wind speed, this is determined by:

$$DEL = \left(\frac{1}{N_{eq}} \sum_i n_i \left(\Delta M_{X,i} \right)^m \right)^{1/m} \tag{6.25}$$

where n_i is the number of load cycles with range $\Delta M_{X,i}$; i is the fatigue cycle index; and N_{eq} is the equivalent number of load cycles, typically 10^7 cycles. The Wöhler exponent is $m = 10$ for a composite structure.

TABLE 6.5
Configuration of the Finite Element–Based Simulators of the Blade

Simulator	**1**	**2**	**3**	**4**	**5**	**6**	**7**	**8**	**9**	**10**
Dimensions	1D	1D	3D	3D	2D	2D	3D	3D	3D	3D
Element type	Euler–Bernoulli beam	Euler–Bernoulli beam	Euler–Bernoulli beam	Timoshenko beam	Plane stress (2D elas.)	Plane stress (2D elas.)	Solid (3D elas.)	Solid (3D elas.)	Solid (3D elas.)	Solid (3D elas.)
Shape function	Linear	Linear	Linear	Quadratic	Linear	Quadratic	Linear	Quadratic	Quadratic	Quadratic
Modes	2	4	4	8	4	8	4	8	10	12
No. of elements	16	48	48	48	16	48	$2 \times 2 \times 96$	$4 \times 4 \times 192$	$8 \times 8 \times 384$	$16 \times 16 \times 768$
Load nodes	6	14	10	14	6	14	6	10	14	28
Nodal force orientation	In-plane projection*	In-plane projection	According to γ**	According to γ	In-plane projection	In-plane projection	In-plane projection	According to γ	According to γ	According to γ
Torsion Degree of Freedom (DOF)	off	off	on	on	off	off	off	off	on	on

* *In-plane projection:* The resultant of the forces is projected onto the $YZ - plane$.
** *According to γ:* The forces are oriented according to the angle γ, which depends on the relative wind speed, the blade pitch, and twist angles.

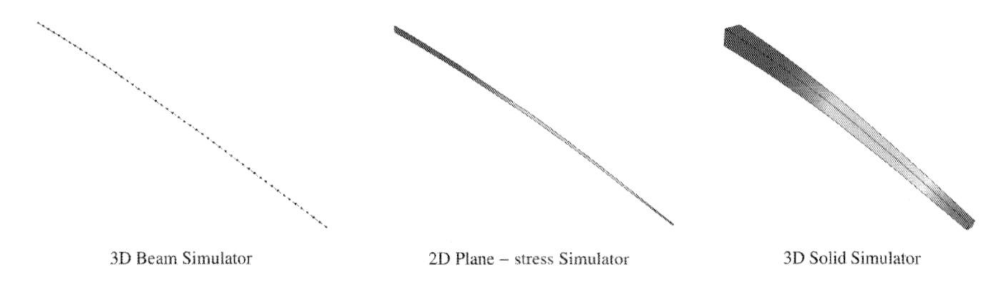

FIGURE 6.13 Illustration of the finite element model (FEM) simulators: 3D beam model (simulator 4), FEM formulation of 2D elasticity (i.e., plane stress/strain; simulator 6), and FEM formulation of 3D elasticity (i.e., 3D elastic solids; simulator 10).

Figure 6.14a compares the DEL estimates, as delivered by the individual FE-based simulators, versus the proposed ULC-BAG scheme, as a function of wind speed. In general, the simulators are performing on par, which is expected due to the aggregate nature of fatigue DELs. This is particularly true for low levels of excitation, where the structural response lies in the linear range. For higher levels of wind speed (faster than 15 *m/s*), simulators 5 and 6 start to significantly diverge. This confirms a main assumption of the proposed ULC-BAG scheme; the performance and quality of the prediction from alternate simulators may be hard to distinguish, especially for certain ranges of the input covariate space, irrespective of their a priori assumed fidelity level. Through aggregation, the ULC-BAG predictor succeeds in narrowing the uncertainty surrounding the delivered estimate as compared to the original dataset, which forms one main strength of the proposed approach.

The outcome of clustering is a map of the joint probability $P\left(\mathcal{S}_i, C_j \middle| \Delta\mathbf{x}\right)$. The explicit accounting for these probabilities when bootstrapping the data shifts the ensemble aggregated predictor toward clusters of higher densities. We observe

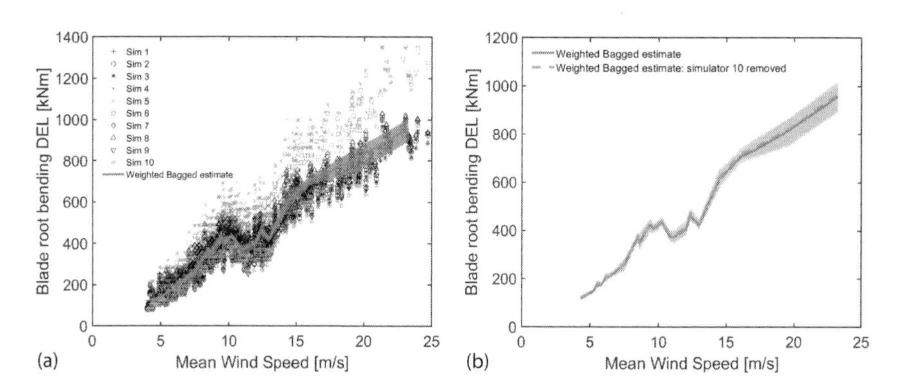

FIGURE 6.14 (a) Comparing the DEL of the blade root bending from 10 FEM simulators, and the local cluster-weighted bagged predictor with the 95% CI. (b) Comparing the local cluster-weighted bagged predictor when the highest-fidelity simulator 10 is removed from the ensemble. (Reproduced from Ref. [30].)

that this is indeed the case for wind speeds higher than 15 *m/s*, where the predictions from simulators 5 and 6 start to deviate from the main cluster.

We further experiment with removing the highest-fidelity FEM simulator (simulator 10) from the ensemble of simulators and rerun the algorithm. According to Figure 6.14b, the local cluster-weighted bagged predictor is largely unaffected, which is computationally attractive since the output of the highest-fidelity FEM simulator (10) is indistinguishable from the cluster of outputs from simulators 1–4 and 7–9; but it takes an order of magnitude more time to run, with no perceptible change in the ensemble aggregate predictive ability. In this respect, we remind the reader that this trait is conditional to the examined quantity of interest. A reliability or failure analysis focusing on extreme responses, or nonlinear effects, could instead require an alternate weighting of the higher-precision simulator.

6.5.3 Comments and Discussion

We proposed an ensemble learning framework based on unsupervised VBGM clustering, and locally weighted bootstrap aggregating of the stochastic outputs from multiple distinct simulators (in the absence of any measurements). We call the method *unsupervised local cluster-weighted bootstrap aggregation* (ULC-BAG). Clustering served the purpose of deriving the probability map (weights). Clustering is carried out on the stochastic output corresponding to the binned input space. Our method does not assign weights to individual simulators per se. Rather, it assigns weights to clusters of output, that is, to the collection of *similar* output data, originating from the various simulators in each local region of the binned input space. The outputs are then combined via the cluster-weighting bootstrap aggregation step. Furthermore, even though the assimilation of measurements is not the main focus, we advance the method, demonstrating how measurement replications may be exploited in order to update the weights. Our insight is that the measurements may be considered as further physical simulators, in the ensemble with other numerical simulators; the proposed algorithm can simply be rerun without mathematical manipulations. We argue that such a way of performing data assimilation is crucial when the correct simulator output is clustered separately from the larger density clusters, for certain regions of the input space. In other words, the derived weights may be inadequate, because the method fails to recognize that certain simulators are more fitting than others, in certain regions of the input space.

6.6 SPATIALLY DISTRIBUTED VIRTUAL SENSING FOR DYNAMIC LOADS ESTIMATION

This section summarizes a hybrid approach, first introduced in Refs. [39, 40]. The method relies on the fusion of data and physics-based models for the purpose of virtual sensing, in the distributed sense, for a monitored wind turbine blade. The virtual-sensing task lies in predicting the response of the structure under unknown, distributed input loads, given sparse vibration measurements. The distributed loading is represented by a GP model, and a space–time filtering approach is proposed for estimating the unknown inputs (loads). A dual Kalman filter (DKF) [41] is then used for the estimation of the blade's response in unmeasured locations. The method is exemplified in a simulated case study of a monitored wind turbine blade, which is excited by a distributed drag and lift load along the length of the blade.

6.6.1 Method

The proposed methodology is a hybrid one, assuming (i) the availability of a physics-based dynamical system model, (ii) that the distributed excitation is an observable and spatially continuous process, and (iii) that a sparse sensor network tracks the distributed excitation at certain spatial points.

6.6.1.1 Dynamical System Formulation

Initiating from the first assumption, we assume the availability of a linear system model, which can be cast as a continuous-time second-order differential equation:

$$\mathbf{M}\ddot{\mathbf{u}}(t) + \mathbf{C}\dot{\mathbf{u}}(t) + \mathbf{K}\mathbf{u}(t) = \mathbf{S}_{\mathrm{p}}\mathbf{p}(t) \tag{6.26}$$

where $\mathbf{u}(t) \in \mathbb{R}^n$ is the system's displacement vector, while $\mathbf{M}$, $\mathbf{C}$, and $\mathbf{K} \in \mathbb{R}^{n \times n}$ are the structural matrices reflecting mass, damping, and stiffness, respectively. The compilation of the former equation necessitates the availability of a numerical model of the system, usually delivered in the form of a FEM. The right-hand side of Eq. (6.27) reflects the applied loading, which is represented here as the product of a selection matrix $\mathbf{S}_{\mathrm{p}} \in \mathbb{R}^{n \times n_{\mathrm{p}}}$ with the force vector $\mathbf{p}(t) \in \mathbb{R}^{n_{\mathrm{p}}}$, where $n_{\mathrm{p}} \leq n$ designates the number of loaded degrees of freedom. It should be noted that the assumption of linearity is convenient, but not necessary. In the case of a nonlinear system model, a nonlinear filtering approach can be adopted.

In this framework, we will pursue a Bayesian filtering approach, which necessitates an observer setup. To this end, we recast the system dynamics into a state–space formulation upon definition of the state vector $\mathbf{x}(t) = \mathrm{vec}\left(\left[\mathbf{u}(t)\dot{\mathbf{u}}(t)\right]\right) \in \mathbb{R}^{2n}$. Since we are dealing with measurements collected at a discrete rate, we further transform the originally continuous state–space form into its discrete counterpart. Lastly, following the Bayesian filtering logic, we assume the presence of uncertainties in the modeling (process) and measurement equations, which are accounted for by corresponding Gaussian noise terms, $\mathbf{v}_k \sim \mathcal{N}(0, \mathbf{Q})$ and $\mathbf{w}_k \sim \mathcal{N}(0, \mathbf{R})$, respectively. The interested reader is referred to Refs. [42, 43] for further details on such a representation. Finally, the discrete state–space form is delivered as follows:

$$\mathbf{x}_{k+1} = \mathbf{A}\mathbf{x}_k + \mathbf{B}\mathbf{p}_k + \mathbf{v}_k \tag{6.27}$$

$$\mathbf{y}_k = \mathbf{G}\mathbf{x}_k + \mathbf{J}\mathbf{p}_k + \mathbf{w}_k \tag{6.28}$$

where the system matrices $\mathbf{A} \in \mathbb{R}^{2n\times 2n}$ and $\mathbf{B} \in \mathbb{R}^{2n\times n_p}$ occur through discretization, while the output and direct transmission matrices $\mathbf{G} \in \mathbb{R}^{n_y\times 2n}$ and $\mathbf{J} \in \mathbb{R}^{n_y\times n_p}$ are formulated so as to extract the measured quantities contained in vector $\mathbf{y}_k \in \mathbb{R}^{n_y}$. Vibration-based measurements can reflect different types of dynamic response, with acceleration and strain forming typical in such instances.

The virtual-sensing task, in this case, falls in the class of joint input–state estimation, which consists of recursively estimating both the dynamic response (state) $\mathbf{x}_k$ and distributed load (input) $\mathbf{p}_k$ under the availability of sparse response measurements $\mathbf{y}_k$. The novelty of the suggested approach lies in its allowance for distributed input loads, as opposed to the task of identification of concentrated loads, which has been more commonly tackled in existing literature, such as Ref. [41]. The distributed nature of the applied loading is represented by a GP model.

6.6.1.2 Spatiotemporal Filtering

A spatiotemporal process $p(\mathbf{s},t)$ is assumed to describe the distributed excitation that is acting onto the monitored dynamical system,

$$p(\mathbf{s},t) = \mu(\mathbf{s},t) + v(\mathbf{s},t) \tag{6.29}$$

with $\mathbf{s} \in \mathbb{S}$ designating the spatial variable, which is defined to lie within a domain $\mathbb{S}$. $\mu(\mathbf{s},t)$ represents the mean value, or trend, of the assumed process, while $v(\mathbf{s},t)$ is a random GP that accounts for variability.

We further assume a dynamic evolution model, which governs the evolution of $\mu(\mathbf{s},t)$ as:

$$\mu(\mathbf{s},t) = \int_S w(\mathbf{z})\mu(\mathbf{z},t-1)d\mathbf{z} + \xi(\mathbf{s},t) \tag{6.30}$$

In the above formulation, $\xi(\mathbf{s},t)$ comprises a spatially colored noise process, while $\mu(\mathbf{s},t)$ and $w(\mathbf{z})$ are approximated by means of regression on an orthogonal basis $f_i(\mathbf{s})$ $(i = 1,2,\ldots,p)$, with $\int_S f_i(\mathbf{s})f_j(\mathbf{s})d\mathbf{s} = \delta_{ij}$,

$$\mu(\mathbf{s},t) = b_1(t)f_1(\mathbf{s}) + b_2(t)f_2(\mathbf{s}) + \ldots + b_p(t)f_p(\mathbf{s}) = \mathbf{f}^{\mathrm{T}}(\mathbf{s})\mathbf{b}(t) \tag{6.31}$$

$$w(\mathbf{z}) = c_1(\mathbf{s})f_1(\mathbf{z}) + c_2(\mathbf{s})f_2(\mathbf{z}) + \ldots + c_p(\mathbf{s})f_p(\mathbf{z}) = \mathbf{c}^{\mathrm{T}}(\mathbf{s})\mathbf{f}(\mathbf{z}) \tag{6.32}$$

with $b_i(t)$ and $c_i(t)(i = 1,2,\ldots,p)$ comprising the zero-mean regression coefficients. The exploitation of the orthogonality property and substitution of the regression formulas into the temporal evolution model of Eq. (6.31) result in the following expression for the model of the trend of the spatiotemporal stochastic process $p(\mathbf{s}, t)$:

$$\mu(\mathbf{s},t) = \mathbf{c}^{\mathrm{T}}(\mathbf{s})\mathbf{b}(t-1) + \xi(\mathbf{s},t) \overset{(31)}{=} \mathbf{f}^{\mathrm{T}}(\mathbf{s})\mathbf{b}(t) \tag{6.33}$$

where $\mathbf{c}(\mathbf{s}) = \left[c_1(\mathbf{s})c_2(\mathbf{s})\ldots c_p(\mathbf{s})\right]^{\mathrm{T}}$, and $\mathbf{b}(t-1) = \left[b_1(t)b_2(t)\ldots b_p(t)\right]^{\mathrm{T}}$. The previous equation defines an evolution model for the regression coefficients $\mathbf{b}(t)$, which can be evaluated at the n_{s} available measurement locations (sensor channels) $\left[\mathbf{s}_1^{\mathrm{m}}, \mathbf{s}_2^{\mathrm{m}}, \ldots, \mathbf{s}_{n_{\mathrm{s}}}^{\mathrm{m}}\right]$, rendering the following equation:

$$\mathbf{F}^{\mathrm{T}}\mathbf{b}(t) = \mathbf{C}\mathbf{b}(t-1) + \Xi(t) \tag{6.34}$$

where $\mathbf{F} = [\mathbf{f}(\mathbf{s}_1^{\mathrm{m}})\;\; \mathbf{f}(\mathbf{s}_2^{\mathrm{m}})\;\; \ldots\;\; \mathbf{f}(\mathbf{s}_{n_{\mathrm{s}}}^{\mathrm{m}})] \in \mathbb{R}^{p \times n_{\mathrm{s}}}$, $\mathbf{C} = [\mathbf{c}(\mathbf{s}_1^{\mathrm{m}})\;\; \mathbf{c}(\mathbf{s}_2^{\mathrm{m}})\;\; \ldots\;\; \mathbf{c}(\mathbf{s}_{n_{\mathrm{s}}}^{\mathrm{m}})]^{\mathrm{T}} \in \mathbb{R}^{n_{\mathrm{s}} \times p}$, and $\Xi(t) = [\xi(\mathbf{s}_1^{\mathrm{m}}, t)\;\; \xi(\mathbf{s}_2^{\mathrm{m}}, t)\;\; \ldots\;\; \xi(\mathbf{s}_{n_{\mathrm{s}}}^{\mathrm{m}}, t)]^{\mathrm{T}} \in \mathbb{R}^{n_{\mathrm{s}}}$.

Under the assumption that $n_{\mathrm{s}} \geq p$ and that $\left(\mathbf{F}^{\mathrm{T}}\mathbf{F}\right)^{-1}$ is non-singular, the least-squares solution in terms of the regression coefficients $\mathbf{b}(t)$ is obtained as:

$$\mathbf{b}(t) = \mathbf{L}\mathbf{b}(t-1) + \mathbf{N}\Xi(t), \quad \text{with} \quad \mathbf{L} = \mathbf{N}\mathbf{C} \in \mathbb{R}^{p \times p} \quad \& \quad \mathbf{N} = \left(\mathbf{F}^{\mathrm{T}}\mathbf{F}\right)^{-1}\mathbf{F}^{\mathrm{T}} \in \mathbb{R}^{p \times p} \tag{6.35}$$

6.6.1.3 Input–State Estimation

The parameterization of the loading process in terms of the regression coefficient vector $\mathbf{b}(t)$ allows us to tackle the virtual-sensing task by means of a joint input–state estimation approach. In this case, we opt for the DKF formulation [41], which exploits Eq. (6.35) as the required state–space model that describes the evolution of the load (input) coefficients. With the state evolution and measurement equations described by Eqs. (6.27) and (6.28), respectively, the DKF update equations for the input are summarized as follows:

Input prediction:

$$\mathbf{b}_{k|k-1} = \mathbf{L}\mathbf{b}_{k-1|k-1} \tag{6.36}$$

$$\mathbf{P}_{k|k-1}^{\mathrm{b}} = \mathbf{L}\mathbf{P}_{k-1|k-1}^{\mathrm{b}}\mathbf{L}^{\mathrm{T}} + \mathbf{N}\mathbf{Q}\mathbf{N}^{\mathrm{T}}, \quad \mathbf{Q} = \mathrm{var}(\xi) \tag{6.37}$$

Input update:

$$\mathbf{K}_k^{\mathrm{b}} = \mathbf{P}_{k|k-1}^{\mathrm{b}}\Phi\mathbf{J}^{\mathrm{T}}\left(\mathbf{J}\Phi^{\mathrm{T}}\mathbf{P}_{k|k-1}^{\mathrm{b}}\Phi\mathbf{J}^{\mathrm{T}} + \mathbf{R}\right)^{-1} \tag{6.38}$$

$$\mathbf{b}_{k|k} = \mathbf{b}_{k|k-1} + \mathbf{K}_k^{\mathrm{b}}\left(\mathbf{y}_k - \mathbf{G}\mathbf{x}_{k-1} - \mathbf{J}\Phi^{\mathrm{T}}\mathbf{b}_{k|k-1}\right) \tag{6.39}$$

$$\mathbf{P}_{k|k}^{\mathrm{b}} = \mathbf{P}_{k|k-1}^{\mathrm{b}} - \mathbf{K}_k^{\mathrm{b}}\mathbf{J}\Phi^{\mathrm{T}}\mathbf{P}_{k|k-1}^{\mathrm{b}} \tag{6.40}$$

where $\mathbf{b}_k = \mathbf{b}(k\Delta t)$, and $1/\Delta t$ is the employed sampling rate. $\Phi = \left[\begin{array}{cccc} \mathbf{f}(\mathbf{s}_1) & \mathbf{f}(\mathbf{s}_2) & \ldots & \mathbf{f}(\mathbf{s}_{n_{\mathrm{p}}}) \end{array}\right] \in \mathbb{R}^{p \times n_{\mathrm{p}}}$, with the basis functions evaluated along the degrees of freedom on which loading is exerted $\mathbf{s}_i (i = 1, 2, \ldots, n_{\mathrm{p}})$.

The a posteriori estimate of the distributed load $p_k(\mathbf{s})$ is then reconstructed on the basis of the posterior estimate of the input coefficients:

$$p_{k|k}(\mathbf{s}) = \mathbf{f}^{\mathrm{T}}(\mathbf{s})\mathbf{b}_{k|k} + c_{v}^{\mathrm{T}}(\mathbf{s})\left(\mathbf{C}_0^p\right)^{-1}\hat{\mathbf{p}}_k^{\mathrm{m}} \tag{6.41}$$

where $\mathbf{C}_0^p = \mathrm{cov}\left[\hat{\mathbf{p}}_k^{\mathrm{m}}\hat{\mathbf{p}}_k^{\mathrm{m}}\right]$, $\hat{\mathbf{p}}_k^{\mathrm{m}}$ is the predicted input at the measurement locations, and $\mathbf{c}_v(\mathbf{s}) = \mathbb{E}\left[v(\mathbf{s},t)v(t)\right]$, which is obtained upon fitting a spatially stationary and isotropic model on a set of simulated input data [44]. Thereafter, the input at the physical degrees of freedom of the model can be obtained by evaluating the process $p_{k|k}(\mathbf{s})$ at the nodes of the model as follows:

$$\mathbf{p}_{k|k} = \left[\begin{array}{cccc} p_{k|k}(\mathbf{s}_1) & p_{k|k}(\mathbf{s}_2) & \ldots & p_{k|k}(\mathbf{s}_{n_p}) \end{array}\right]^{\mathrm{T}} \tag{6.42}$$

Once the input has been defined on all loaded nodes of the employed dynamical model, it is assumed as known for the problem formulation of Eqs. (6.27) and (6.28). This allows for a state estimation problem to be solved on the basis of a second Kalman filter, which serves for estimating the state in all model locations, including those that are unmeasured. It should be noted that the effect of distributed loads can be alternatively calculated in a reduced space [45], instead of the physical, in the form of generalized forcing terms. The Bayesian filter equations follow the formulation of the standard Kalman filter as follows:

State prediction:

$$\mathbf{x}_{k|k-1} = \mathbf{A}\mathbf{x}_{k-1|k-1} + \mathbf{B}\mathbf{p}_{k|k} \tag{6.43}$$

$$\mathbf{P}_{k|k-1} = \mathbf{A}\mathbf{P}_{k-1|k-1}\mathbf{A}^{\mathrm{T}} + \mathbf{Q} \tag{6.44}$$

State update:

$$\mathbf{K}_k = \mathbf{P}_{k|k-1}\mathbf{G}^{\mathrm{T}}\left(\mathbf{G}\mathbf{P}_{k|k-1}\mathbf{G}^{\mathrm{T}} + \mathbf{R}\right)^{-1} \tag{6.45}$$

$$\mathbf{x}_{k|k} = \mathbf{x}_{k|k-1} + \mathbf{K}_k\left(\mathbf{y}_k - \mathbf{G}\mathbf{x}_{k|k-1} - \mathbf{J}\mathbf{p}_{k|k}\right) \tag{6.46}$$

$$\mathbf{P}_{k|k} = \mathbf{P}_{k|k-1} - \mathbf{K}_k\mathbf{G}\mathbf{P}_{k|k-1} \tag{6.47}$$

6.6.1.4 Illustrative Example

We illustrate the previous framework on the response estimation of a simulation of the blade component of the 5 MW NREL reference wind turbine [46]. The numerical model employed for each simulation adopts 50 Euler–Bernoulli beam elements for the approximation of the system, which is illustrated in Figure 6.15. The blade loading is simulated in the form of drag dF_D and lift dF_L forces that are acting as distributed loads along the length and can be analyzed into the tangential dF_T and normal dF_N forces. The blade simulation is carried out via the use of a turbulent wind field of 600 sec duration, implemented in TurbSim [46]. The blade

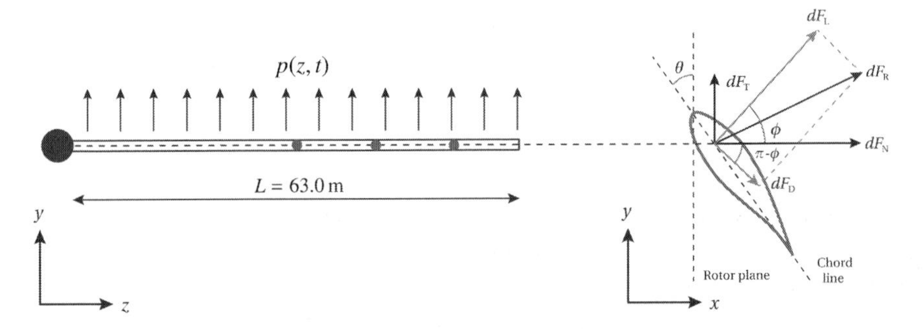

FIGURE 6.15 (Left) The simulated wind turbine (WT) blade, with the measurement points located at 30, 40, and 50 m from the root; and (right) a schematic of the assumed distributed loading in the tangential direction.

element momentum (BEM) theory [47] and Runge–Kutta integration are used to compute the aerodynamic load.

Measurements are assumed to be derived in the form of accelerations and strains at three positions along the blade, indicated as red dots in Figure 6.15. The acceleration measurements capture the response in both x and y directions, while the strains are measured only in the x direction. Both types of measurements are collected with a sampling frequency $F_s = 100$ Hz. The simulated measurements $\mathbf{y}_k$ are artificially polluted with a 2% white Gaussian noise, whose level is equal to 2% of the RMS ratio, simulating sensor imprecision, prior to being fed into the previously outlined Bayesian virtual-sensing Scheme 5.1.2.

A training phase is required for deriving the GP basis functions, which are assumed to correspond to the first 200 sec of the simulated time series. Snapshots are drawn from the simulated input distribution. The empirical orthogonal functions are extracted by applying a singular value decomposition to the pool of snapshots and retaining only the first three components (because the latter is bounded by the number of response measurement points). All unknown variables (state variables and regression coefficients) are initiated with a zero value, while the corresponding values of the initial covariance matrices are set to $10^{-3} \times \mathbf{I}_3$ and $10^{-10} \times \mathbf{I}_{2n}$, respectively. The process noise covariance matrix is selected as $\mathbf{Q} = 10^{-12} \times \mathbf{I}_{2n}$, and the measurement noise is set to $\mathbf{R} = 10^{-6} \times \mathbf{I}_{n_y}$.

Figure 6.16 displays snapshots of the distributed normal force dF_N along the length of the blade. The snapshots are extracted at randomly selected time steps during the 600 sec simulation. The actual versus estimated distribution of the normal force is presented in Figure 6.17 for two representative time instants, demonstrating a good approximation. The estimation is particularly accurate in the vicinity of available measurements, with larger deviations observed at the first half of the length, which is more remote from the sensed region (red markups in Figure 6.15). In order to demonstrate the second, and perhaps the primary, target of the virtual-sensing task, we further contrast the actual versus estimated displacement response at an unmeasured location. The displacement along the x direction at the tip of the blade is plotted in Figure 6.18, revealing successful tracking.

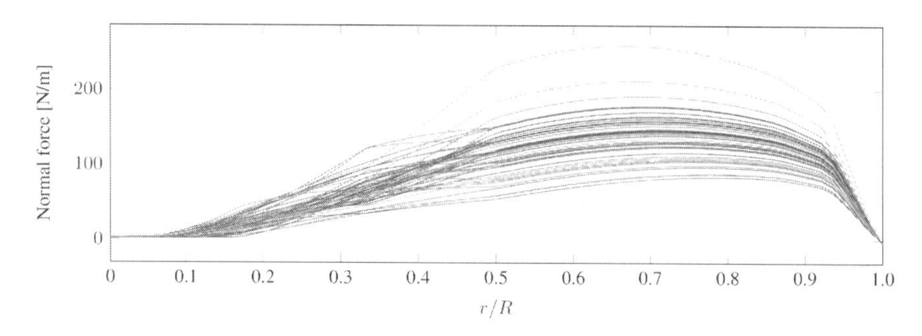

FIGURE 6.16 Snapshots of the distributed load profile along the length of the blade. (Reproduced from Ref. [40].)

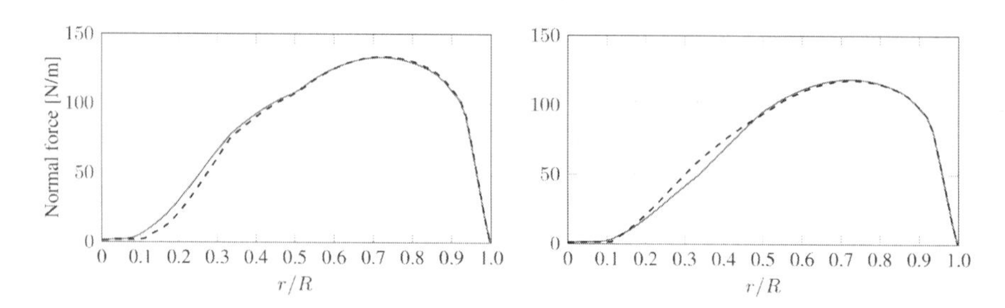

FIGURE 6.17 Actual (black) versus estimated (red) distribution of the axial force along the length of the blade at two different time instants. (Reproduced from Ref. [40].)

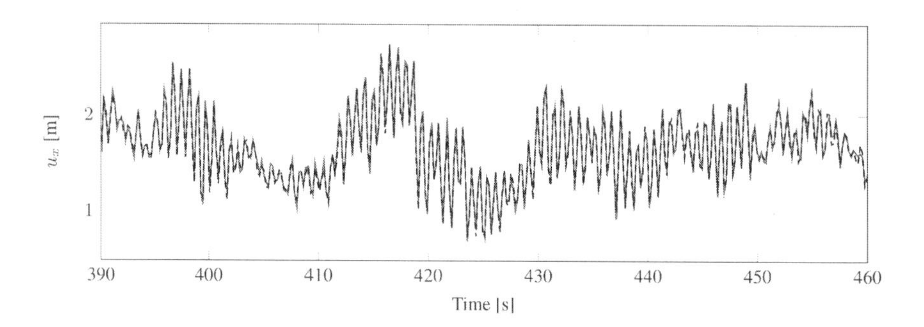

FIGURE 6.18 Actual (black) versus estimated (red) displacement at the tip of the blade in the x direction. (Reproduced from Ref. [40].)

6.6.2 Comments and Discussion

This section focused on the illustration of a spatiotemporal filtering approach for virtual sensing. Contrary to joint input–state identification schemes that regard the inputs as independent point loads, a spatial interpolation scheme is exploited here for representing the input. The scheme is particularly relevant for the task of virtual sensing in the context of monitoring wind energy structures, where aerodynamic forces admit such a distributed loading assumption. We demonstrate the principle via estimation of the distributed drag and lift forces on a wind turbine blade. A GP model is employed for interpolation in the spatial sense. A Bayesian filtering step is subsequently applied for estimating the dynamic response of the system by means of a dual input and state estimation scheme. The concept is verified on a simulated blade case study, under the assumption of sparse output-only vibration measurements.

6.7 SYSTEMS-LEVEL MODELING OF WIND FARM DATA

A systems-level approach is presented for wind farm monitoring, which brings together several existing methods [48, 49]—including data storage and extraction and the statistical analysis of turbine population data (SCADA data). We present work toward a standardized data-storage methodology, considering channel data recorded from wind farms (regardless of the asset owner, turbine type, or monitoring system). A combined inference is then summarized, which should allow information to be shared between similar measurement channels from collected wind farms using (i) mixture models and (ii) multilevel modeling. Each method improves predictors (power curves) in wind farms by considering data recorded from the collected population, rather than individual systems. In each example, a set of functions (or *tasks*) are learnt over the population of assets, which share information via data-pooling (in various forms). We favor explainability and move toward a parameterized, multilevel view of the wind farm data. The resultant methodology allows the practitioner to interpret which groups of turbines share correlated information, for which effects or parameters. In the resultant model, turbine groups with sparse data (automatically) borrow statistical strength from those that are *data-rich*.

6.7.1 A DATABASE FOR SCADA MONITORING

Traditionally, each asset owner will have their own format for storing monitoring data from a wind farm. When moving to a general approach where all data can be considered across multiple structures (or even multiple types of structures), a standardized method for data storage is required. We summarize a NoSQL database via the *PBSHM Schema*, originally presented in Ref. [48]. The database restructures SCADA data by grouping measurements according to structure name and time of data collection. Figure 6.19 outlines the composition of a channel document within the PBSHM Schema.

For instance, a structure called *turbine 1*, which has only one temperature sensor, would record the SCADA values for 1 January 2020 as follows:

```
{
        "name": "turbine-1",
        "population": "wind farm-name",
        "timestamp": 1577836800000000000,
        "channels": [
```

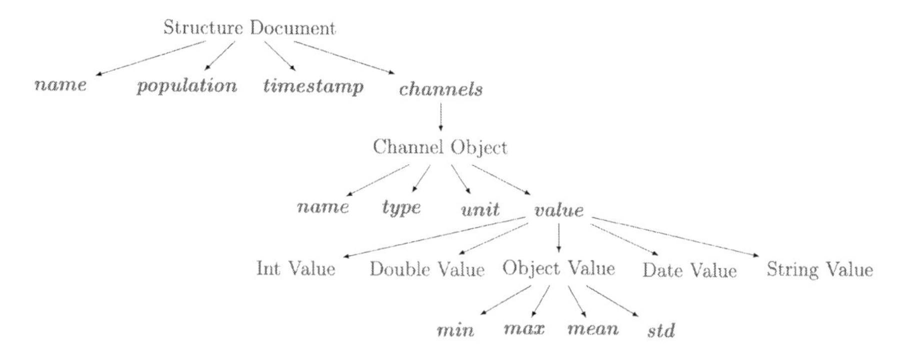

FIGURE 6.19 PBSHM Schema hierarchical structure.

```
                {
                        "name": "sensor-1",
                        "type": "temperature",
                        "unit": "C",
                        "value": {
                                "min": 9.5,
                                "max": 10.0,
                                "mean": 9.75,
                                "std": 0
                        }
                }
        ]
}
```

Once data is stored in a common format, this enables the development of generic algorithms that can function across the varying datasets, regardless of the turbine owner or type. One of the core features of a database is the ability to extract data. For instance, if we wish to view temperature data across a variety of turbines, one would execute a general *find* command upon an active connection to the database. The following code demonstrates a *find* command in Python to retrieve all temperature channel data, regardless of structure, within the selected collection:

```
# Retrieve Temperature Channels
for document in collection.find(
     {"channels.type": "temperature"},
     {
           "_id": 0, "name": 1, "population": 1,
"timestamp": 1,
           "channels": {"$elemMatch": {"type":
"temperature"}}
     }
):
     print(document)
```

6.7.2 Toward a Multilevel View of Wind Farm Data

Following a procedure to organize and extract measurements from a wind farm database, we explore methods to model the associated channel data across a variety of turbines—and the potential to share information between assets, to extend the value of data. Here, we restrict ourselves to modeling the relationship between two channels (input $\mathbf{x}_k$ and output $\mathbf{y}_k$) for K related datasets:

$$\left\{\mathbf{x}_k, \mathbf{y}_k\right\}_{k=1}^{K} = \left\{\left\{x_{ik}, y_{ik}\right\}_{i=1}^{N_k}\right\}_{k=1}^{K}$$

where the $k-$ index associates observations with specific labels (e.g., the turbine identifier or operating condition), while N_k is the number of observations associated with that subset. With continuous variables, the population data can be assumed to represent samples from a set of regression *tasks* f_k:

$$\left\{\mathbf{y}_k = f_k\left(\mathbf{x}_k\right) + \epsilon_k\right\}_{k=1}^{K} \tag{6.48}$$

In the examples here, the tasks are *power curves*, which capture the relationship between power ($\mathbf{y}_k$) and wind-speed ($\mathbf{x}_k$) measurement channels—Figure 6.20 is an example of such in-the-field measurements (normalized in view of data confidentiality). With this data, k associates each observation with one of three underlying trends: (i) normal operation, limiting at unity; (ii) 50% curtailment; or (iii) zero power.

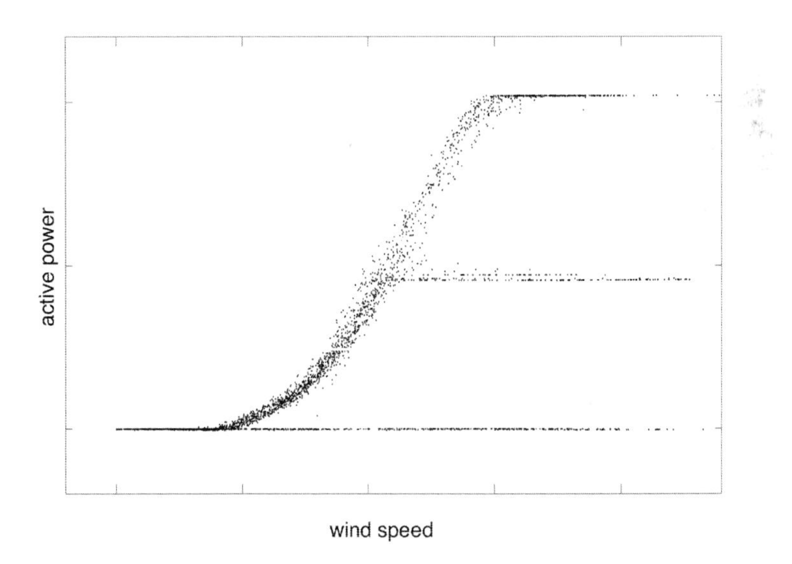

FIGURE 6.20 Data $\left\{\mathrm{x}_k, \mathrm{y}_k\right\}_{k=1}^{K}$, including $K = 3$ power relationships f_k (i.e., tasks). (i) The ideal power curve, limiting at unity; (ii) $\approx 50\%$-limited output; and (iii) zero-limited output. (Reproduced from Ref. [50].)

Considering such multi-task population data, we review two approaches recently presented in the literature:

- Section 6.7.3: Overlapping mixture models [50]
- Section 6.7.4: Multilevel (or hierarchical) modeling [49]

Both utilize *data pooling* to improve task predictions—that is, combining measurements from multiple domains (or sources $k \in \{1,2,\ldots,K\}$) to effectively extend the data available for training. Throughout, when discussing each method, we use $\boldsymbol{\theta}$ (and variants, e.g., $\boldsymbol{\theta}'$ and $\boldsymbol{\theta}_l$) to refer to the collected parameters over wind farm predictors.

Overlapping mixture models: Applied to the data in Figure 6.20, mixture models approximate tasks f_k, associated with different operating conditions. For a given operating condition, the data from all turbines is assumed to be sampled from a *single* underlying function (complete pooling), and the models of each condition are assumed independent (i.e., the model-specific parameters $\boldsymbol{\theta}_k$ are independent). In turn, while the training data are extended (for each task), there is no means for similar tasks to borrow statistical strengths from each other—and the inter-turbine variances are not quantified by the model. This approach is visualized in Figure 6.21a, where the tasks over the population can share information

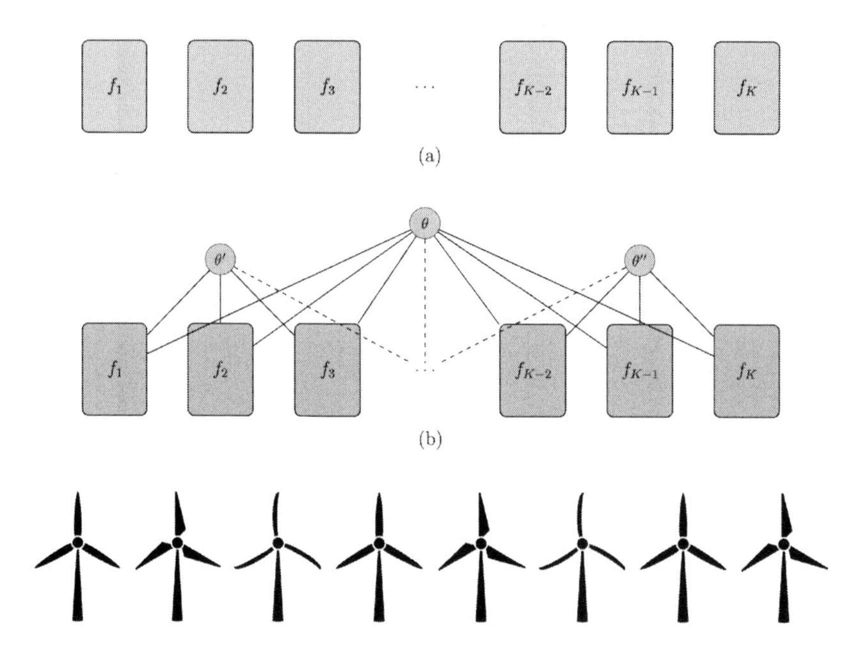

FIGURE 6.21 Visualization of (a) collected, independent models learnt from wind farm population data—information is only shared at the data level, and there are no interdependencies between tasks; and (b) hierarchical modeling to represent nested groups (e.g., relating to turbine or operating condition) around parameterized models. Here, $\boldsymbol{\theta}$ is the set of (shared) population-level parameters, while $\boldsymbol{\theta}'$ and $\boldsymbol{\theta}''$ are subgroup-specific parameters.

in terms of data, but the tasks themselves are not interconnected (by statistical dependencies). An advantage of the mixture model formulation here, however, is that the label indices k (associating data to tasks) can be more conveniently treated as unknown and inferred from the data.

Multilevel (or hierarchical) models: We then review multilevel modeling of the task-set as multiple, nested groups, relating to (i) turbine label, and (ii) operating condition. This approach allows us to capture the inter-task correlations (between turbines and operating conditions), such that data can be shared (in this case, *partially pooled*) via correlated or tied parameters in an interpretable model. As such, the covariance structure of $\boldsymbol{\theta}$ is captured by the inference. By negating the assumptions of independence (between f_k), sparse features borrow statistical strength from those that are data-rich. Moreover, the covariance structure of the model can inform which tasks share information, for which interpretable parameter. Figure 6.21b visualizes the interdependencies of such a model, with population-level $\boldsymbol{\theta}$ and subgroup-specific $\{\boldsymbol{\theta}', \boldsymbol{\theta}''\}$ parameter sets.

6.7.3 Modeling Power Trends via Mixture Models

This work considers SCADA data recorded from an operational wind farm owned by Vattenfall, originally presented in Ref. [51]. For confidentiality reasons, information regarding the type, location, and number of turbines cannot be disclosed. The data was recorded from a farm containing the same model of turbine, over a period of 125 weeks [51, 52].

In this example, the data is *unlabeled* in terms of the subgroup indices k (i.e., records of the operating condition are unavailable). Referring to Figure 6.20, this implies that there is no ground truth to associate samples with operating conditions (tasks): (i) normal, (ii) ≈50% curtailed, or (iii) zero-power. In turn, labels to associate data with tasks (i)–(iii) are unobserved and must be represented as latent variables. Importantly, if labels were available (e.g., in a control log), they should be treated as *observations* in the model. This scenario is presented in the second method in Section 6.7.4, and it brings several advantages since a more complex (correlated) inference becomes feasible. However, without specified labeling for tasks (i)–(iii), the model must categorize observations in an *unsupervised* manner, which can be problematic in practice [52].

6.7.3.1 An Overlapping Mixture of Gaussian Processes

An overlapping mixture of Gaussian processes (OMGP) [53, 54] is used to learn the set of power curve tasks. Rather than reintroduce GPR, the reader is referred to Section 6.3.1.2, where we follow a similar notation. Here, however, we assume K latent functions represent the wind farm data $\{\mathbf{y}_k = f_k(\mathbf{x}_k) + \epsilon_k\}_{k=1}^{K}$, and, in turn, K GPs are required to approximate the trends. The mixture is overlapping since each observation is assumed to be sampled from one of the K GP components across the input. As discussed, labels to assign observations to tasks (i)–(iii) are unknown; this introduces the latent variable $\mathbf{Z}$—a binary indicator matrix, such that $\mathbf{Z}[i,k] \neq 0$ indicates that observation i was generated by function k. There is only one non-zero entry per row in $\mathbf{Z}$ (each observation generated by one task). Note that, as the k-indices are unknown, they are removed from the $\mathbf{x}$, $\mathbf{y}$ notation for the remainder of this section; instead, the (probabilistic) categorization into tasks is represented by rows of $\mathbf{Z}$.

The likelihood of the OMGP is therefore [53]:

$$p\left(\mathbf{y} \mid \{\mathbf{f}_k\}_{k=1}^{K}, \mathbf{Z}, x\right) = \prod_{i,k=1}^{N,K} p\left(y_i \mid f_k(x_i)\right)^{\mathbf{Z}[i,k]} \tag{6.49}$$

Prior distributions are placed over each latent function and variable from the set of K tasks in Eq. (6.48):

$$P(\mathbf{Z}) = \prod_{i,k=1}^{N,K} \Pi[i,k]^{\mathbf{Z}[i,k]} \tag{6.50}$$

$$f_k(\mathbf{x}) \sim \mathcal{N}\left(\mu_k(\mathbf{x}), \mathbf{K}_{\mathbf{xx}}^{(k)}\right) \tag{6.51}$$

$$\epsilon_{ki} \overset{iid}{\sim} \mathcal{N}(0, \sigma^2) \tag{6.52}$$

Eq. (6.50) is a multinomial-distributed prior over the indicator matrix, where $\Pi[i,:]$ is a histogram over the K components for the ith observation and[1] $\sum_{k=1}^{K} \Pi[i,k] = 1$. Eq. (6.51) corresponds to a GP prior over each latent function f_k, with distinct mean and kernel functions $\left(\mu_k(x_i), k_k(x_i, x_j)\right)$. The kernel function populates the covariance matrix $\mathbf{K}_{\mathbf{xx}}[i,j] = k(x_i, x_j)\ \forall i, j \in \{1, \ldots, N\}$, as outlined in Section 6.3.1.2. For now, the noise variance prior is defined by a *shared* hyperparameter σ (this is modified later).

The hyperparameters for the model can be collected as $\theta = \left\{\{\theta_k\}_{k=1}^{K}, \Pi\right\}$, where subscripted $\boldsymbol{\theta}_k$ denotes a separate set of mean/kernel hyperparameters for the kth component (including the additive noise in Eq. (6.52)). Referring back to the channel data in Figure 6.20, it is possible to postulate prior distributions from domain expertise, by specifying the associated mean $\mu_k(x_i)$ and covariance $k_k\left(x_i, x_j\right)$ functions. Unlike Section 6.3.1.2, here, we use a non-zero (parameterized) mean, while the covariance functions remain the same—the squared exponential (6.2).

The mean functions can be specified, since there are several established parameterizations for power curves [55]. Here, we use the following set of assumptions to define the model:

- It should be clear from the data that three latent functions will be representative, such that $K = 3$ (alternatively, K can be selected by cross-validation [50]).
- For the zero-power relationship (iii), a linear regression (with a constant kernel) should be representative.
- For the remaining functions ((i) ideal and (ii) curtailed), the soft-clip [56] appropriately describes the expected relationships. The scaled

soft-clip exhibits near-linearity within bounds (cut-in and cut-out wind speeds) and horizontal asymptotes (regarding min/max power) for high and low inputs:

$$\mu_k(x_i;\beta,\alpha)=\frac{\alpha_1}{\beta}\log\left\{\frac{1+e^{\beta v}}{1+e^{\beta(v-1)}}\right\}$$

$$v \triangleq \alpha_2 x_i+\alpha_3$$

$$\alpha \triangleq \{\alpha_1,\alpha_2,\alpha_2\}$$

where, relating to power curves, the hyperparameters $\{\beta,\boldsymbol{\alpha}\}$ are interpretable; α_1 determines the value of the horizontal (non-zero) asymptote, which corresponds to the maximum (or limited) power; β controls the *rate* at which the near-linear section tends to the asymptotic values (around the cut-in and cut-out wind speeds); and, finally, α_2 scales and α_3 translates the function with respect to the input. Figure 6.22 illustrates the effects of $\{\beta,\boldsymbol{\alpha}\}$.

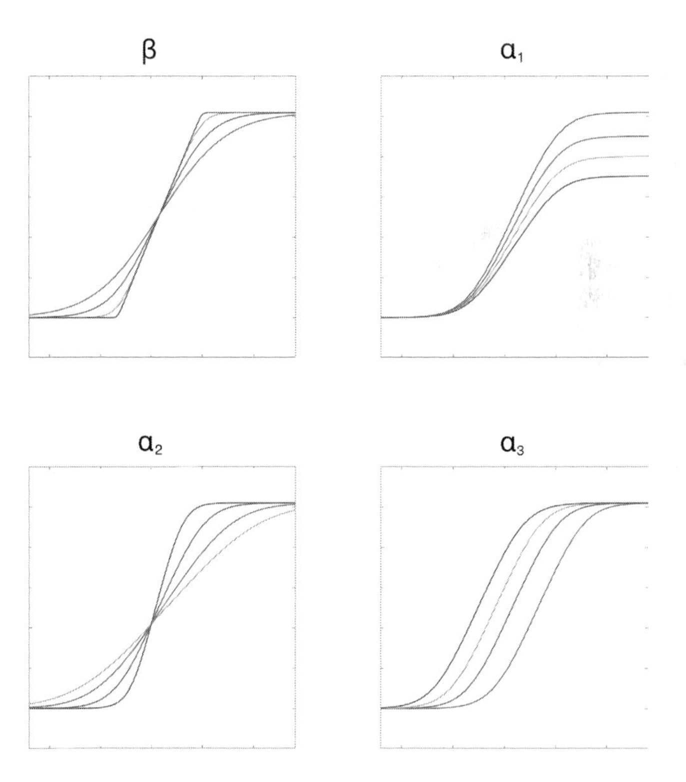

FIGURE 6.22 Effects of the hyperparameters on the mean function $m(x_i;\beta,\alpha)$. (Reproduced from Ref. [50].)

To summarize, the mixture includes two independent GPs with a soft-clip mean and squared-exponential kernel (6.2) function. These priors correspond to the ideal and curtailed curves. For the final component, a constant kernel is selected $k^{(3)}(x_i, x_j) = c$. This assumption reduces the latent function to a (zero-gradient) linear regression, to approximate the zero-power trend. The collected hyperparameters of the model are then: $\theta_k = \{\beta_k, \alpha_k, \sigma_f^{(k)}, l_k, \sigma\}_{k=1}^{2}$ and $\theta_3 = \{c, \sigma\}$.

6.7.3.2 Inference for the OMGP

Since both $\{\mathbf{f}_k\}$ and $\mathbf{Z}$ are unknown, computation of the posterior distribution $p(\{\mathbf{f}_k\}_{k=1}^{K}, \mathbf{Z}|\mathbf{x}, \mathbf{y})$ is intractable, unlike standard GPR (type II maximum likelihood [6]). As such, VI [57] is implemented as an approximate procedure. VI involves the selection of an approximate density family $q \in \mathcal{Q}$ over the target conditional $p(\Theta|\mathbf{y})$. The best candidate $\hat{q}(\Theta)$ in the family $q(\Theta) \in \mathcal{Q}$ is the one that is *closest* to the intractable target $p(\Theta|\mathbf{y})$ in terms of the KL divergence:

$$\hat{q}(\Theta) = \underset{q \in \mathcal{Q}}{argmin}\, KL\left(q(\Theta) \| p(\Theta|\mathbf{y})\right) \tag{6.53}$$

In this case, $\Theta \triangleq \{\{\mathbf{f}_k\}, \mathbf{Z}\}$. The KL divergence for Eq. (6.54) can then be defined:

$$KL\left(q(\Theta) \| p(\Theta|\mathbf{y})\right) = \mathbb{E}_{q(\Theta)}[\log q(\Theta)] - \mathbb{E}_{q(\Theta)}[\log p(\Theta|\mathbf{y})] \tag{6.54}$$

$$= \mathbb{E}_{q(\Theta)}[\log q(\Theta)] - \mathbb{E}_{q(\Theta)}[\log p(\Theta, \mathbf{y})] + \log p(\mathbf{y}) \tag{6.55}$$

The above reveals a dependence on $p(\mathbf{y})$, which is intractable, and why VI is needed in the first place [57]. So, rather than the KL divergence (Eq. (6.55)), an alternative object is optimized that is equivalent to the (negative) KL divergence up to the term $\log p(\mathbf{y})$, which is a constant with respect to Θ:

$$\mathcal{L}_b(\Theta) = \mathbb{E}_{q(\Theta)}[\log p(\Theta, \mathbf{y})] - \mathbb{E}_{q(\Theta)}[\log q(\Theta)] \tag{6.56}$$

$$= \int q(\Theta) \log \frac{p(\Theta, \mathbf{y})}{q(\Theta)} d\Theta \tag{6.57}$$

This quantity is referred to as the ELBO. From Eq. (6.55), it can be seen that *maximizing* this object will *minimize* the KL divergence between $q(\Theta)$ and $p(\Theta \mid \mathbf{y})$. Conveniently, Eq. (6.57) can be used to construct a lower bound on the marginal likelihood $p(\mathbf{y})$, by rearranging Eq. (6.55) and substituting in Eq. (6.56):

$$\log p(\mathbf{y}) = KL\left(q(\Theta) \| p(\Theta \mid \mathbf{y})\right) + \mathcal{L}_b \tag{6.58}$$

Since $KL(\cdot) \geq 0$ [58], it follows that the evidence is lower-bounded by the ELBO, in other words, $\log p(\mathbf{y}) \geq \mathcal{L}_b$. This inequality is useful, as it shows that $\mathcal{L}_b$

can be used to monitor the marginal likelihood during inference or optimization of the hyperparameters $\boldsymbol{\theta}$ (as with the conventional GP; see Section 6.3.1.2).

A family $q \in \mathcal{Q}$ must be chosen for the variational approximation of the full posterior distribution $p(\{\mathbf{f}^{(k)}\}, \mathbf{Z} \mid \mathbf{y})$. We adopt a *mean-field* assumption, such that q factorizes, $q(\{\mathbf{f}^{(k)}\}, \mathbf{Z}) = q(\{\mathbf{f}^{(k)}\})q(\mathbf{Z})$ [57], implying that each variable (associated with each factor) is statistically independent. In consequence, by utilizing conjugate prior distributions, it is possible to analytically update each latent variable in turn (presented in Ref. [53]) while keeping the others fixed, such that the bound $\mathcal{L}_b$ is maximized, with respect to that variable. Updates for each factor are iterated until convergence in the lower bound $\mathcal{L}_b$—in practice, however, a corrected lower bound is used $(\mathcal{L}_{bc})$, which is more stable to implement (again, see Ref. [53] for details).

In brief terms, the inference scheme alternates between updating the approximated (factorized) posterior and then optimizing the hyperparameters of the model, while the improved lower bound $\mathcal{L}_{bc}$ on the marginal likelihood is maximized. The following expectation maximization (EM) steps are alternated:

1. *E-step*: Iterate mean-field updates of $\{\mathbf{f}_k\}$ and $\mathbf{Z}$ [53] until convergence in $\mathcal{L}_{bc}$ (fix hyperparameters).
2. *M-step*: Optimize the lower bound $\mathcal{L}_{bc}$ with respect to all hyperparameters until convergence:

$$\left\{\left\{\hat{\theta}_k\right\}_{k=1}^{K}, \hat{\Pi}\right\} = \underset{\left\{\{\theta_k\}_{k=1}^{K}, \Pi\right\}}{\operatorname{argmax}} \{\mathcal{L}_{bc}\}$$

The distribution $q(\mathbf{Z})$ is kept fixed.

Having initialized each component from the prior, steps 1 and 2 are iterated until convergence in $\mathcal{L}_{bc}$.

6.7.3.3 A Note on Model Dependence

While mean-field VI enables elegant inference via conjugate updates, it fails to capture the inter-task relationships *in this case*. More generally, mean-field families do not capture the correlation between any factors of $\mathcal{Q}$. Figure 6.23 visualizes this effect, reproducing the 2D example from Ref. [57], which approximates a highly correlated, Gaussian-distributed posterior—the target is shown by the elongated (off-axis) teal ellipse. Crucially, the VI approximation (red ellipse) cannot represent correlations between θ_1 and θ_2, since they are specified as independent factors in the mean-field family.

The importance of cross-correlation depends on the application. When the expected parameter values (ellipse centers) are a priority, a mean-field approximation can be sufficient, since the expectations are typically similar [57]. We argue, however, that in most applications of population monitoring, the relationship between task parameters is highly informative. Modeling these correlations extends insights into inter-turbine and wind farm behavior, rather than pooling

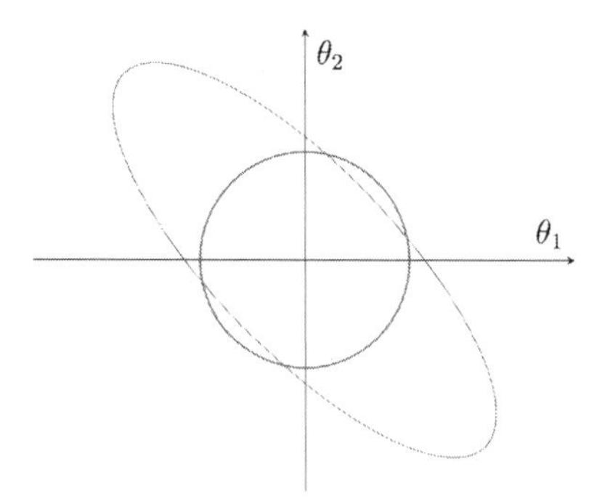

FIGURE 6.23 Visualization of a mean-field variational inference (VI) approximation $\hat{q}(\Theta)$ (red ellipse) for a two-dimensional Gaussian distributed posterior, compared to the ground truth $p(\Theta \mid \mathbf{y})$ (teal ellipse).

where the effective size of training data is extended only. As discussed, this information can be used to postulate which groups of systems and tasks share information, for which interpretable effect; this motivates the methodology presented in Section 6.5.

6.7.3.4 Prediction

The OMGP can be used to predict outputs that are yet to be observed $\mathbf{y}_*$. The posterior predictive likelihood for new inputs $\mathbf{x}_*$ is:

$$p(\mathbf{y}_* | \mathbf{x}_*, \mathcal{D}) \propto \sum_{k=1}^{K} \mathcal{N}\left(\mathbf{y}_* | \bar{\mathbf{f}}_*^{(k)}, \mathbf{Q}_*^{(k)}\right) \tag{6.59}$$

$$\begin{aligned}
\bar{\mathbf{f}}_*^{(k)} &\triangleq \mu_*^{(k)} + \mathbf{K}_{\mathbf{x}_*\mathbf{x}}^{(k)} \left(\mathbf{K}_{\mathbf{xx}}^{(k)} + \mathbf{B}^{(k)-1}\right)^{-1} \left(\mathbf{y} - \mu^{(k)}\right) \\
\mathbf{Q}_*^{(k)} &\triangleq \mathbf{K}_{\mathbf{x}_*\mathbf{x}_*}^{(k)} - \mathbf{K}_{\mathbf{x}_*\mathbf{x}}^{(k)} \left(\mathbf{K}_{\mathbf{xx}}^{(k)} + \mathbf{B}^{(k)-1}\right)^{-1} \mathbf{K}_{\mathbf{x}_*}^{(k)} + \mathbf{R}_*^{(k)} \\
\mathbf{R}_*^{(k)} &\triangleq \sigma^2 \mathbf{I}_M \\
\mathbf{B}^{(k)} &= \text{diag}\left(\left\{\frac{\hat{\Pi}[1,k]}{\sigma^2}, \ldots, \frac{[\hat{\Pi}[N,k]]}{\sigma^2}\right\}\right)
\end{aligned} \tag{6.60}$$

We vectorize prediction, such that $\mathbf{K}_{\mathbf{xx}'}[i,j] = k(x_i, x'_j)\ \forall i,j$, and so on. In the mixture, the prior mixing proportion for new observations weights each power curve equally, such that the associated constant is ignored in Eq. (6.59). Interestingly, the predictive equations for the OMGP are similar to the conventional GP (Eq. (6.4)). The noise component differs $\left(\mathbf{B}^{(k)-1}\right)$, as it is scaled according to $\hat{\Pi}[i,k]^{-1}$ [53], which *weights* the contribution of each training observation between the K tasks.

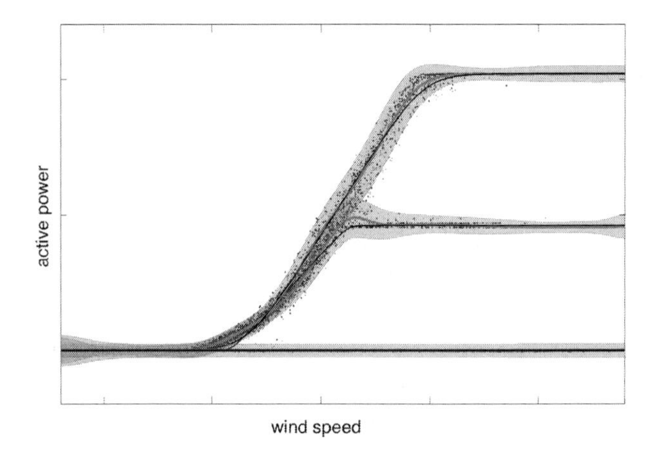

FIGURE 6.24 OMGP regression of wind farm channel data. Black lines show the soft-clip mean functions of the prior. Gray lines show the predictive mean $\overline{\mathbf{f}}_*^{(k)}$, and shaded regions show three-sigma of the predictive variance $diag(\mathbf{Q}_*^{(k)})$. Small · markers show the test set, and larger • markers show the training set. (Reproduced from Ref. [50].)

6.7.3.5 Results and Performance-Monitoring Examples

In total, 8900 observations were sampled from the wind farm data, corresponding to a (selected) subset of seven operational turbines over nine weeks. OMGP regression of the curtailed data is shown in Figure 6.24. Given the training observations (larger • markers), the model has inferred the multivalued behavior in an unsupervised manner, including the ideal curve, ≈50% curtailment, and the zero-power behavior. Figure 6.25 shows heteroskedastic (input-dependent) noise over

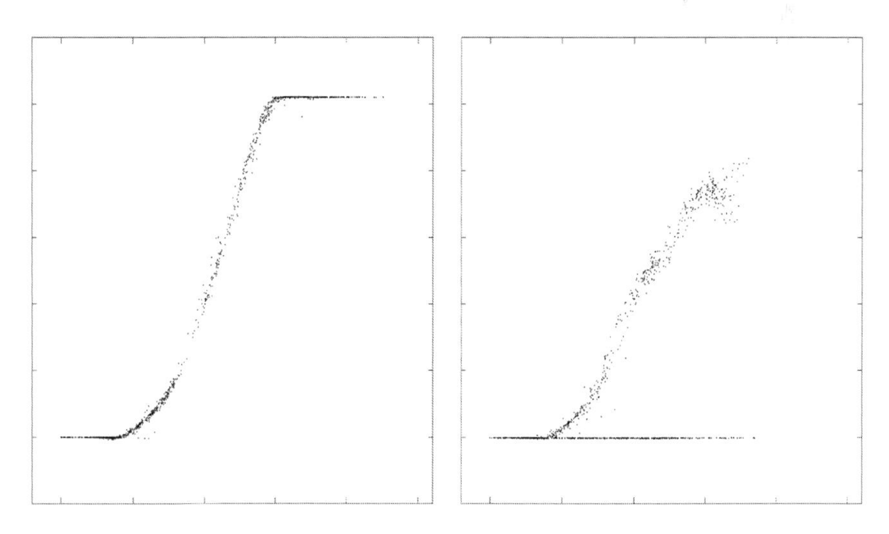

FIGURE 6.25 Weekly datasets, compared to the mixture model. (Reproduced from Ref. [60].)

the posterior predictive covariance $\mathbf{Q}_*^{(k)}$. For details of how input-dependent noise is estimated for this example, we direct the reader to Ref. [59].

Such a wind farm model is useful in a variety of contexts, such as forecasting, monitoring, and control. Here, we show a performance-monitoring example, where OMGP is compared to future (test) data from *all* turbines in the population. This was originally presented in Ref. [60], where a similar power curve model was used to inform outlier analysis by measuring the combined predictive-likelihood of future data under the OMGP. Examples of inlying or outlying (weekly) data subsets (from across the wind farm) are shown in Figure 6.25.

The functional outlier analysis is useful when monitoring, since the outlier in Figure 6.25b resembles a typical *suboptimal* power curve [52], while the inlying example in Figure 6.25a resembles one of the K permitted normal conditions.

6.7.4 Multilevel Models for Partial Pooling

Multilevel or *hierarchical* models offer another method to approximate collected power curve data.[2] As alluded to in Section 6.7.3, a hierarchical model structure can capture correlations between tasks, where the data provides evidence of task similarity (this information was not represented in the previous model). In turn, the wind farm model not only extends the training data to improve prediction but offers further insights into population-wide behavior. For example:

- Which turbines are correlated, for which (ideally interpretable) parameter
- A formal quantification of how parameters vary over the farm
- The ability to learn (or encode as prior information) shared parameters or turbine-specific parameters

For demonstration, we introduce standard hierarchical linear regression, which is developed into a piecewise linear model for power curve prediction—originally presented in Ref. [49].

6.7.4.1 Introduction to Multilevel Models as Linear Regression

We initially consider the set of tasks from Eq. (6.48) as K nested linear regressions. Building a hierarchical structure allows us to learn shared or task-specific *effects* for different turbine groupings (consider the visualization in Figure 6.21). In this application, the parameters of the tasks themselves f_{kl} will be learnt at the:

- System (turbine) level $\left(k \in \{1,\ldots,K\}\right)$;
- Subgroup level, relating to operating conditions $\left(l \in \{1,\ldots,L\}\right)$; or
- Whole-population level (no index).

In the introductory example, this will be parameterized as follows:

$$\left\{\left\{\mathbf{y}_{kl} = \underbrace{\Phi_{kl}\alpha_k}_{\text{per-turbine}} + \underbrace{\Psi_{kl}\beta_l}_{\text{per-condition}} + \epsilon_{kl}\right\}_{l=1}^{L_k}\right\}_{k=1}^{K} \tag{6.61}$$

where Φ_{kl} and Ψ_{kl} are the *design* matrices, containing (transformations of) measurements from each turbine. A typical example would set $\Phi_{kl} = [\mathbf{1}, \mathbf{x}_k]$, such that the first term in Eq. (6.61) is simple linear regression, with a *slope* and *intercept*; while Ψ_{kl} might correspond to the observations transformed by some nonparametric basis functions (i.e., splines). The matrices correspond to turbine-specific $\left(k \in \{1,\ldots,K\}\right)$ and condition-specific $\left(l \in \{1,\ldots,L_k\}\right)$ *weight* vectors $\left\{\alpha_k \text{ and } \beta_l\right\}$, respectively. In the above example, there would be a linear effect that is modeled for each turbine k, and some nonparametric component that is shared between the operating conditions l. The noise vector is normally distributed for all tasks $\epsilon_{kl} \sim N\left(0, \sigma^2\mathbf{I}\right)$—more on this assumption later.

Keeping a general notation, the likelihood of the response becomes:

$$\mathbf{y}_{kl} \mid \mathbf{x}_{kl} \sim \mathrm{N}\left(\Phi_{kl}\alpha_k + \Psi_{kl}\beta_l, \sigma^2\mathbf{I}\right) \tag{6.62}$$

In a Bayesian manner, we can set a hierarchy of prior distributions over the weights $\{\alpha_k, \beta_l\}$ for each group $\{k, l\}$ to encode our domain expertise of wind farm interdependence, in terms of shared (tied) or correlated (varying) parameters. A standard example might be:

$$\{\alpha_k\}_{k=1}^{K} \overset{i.i.d}{\sim} N\left(\mu_\alpha, diag\left\{\sigma_\alpha^2\right\}\right) \tag{6.63}$$

$$\mu_\alpha \sim N\left(m_\alpha, \mathrm{diag}\{s_\alpha\}\right) \tag{6.64}$$

$$\sigma_\alpha \overset{\text{i.i.d}}{\sim} \mathrm{IG}(a, b) \tag{6.65}$$

$$\beta_l \sim N\left(\mu_\beta, \mathrm{diag}\left\{\sigma_\beta^2\right\}\right) \tag{6.66}$$

$$\sigma \sim \mathrm{IG}(a_\sigma, b_\sigma) \tag{6.67}$$

This model and a prior formulation are represented in Figure 6.26. In words, Eq. (6.64) assumes that the weights $\{\alpha_k\}_{k=1}^{K}$ are normally distributed $N(\cdot)$ with mean μ_α and covariance[3] $diag\left\{\sigma_\alpha^2\right\}$. Similarly, Eq. (6.65) states that the prior expectation of the weights α_k is normally distributed with mean $\boldsymbol{m}_\alpha$ and covariance $diag\{s_\alpha\}$; lastly, Eq. (6.66) states that the deviation of the slope and intercept is inverse-gamma distributed a priori $\mathrm{IG}(\cdot)$, with shape and scale parameters a and b, respectively.

The effect of this hierarchy is that task-specific weight vectors (α_k) are learnt for each turbine. These weights α_k are conditionally dependent on the common

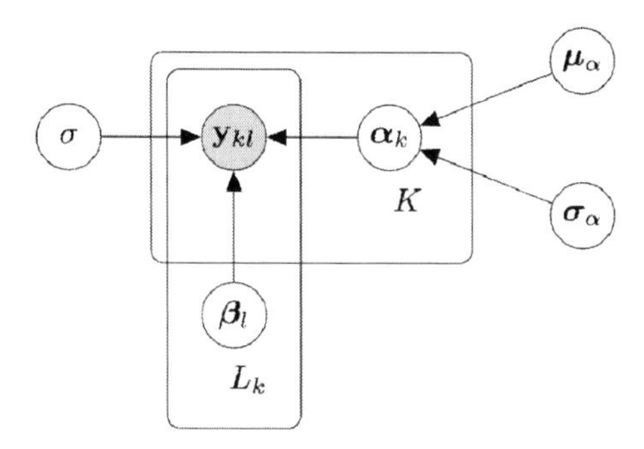

FIGURE 6.26 Directed graphical model of hierarchical linear regression with nested groups. The total number of tasks is $\sum_{k=1}^{K} L_k$.

latent variables $\{\mu_\alpha, \sigma_\alpha^2\}$ (i.e., parent nodes in Figure 6.26). In turn, sparse domains borrow statistical strength from data-rich domains since the parent nodes $\{\mu_\alpha, \sigma_\alpha^2\}$—which influence all tasks—are inferred from the collected population data. On the other hand, the parameters from β_l are *not coupled* via parent nodes. Instead, these weights are tied between each operating condition $l \in \{1, \dots, L_k\}$ for all turbines. This is highlighted by an l-index (and absent parent nodes) in the directed graphical model of Figure 6.26. As such, the parameters in β_l are shared between all turbines (k) but separate for each operating condition (l), similar to the OMGP. Finally, the variance of the additive noise σ has no index—it is learnt wholly at the population level since the noise is assumed consistent between all tasks (turbines and operating conditions).

Selecting appropriate prior distributions, and their associated hyperparameters $\{\mathbf{m}_\alpha, \mathbf{s}_\alpha, a, b\}$, is essential to the success of hierarchical models—which can be notoriously unstable when implemented incorrectly [61]. We now justify these decisions in an example by encoding engineering knowledge as weakly informative priors [62] for a set of power curves.

6.7.4.2 Model and Prior Formulation: Segmented Linear Power Curves

Figure 6.27 shows the multi-task power curve data used in this example, provided by Visualwind and recorded from three operational turbines. The turbines are the same make and model but in different locations. Unlike the mixture model example, where each function f_k was assumed independent, knowledge transfer is enabled by correlating and tying the parameters using a hierarchical formulation.

Figure 6.27 shows 10,581 (SCADA) observations in total. The data was labeled in weekly subsets, according to turbine $k \in \{1, 2, 3\}$ and operational condition (normal or curtailed) $l \in \{1, 2\}$. As before, each point corresponds to a 10-min average of power y_{ikl} and wind speed x_{ikl}. The first turbine has two weeks of data, the second has four weeks, and the third has 11.5 weeks. Missing values and very sparse outliers were removed from the dataset. For details of data pre-processing, refer to the original paper in Ref. [49].

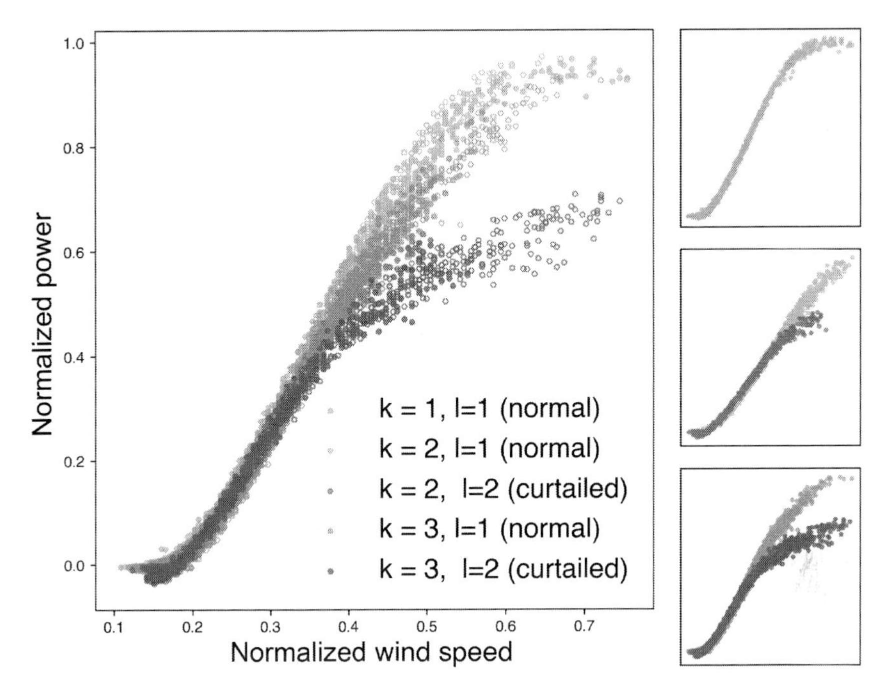

FIGURE 6.27 Power curve data from three $k \in \{1,2,3\}$ wind turbines of the same make and model. Relationships correspond to normal $l = 1$ (light colors) and curtailed $l = 2$ (dark colors) operation. (Reproduced from Ref. [49].)

Since the first turbine presents a normal power curve only $(l = 1)$, there are five tasks altogether, $\sum_{l=1}^{L} L_k = 1 + 2 + 2 = 5$. Specific tasks are subsampled to own less data than others: in particular, referring to Figure 6.27, the data-rich normal task from the first turbine ($k = 1, l = 1$: light blue) should support the sparse normal tasks ($k \in \{2,3\}$: light green and red); while the data-rich curtailment from the third turbine (dark red) should support the curtailed relationship of the second turbine (dark green).

A standard power curve model assumes segmented-linear regression [55]. A similar formulation is adopted here:

$$P(x_i) = \begin{cases} 0 & x_i < p \\ m_1(x_i - p) & p < x_i < q \\ m_2(x_i - q) + m_1(q - p) & q < x_i < r \\ P_m & x_i > r \end{cases} \tag{6.68}$$

$$m_2 \triangleq \frac{P_m - m_1(q - p)}{(r - q)}$$

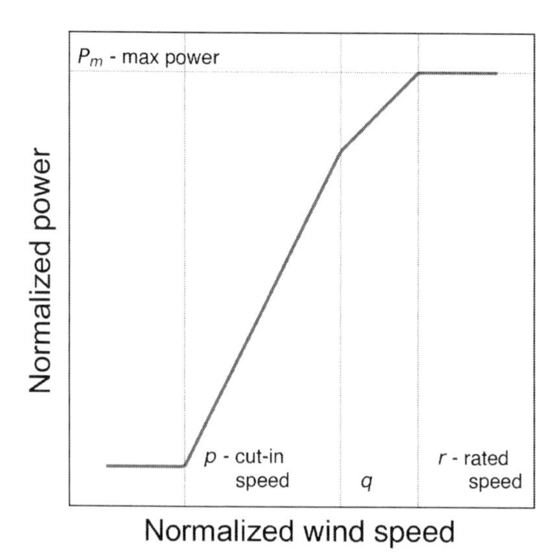

FIGURE 6.28 The segmented linear power curve model, indicating interpretable parameters $\{p,q,r,P_m\}$. (Reproduced from Ref. [49].)

Although simple, Eq. (6.68) presents interpretable parameters—visualized in Figure 6.28. p is the cut-in speed, and r is the rated speed (for normal operation); the change-point q corresponds to the initiation of the limit to maximum power P_m (where $p < q < r$). The gradients m_1 and m_2 (linearly) approximate the response between p-q and q-r, respectively. The second change point and gradient $\{q,m_2\}$ enable *soft* curtailments, rather than a hard limit at maximum power P_m.

Each segment of the segmented regression could follow a similar structure to Eq. (6.62), such that each component is a varying intercepts–slope model [62]. Herein, we avoid matrix notation to present the model around the interpretable parameters $\{P_m,m_1,m_2,p,q,r\}$. This makes formulating the structure of the population model and prior distributions much more intuitive, to naturally encode domain expertise.

6.7.4.3 Encoding Multilevel Domain Expertise

From knowledge of turbine operation, we expect the power before cut-in to be zero for all turbines (i.e., a tied parameter). The cut-in speed p can also be tied and inferred at the population level since all turbines have the same design. Considering groups of operating condition, the max power P_m should be tied between operational labels $l \in \{1,2\}$, such that one parameter is learnt for the normal tasks $(l=1)$ and one for the curtailed tasks $(l=2)$. On the other hand, the change-points $\{q,r\}$ and gradients $\{m_1,m_2\}$ are assumed to be distinct but correlated between all tasks via the parent nodes—since these variables are more affected by the environment and use-type of the turbine. Within the covariance structure of the model, we expect the curtailed relationships $(l=2)$ to be more correlated than the normal relationships $(l=1)$, and vice versa.

The (expected) task set can then be summarized as hierarchical segmented linear models:

$$\left\{\left\{\hat{y}_i^{(kl)} = \begin{cases} 0 & x_i < p \\ m_1^{(kl)}(x_i - p) & p < x_i < q^{(kl)} \\ m_2^{(kl)}(x_i - q^{(kl)}) + m_1^{(kl)}(q^{(kl)} - p) & q^{(kl)} < x_i < r^{(kl)} \\ P_m^{(l)} & q^{(kl)} < x_i < r^{(kl)} \end{cases}\right\}_{k=1}^{K_l}\right\}_{l=1}^{L} \tag{6.69}$$

$$m_2^{(kl)} \triangleq \frac{P_m^{(l)} - m_1^{(kl)}(q^{(kl)} - p)}{(r^{(kl)} - q^{(kl)})} \tag{6.70}$$

To reiterate, the parameters that vary between turbines have a k-index (subscript or superscript), the parameters that vary between operating condition have an l-index, and population-level parameters have no index. The hierarchy of parameter dependencies is visualized with a directed graphical model in Figure 6.29, where plates collect parameters from a total of $\sum_{k=1}^{K} L_k$ tasks relating to operating condition l or turbine labeling k. (Parent nodes are discussed below.) Figure 6.29 is useful since it becomes clear which parameters are learnt at which groupings (i.e., which nodes are enclosed by which plates).

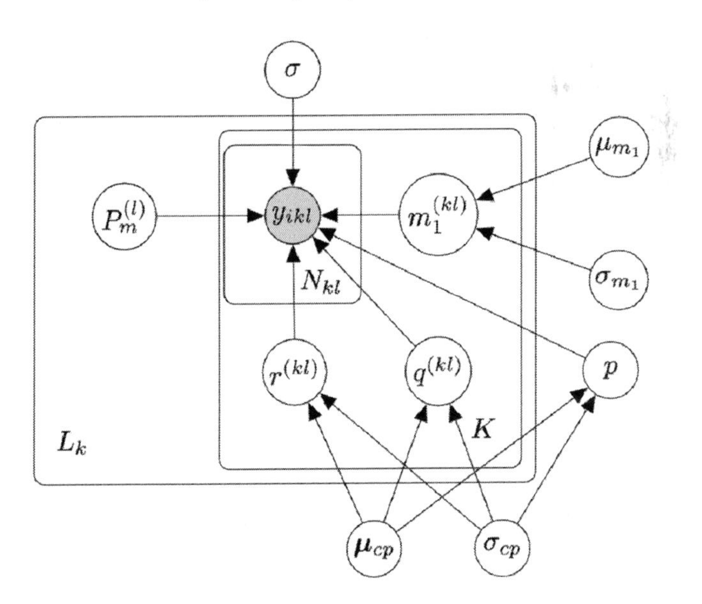

FIGURE 6.29 Directed graphical model of the hierarchical segmented linear power curve for collected wind turbine data. The total number of tasks is $\sum_{k=1}^{K} L_k$, for $k \in \{1, \ldots, K\}$ turbines and $l \in \{1, \ldots, L_K\}$ operating conditions.

The likelihood of the response can be specified using Eq. (6.72):

$$y_{ikl} \mid x_{ikl}, \boldsymbol{\theta}_{kl} \sim \mathrm{N}\left(\hat{y}_i^{(kl)}, \sigma^2\right) \tag{6.71}$$

where $\boldsymbol{\theta}_{kl}$ is the parameter set relevant to turbine k and curtailment l, that is, $\boldsymbol{\theta}_{kl} = \left\{P_m^{(l)}, m_1^{(kl)}, p, q^{(kl)}, r^{(kl)}\right\}$. Since we have a clear interpretation of each parameter, weakly informative prior distributions [62] are postulated. For the change points:

$$\begin{gathered} p \sim \mathrm{N}(\mu_p, \sigma_{cp}^2), \quad q^{(kl)} \sim \mathrm{N}(\mu_q, \sigma_{cp}^2), \quad r^{(kl)} \sim \mathrm{N}(\mu_r, \sigma_{cp}^2) \\ \mu_{cp} \sim \mathrm{N}([.2, .4, .6], .5 \times \mathbf{I}), \quad \sigma_{cp} \sim \mathrm{IG}(1,1) \end{gathered} \tag{6.72}$$

These distributions encode our belief that the change points are expected to occur at regular intervals across the input with high variance (respective to a normalized scale). The expected change points have been collected in one vector $\mu_{cp} = \left\{\mu_p, \mu_q, \mu_r\right\}$. The priors for gradient and maximum power are also based on domain expertise:

$$\begin{gathered} m_1^{(kl)} \sim \mathrm{N}(\mu_{m_1}, \sigma_{m_1}^2) \\ \mu_{m_1} \sim \mathrm{N}(2.5, .5), \quad \sigma_{m_1} \sim \mathrm{IG}(1,1) \end{gathered} \tag{6.73}$$

$$P_m^{(1)} \sim \mathrm{N}(1, .1), \qquad P_m^{(2)} \sim \mathrm{N}(.8, .1) \tag{6.74}$$

These distributions present the expected gradient m_2 in a normalized space, a unit max power $P_m^{(1)}$ for normal operation, and typical 80% curtailment [50] for the limited output $P_m^{(2)}$. No prior is required for m_2, since it is specified by $\left\{P_m, m_1, p, q, r\right\}$ in Eq. (6.71). The $\mathrm{IG}(1,1)$ distributions encourage weak inter-task correlations, such that task dependencies are mostly informed by the evidence in the data. The hierarchical, segmented structure of the model means that the posterior is intractable and inferred with MCMC. Here, this is implemented in the probabilistic programming language Stan [63], which utilizes the no-U-turn implementation of Hamiltonian Monte Carlo [64]. The burn-in period is 1000 iterations, and 2000 iterations are used for inference.

6.7.4.4 Results and Interpreting the Wind Farm Model

The wind farm inference is compared to two benchmarks, each with the same hierarchical prior distributions and hyperparameters (for single or independent models):

- *Single-task learning (STL)*: The task model learnt from each domain independently
- *Complete pooling (CP)*: The model learnt for normal or curtailed power curves, considering the data from all turbines as a single task

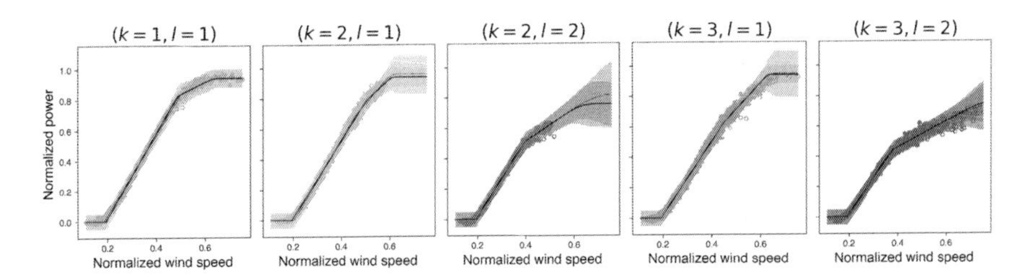

FIGURE 6.30 Posterior predictive distribution. The mean and three-sigma deviation for: (Light shading, dashed line) K independent power curve models, and (dark shading, solid line) multitask learning via hierarchical Bayes. (Reproduced from Ref. [49].)

Figure 6.30 shows posterior predictive distributions compared to STL. Although the differences are small, variance reductions are (intuitively) more obvious in domains with less data, or data that only corresponds to a subset of the input domain (greens and light red).

Table 6.6 quantifies changes in task-wise predictions compared to the benchmarks. The combined predictive likelihood $\mathcal{L}$ is increased when population modeling, compared to single-task learning, from 8229 to 8258. Task-wise, there is a likelihood increase in all domains other than $(k=2, l=1)$ and $(k=3, l=2)$. We believe these reductions occur because the inference maximizes the overall likelihood, $\mathcal{L}$. In turn, the performance in data-rich domains is reduced in a trade-off, as our prior belief is best suited to data-rich tasks—when the prior becomes more informed by data, it becomes less suitable in data-rich domains and represents the population instead. To combat this, less informative or uninformative priors should be considered [62].

As an example, Figure 6.31 presents insights relating to maximum power estimates P_m; for a discussion around the other parameters, see the original publication in Ref. [49]. Intuitively, when the normal maximum $P_m^{(k,1)}$ is tied between turbines, the estimate moves toward the value evidenced by the data-rich domain (blue), while the curtailed maximum $P_m^{(k,2)}$ moves toward an average of the relevant tasks (where $l=2$). In both cases, parameter tying enables a posterior distribution with a clear expected value, rather than a relatively vague, *flat* distribution, which does not provide much insight beyond our prior belief. Predictive capabilities are improved, since the population-level model allows the data to be extended, and

TABLE 6.6
Predictive Log-Likelihood: l Corresponds to the Operating Condition (Normal $l=1$, or Curtailed $l=2$), and k Is the Turbine Identifier

Method	$k=1, l=1$	$k=2, l=1$	$k=3, l=1$	$k=2, l=2$	$k=3, l=2$	$\mathcal{L}$
Complete pooling	−168	1555	4681	594	452	7114
Single-task learning	202	**1619**	5147	538	**722**	8229
Multi-task learning	**218**	1599	**5206**	**549**	686	**8258**

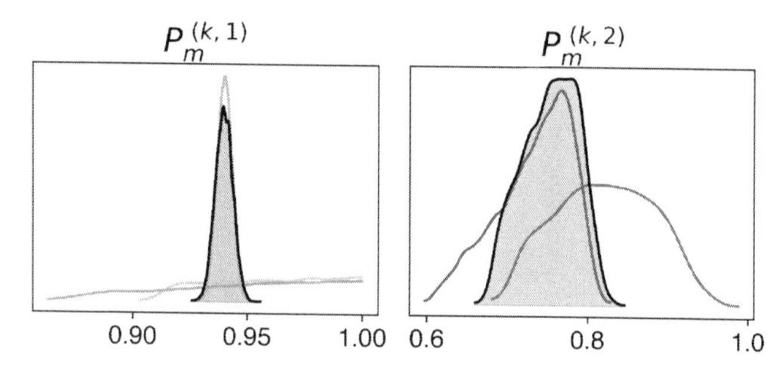

FIGURE 6.31 Changes in the posterior distribution of the independent models (hollow), compared to population-level modeling (shaded). (Reproduced from Ref. [49].)

parameters are automatically learnt from domains that have data to describe the associated effect—in this case, maximum power.

To demonstrate how the covariance structure is also insightful, Figure 6.32 plots the Pearson correlation coefficient of the pair-wise conditionals for q between tasks. (We present q since it is the most structured and insightful.) The heatmap visualizes how hierarchical modeling can capture the correlation between related tasks, with two distinct blocks associated with the normal and curtailed groups.

The resultant covariance structure should demonstrate the potential for a combined and interdependent wind farm analysis—that is, indicating which tasks (turbine, operating condition, or environmental effect) are correlated for which effect (interpretable parameter). This knowledge should allow data-centric monitoring to be more informed, considering how individual turbines constitute collective wind farm behavior.

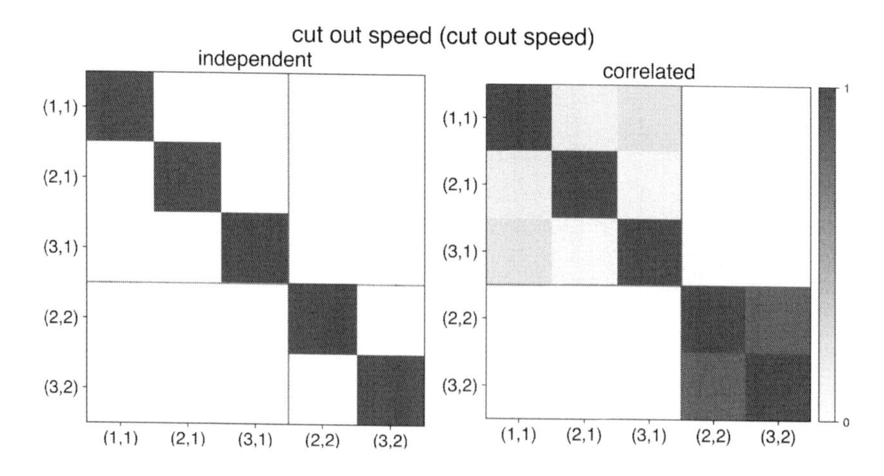

FIGURE 6.32 Pearson correlation coefficient of the conditional posterior distribution of q. Tick labels correspond to (k,l). Purple lines separate the normal $(l=1)$ from the curtailed $(l=2)$ task parameters. (Reproduced from Ref. [49].)

6.7.5 Comments and Discussion

This section has brought together several recent advances in data-centric monitoring of wind farm populations [48–50, 60], suggesting a workflow to monitor the combined data recorded from a large number of turbines (focusing here on SCADA measurements). For data management, we presented a standardized procedure (the *PBSHM Schema*) of storage, extraction, and manipulation of data—in the form of a NoSQL database. The data format is agnostic to the acquisition system, asset owner, and turbine type. The generalized method of data management makes the handling of measurements from large collected groups of turbines (and wind farms) more efficient before any statistical pattern recognition is implemented.

When modeling the extracted measurements, we summarize two methods of approximating the population or wind farm data: (i) mixture modeling, and (ii) hierarchical or multilevel modeling. Each method effectively extends the number of training data by considering that turbines constitute a *whole* (the wind farm)—this is achieved by the *pooling* of data in various forms. In both examples, the predictive capability of power models is improved by considering a collected population of turbines, rather than individual systems, as information can be shared between turbine predictors. Multilevel models take this a step further, however, since they enable additional insights into population-wide behavior, as well as predictive improvements. Insights include a formal quantification of how (interpretable) parameters vary over the farm and an identification of which turbines are similar, for which interpretable effect. This information has the potential for high impact in practical applications since it can inform downstream analysis and decision making (under uncertainty) at the population level, in view of the combined wind farm performance.

6.8 CONCLUDING REMARKS

This chapter has collected recent works in assimilating knowledge from multiple sources of information to improve wind farm monitoring procedures—considering applications of design, control, and maintenance. Each section presents methods to formally combine predictions, data, and knowledge from multiple resources, focusing on:

- Multiple turbines or operating conditions in a wind farm
- An ensemble or population of FEM simulations
- The augmentation of physics-based models with machine learning

One overarching theme is to monitor wind farms collectively and capture the associated interactions between turbines. Information is then shared between predictors, and the value of measurements can be extended. For example, Section 6.3 utilizes a highly instrumented turbine to predict the loads on downwind neighbors (in a simulated study), while Section 6.4 utilizes graph neural networks to learn relational structure in wind farm data—informing how operational characteristics are affected by wind farm design (e.g., turbine separation). Section 6.7 continues this theme by learning large (correlated) power models, considering subgroups of turbines and operating conditions as nested groups in a multilevel Bayesian model.

A secondary theme considers the exploitation of physics-based *models*. Section 6.5 utilizes a cluster-weighted average of an ensemble of predictors—consisting of stochastic finite element simulations. The ensemble diversifies the hypothesis space and improves uncertainty quantification. On the other hand, Section 6.6 augments physics-based filtering methods with a typical machine learning model (Gaussian processes) to enable latent force prediction of distributed loads on wind turbine blades.

Looking forward, while *in-the-field* data is used in several of the cases presented here, a primary focus should consider more *in situ* measurements, especially for applications of structural health monitoring (rather than performance monitoring) where SCADA data does not suffice. Another focus should look to downstream analysis and decision making at the systems level, including inspection planning, and control interactions. These procedures should ensure that the emerging models can inform actionable insights for engineers in practice—this will likely involve decision making under uncertainty alongside work toward increased interpretability and explainability.

ACKNOWLEDGMENTS

The authors gratefully acknowledge the support of the UK Engineering and Physical Sciences Research Council (EPSRC) through grant reference numbers EP/R003645/1, EP/R004900/1, EP/R006768/1, and EP/W005816/1. A.B. Duncan and L.A. Bull were supported by Wave 1 of The UKRI Strategic Priorities Fund under EPSRC Grant EP/W006022/1, particularly the *Ecosystems of Digital Twins* theme within that grant, and The Alan Turing Institute. E. Chatzi and I. Abdallah were supported by the SNSF Bridge Discovery project—AeroSense (40B2-0_187087), a novel microelectromechanical systems (MEMS)-based surface pressure and acoustic internet-of-things (IoT) measurement system for wind turbines. This research was supported by Vattenfall (via A.E. Maguire and C. Campos) and also Visualwind (UK).

ACRONYMS

CM	condition monitoring
CRVAE	conditional relational variational auto-encoder
DWM	dynamic wake meandering
EM	expectation maximization
FAST	fatigue, aerodynamics, structures, and turbulence (FAST is an aeroelastic simulator.)
GNN	graph neural network
GPR	Gaussian process regression
MH	Metropolis–Hastings

PBSHM	population-based structural health monitoring
RVAE	relational variational auto-encoder
SCADA	supervisory control and data acquisition
SHM	structural health monitoring
ULC-BAG	unsupervised local cluster-weighted bootstrap aggregation
VAE	variational auto-encoder
VBGM	variational Bayesian Gaussian mixture
VBGMC	variational Bayesian Gaussian mixture clustering
WT	wind turbine

NOTES

1. Colon notation indexes entire columns or rows in a matrix.
2. Like Ref. [24], we use the terms *multilevel* and *hierarchical* interchangeably.
3. The operator diag $\{\mathbf{a}\}$ forms a square diagonal matrix with the elements from $\mathbf{a}$ on the main diagonal and zeros elsewhere.

REFERENCES

1. M. Grieves and J. Vickers. Digital twin: Mitigating unpredictable, undesirable emergent behavior in complex systems. *Transdisciplinary Perspectives on Complex Systems: New Findings and Approaches*, 85–113, 2017.
2. Climate Change Act 2008, 2008.
3. B.R. Resor. Definition of a 5mw/61.5 m wind turbine blade reference model. *Albuquerque, New Mexico, USA, Sandia National Laboratories, SAND2013-2569*, 2013, 2013.
4. D. Veldkamp. *Chances in wind energy–a probabilistic approach to wind turbine fatigue design* [PhD thesis]. Delft University of Technology, 2006.
5. A. Gelman, J.B. Carlin, H.S. Stern, D.B. Dunson, A. Vehtari, and D.B. Rubin. *Bayesian Data Analysis, Third Edition.* Chapman and Hall/CRC, 2013.
6. E. Papatheou, N. Dervilis, A.E. Maguire, C. Campos, I. Antoniadou, and K. Worden. Performance monitoring of a wind turbine using extreme function theory. *Renewable Energy*, 113:1490–1502, 2017.
7. C. Tay and C. Laugier. Modelling smooth paths using Gaussian processes. In *Proceedings of the International Conference on Field and Service Robotics*, Chamonix, France, 2007.
8. P.W. Battaglia, J.B. Hamrick, V. Bapst, A. Sanchez-Gonzalez, V. Zambaldi, M. Malinowski, A. Tacchetti, D. Raposo, A. Santoro, R. Faulkner, et al. Relational inductive biases, deep learning, and graph networks. *arXiv preprint arXiv:1806.01261*, 2018.
9. C.E. Rasmussen and C.K.I. Williams. *Gaussian Processes for Machine Learning.* Adaptive computation and machine learning. MIT Press, 2006.
10. NREL. FLORIS. Version 2.2.0, 2020.
11. K.E. Tatsis, V.K. Dertimanis, C. Papadimitriou, E. Lourens, and E.N. Chatzi. A general substructure-based framework for input-state estimation using limited output measurements. *Mechanical Systems and Signal Processing*, 150:107223, 2021.
12. L.A. Bull, P.A. Gardner, T.J. Rogers, N. Dervilis, E.J. Cross, E. Papatheou, A.E. Maguire, C. Campos, and K. Worden. Bayesian modelling of multivalued power curves from an operational wind farm. *Mechanical Systems and Signal Processing*, 108530, 2021.

13. M. Lydia, S.S. Kumar, A.I. Selvakumar, and G.E. Prem Kumar. A comprehensive review on wind turbine power curve modeling techniques. *Renewable and Sustainable Energy Reviews*, 30:452–460, 2014.
14. M.D. Hoffman, A. Gelman, et al. The No-U-Turn sampler: Adaptively setting path lengths in Hamiltonian Monte Carlo. *Journal of Machine Learning Research*, 15(1):1593–1623, 2014.
15. L. Breiman. Bagging predictors. *Machine Learning*, 24:123–140, 1996.
16. N. Kirchner Bossi and F. Porté-Agel. Multi-objective wind farm layout optimization with unconstrained area shape. In *Wind Energy Science Conference 2019 (WESC 2019)*, number POST_TALK, 2019.
17. C. Mylonas, I. Abdallah, and E. Chatzi. Relational VAE: A continuous latent variable model for graph structured data, 2021.
18. O. Ivanov, M. Figurnov, and D. Vetrov. Variational autoencoder with arbitrary conditioning. *arXiv preprint arXiv:1806.02382*, 2018.
19. D. Rezende and S. Mohamed. Variational inference with normalizing flows. In F. Bach and D. Blei, editors, *Proceedings of the 32nd International Conference on Machine Learning, Volume 37* of Proceedings of Machine Learning Research, pages 1530–1538, Lille, France, 07–09 Jul 2015. PMLR.
20. S.E Azam, E. Chatzi, C. Papadimitriou, and A. Smyth. Experimental validation of the dual Kalman filter for online and real-time state and input estimation. In *Model Validation and Uncertainty Quantification, Volume 3*, pages 1–13. Springer, 2015.
21. M. Lázaro-Gredilla, S. Ṽan Vaerenbergh, and N.D. Lawrence. Overlapping mixtures of Gaussian processes for the data association problem. *Pattern Recognition*, 45:1386–1395, 2012.
22. D.P. Kingma, T. Salimans, R. Jozefowicz, X. Chen, I. Sutskever, and M. Welling. Improved variational inference with inverse autoregressive flow. In *30th Conference on Neural Information Processing Systems, Barcelona, Spain*, 2016.
23. C. Bishop. *Pattern Recognition and Machine Learning*. Springer, ISBN 978-0-387-31073-2, 2006.
24. L. Dinh, J. Sohl-Dickstein, and S. Bengio. Density estimation using real NVP. *arXiv preprint arXiv:1605.08803*, 2016.
25. J. Jonkman, S. Butterfield, W. Musial, and G. Scott. Definition of a 5-MW reference wind turbine for offshore system development. Technical report, National Renewable Energy Laboratory, NREL/EL-500-38060, 2009.
26. E. Papatheou, N. Dervilis, A.E. Maguire, I. Antoniadou, and K. Worden. A performance monitoring approach for the novel Lillgrund offshore wind farm. *IEEE Transactions on Industrial Electronics*, 62:6636–6644, 2015.
27. K. Tatsis, E. Chatzi, and E.-M. Lourens. Reliability prediction of fatigue damage accumulation on wind turbines support structures. In *Proceedings of the 2nd International Conference on Uncertainty Quantification in Computational Sciences and Engineering*, pages 76–89, 2017.
28. Wind Turbines, Part 1 Design Requirements. Technical Report IEC 61400-1:2005(E), International Electrotechnical Commission, 2005.
29. I. Abdallah, K. Tatsis, and E. Chatzi. Unsupervised local cluster-weighted bootstrap aggregating the output from multiple stochastic simulators. *Reliability Engineering & System Safety*, 199:106876, 2020.
30. L.D. Avendaño-Valencia, I. Abdallah, and E. Chatzi. Virtual fatigue diagnostics of wake-affected wind turbine via Gaussian process regression. *Renewable Energy*, 170:539–561, 2021.
31. C.M. Bishop. *Pattern Recognition and Machine Learning (Information Science and Statistics)*. Springer, 1 edition, 2007.

32. J. Liu, A. Kumar, J. Ba, J. Kiros, and K. Swersky. Graph normalizing flows. *arXiv preprint arXiv:1905.13177*, 2019.
33. L.A. Bull, D. Di Francesco, M. Dhada, O. Steinert, T. Lindgren, A.K. Parlikad, A.B. Duncan, and M. Girolami. Hierarchical Bayesian modeling for knowledge transfer across engineering fleets via multitask learning. *Computer-Aided Civil and Infrastructure Engineering.*
34. A.P. Dempster, N.M. Laird, and D.B. Rubin. Maximum likelihood from incomplete data via the EM algorithm. *Journal of the Royal Statistical Society*, 39(1):1–38, 1977.
35. P. Bühlmann. Bagging, boosting and ensemble methods. In J. Gentle and Y. Mori, editors, *Handbook of Comp. Stat.*, chapter 33, pages 985–1022. Springer, 2012.
36. T. Pfaff, M. Fortunato, A. Sanchez-Gonzalez, and P.W. Battaglia. Learning mesh-based simulation with graph networks. *arXiv preprint arXiv:2010.03409*, 2020.
37. V. Zambaldi, D. Raposo, A. Santoro, V. Bapst, Y. Li, I. Babuschkin, K. Tuyls, D. Reichert, T. Lillicrap, E. Lockhart, et al. Relational deep reinforcement learning. *arXiv preprint arXiv:1806.01830*, 2018.
38. BBC News: Wind farms could power every home by 2030, Oct 2020.
39. D.J.C. MacKay and D.J.C. Mac Kay. *Information Theory, Inference and Learning Algorithms*. Cambridge University Press, 2003.
40. K. Kersting, C. Plagemann, P. Pfaff, and W. Burgard. Most likely heteroscedastic Gaussian process regression. In *Proceedings of the 24th International Conference on Machine learning*, pages 393–400, 2007.
41. D.P. Kingma and J. Ba. Adam: A method for stochastic optimization. *arXiv preprint arXiv:1412.6980*, 2014.
42. D.M. Blei, A. Kucukelbir, and J.D. McAuliffe. Variational inference: A review for statisticians. *Journal of the American Statistical Association*, 112(518):859–877, 2017.
43. L.A. Bull, P.A. Gardner, J. Gosliga, T.J. Rogers, N. Dervilis, E.J. Cross, E. Papatheou, A.E. Maguire, C. Campos, and K. Worden. Foundations of population-based SHM, part i: Homogeneous populations and forms. *Mechanical Systems and Signal Processing*, 148:107141, 2021.
44. B. Carpenter, A. Gelman, M.D. Hoffman, D. Lee, B. Goodrich, M. Betancourt, M. Brubaker, J. Guo, P. Li, and A. Riddell. Stan: A probabilistic programming language. *Journal of Statistical Software*, 76(1), 2017.
45. M.D. Klimek and M. Perelstein. Neural network-based approach to phase space integration. *arXiv preprint arXiv:1810.11509*, 2018.
46. W.D. Penny. Variational Bayes for d-dimensional Gaussian mixture models. Technical report, University College London, 2001.
47. J.K. Lundquist, K.K. DuVivier, D. Kaffine, and J.M. Tomaszewski. Costs and consequences of wind turbine wake effects arising from uncoordinated wind energy development. *Nature Energy*, 4(1):26–34, 2019.
48. A. Sanchez-Gonzalez, J. Godwin, T. Pfaff, R. Ying, J. Leskovec, and P. Battaglia. Learning to simulate complex physics with graph networks. In Hal DaumÃ© III and A. Singh, editors, *Proceedings of the 37th International Conference on Machine Learning, Volume 119 of* Proceedings of Machine Learning Research, pages 8459–8468. PMLR, 13–18 Jul 2020.
49. D.P. Kingma and M. Welling. Auto-encoding bariational Bayes. *arXiv preprint arXiv:1312.6114*, 2013.
50. S.R. Bowman, L. Vilnis, O. Vinyals, A.M. Dai, R. Jozefowicz, and S. Bengio. Generating sentences from a continuous space. *arXiv preprint arXiv:1511.06349*, 2015.

51. D.S. Brennan, C.T. Wickramarachchi, E.J. Cross, and K. Worden. Implementation of an organic database structure for population-based structural health monitoring. In K. Grimmelsman, editor, *Dynamics of Civil Structures, Volume 2*, pages 23–41. Springer International Publishing, Cham, 2022. Series Title: Conference Proceedings of the Society for Experimental Mechanics Series.
52. M.O.L. Hansen. *Aerodynamics of Wind Turbines*. Taylor & Francis Ltd, 2015.
53. K. Tatsis, V.K. Dertimanis, T.J. Rogers, E. Cross, K. Worden, and E. Chatzi. A spatiotemporal dual Kalman filter for the estimation of states and distributed inputs in dynamical systems. In W. Desmet, B. Pluymers, D. Moens, and S. Vandemaele, editors, *International Conference on Noise and Vibration Engineering (ISMA 2020) and International Conference on Uncertainty in Structural Dynamics (USD 2020)*, pages 3591–3597, Red Hook, NY, 2021. Curran. 29th International Conference on Noise and Vibration Engineering (ISMA 2020) in conjunction with the 8th International Conference on Uncertainty in Structural Dynamics (USD 2020) (virtual); Conference Location: Leuven, Belgium; Conference Date: September 7-9, 2020; Conference lecture held on September 7, 2020. Due to the Coronavirus (COVID-19) the conference was conducted virtually.
54. M. Betancourt and M. Girolami. Hamiltonian Monte Carlo for hierarchical models. *Current trends in Bayesian Methodology with Applications*, 79(30):2–4, 2015.
55. K.E. Tatsis, V.K. Dertimanis, and E.N. Chatzi. Sequential Bayesian inference for uncertain nonlinear dynamic systems: A tutorial. *Journal of Structural Dynamics*, (1), 2022.
56. K. Tatsis, V.K. Dertimanis, T.J. Rogers, E. Cross, K. Worden, and E. Chatzi. A spatiotemporal dual Kalman filter for the estimation of states and distributed inputs in dynamical systems. In *Proceedings of ISMA-USD 2020*, pages 3591–3597, Heverlee, 2020. KU Leuven, Department Werktuigkunde.
57. J. Gilmer, S.S. Schoenholz, P.F. Riley, O. Vinyals, and G.E. Dahl. Neural message passing for quantum chemistry. In *International Conference on Machine Learning*, pages 1263–1272. PMLR, 2017.
58. C.K. Wikle and N. Cressie. A dimension-reduced approach to space-time Kalman filtering. *Biometrika*, 86(4):815–829, 1999.
59. D.J. Rezende and S. Mohamed. Variational inference with normalizing flows. In *Proceedings of the 32nd International Conference on Machine Learning, Lille, France*, volume 37, 2015.
60. I. Higgins, L. Matthey, A. Pal, C. Burgess, X. Glorot, M. Botvinick, S. Mohamed, and A. Lerchner. *beta*-VAE: Learning basic visual concepts with a constrained variational framework. 2016.
61. C.P. Robert and G. Casella. *Monte Carlo Statistical Methods*. Springer, 2nd edition, 2004.
62. G. Corso, L. Cavalleri, D. Beaini, P. Liò, and P. Veličkovic'. Principal neighbourhood aggregation for graph nets. *arXiv preprint arXiv:2004.05718*, 2020.
63. H. Fu, C. Li, X. Liu, J. Gao, A. Celikyilmaz, and L. Carin. Cyclical annealing schedule: A simple approach to mitigating kl vanishing. *arXiv preprint arXiv:1903.10145*, 2019.
64. M.F. Howland, S.K. Lele, and J.O. Dabiri. Wind farm power optimization through wake steering. *Proceedings of the National Academy of Sciences*, 116(29):14495–14500, 2019.

7 From Structural Health Monitoring to Finite Element Modeling of Heritage Structures

The Medieval Towers of Lucca

Riccardo Mario Azzara, Maria Girardi, Cristina Padovani, and Daniele Pellegrini

7.1 OVERVIEW

This chapter describes the experiments carried out on three medieval masonry towers in the historic center of Lucca, Italy. The towers have been continuously monitored by high-sensitivity seismic stations that record the structures' response to the dynamic actions of the surrounding environment. Special attention is devoted to the Guinigi Tower, one of the most iconic monuments in Lucca, whose monitoring campaign started in 2021. The goal of the chapter is to show the effectiveness of dynamic monitoring as a valuable source of information on the structural properties of the towers and sketch the capabilities of experiment-based finite element (FE) modeling.

7.2 INTRODUCTION

Studying the dynamic effects of the urban environment on buildings is a cutting-edge research topic in heritage preservation. Continuous long-term vibration monitoring is an effective nondestructive technique to investigate the dynamic behavior and check the health status of historical buildings. Data recorded by sensor networks is processed using suitable numerical procedures to determine the structure's dynamic properties, such as frequencies, damping ratios, and mode shapes. This approach, known as operational modal analysis (OMA) (Brincker and Ventura 2015), allows tracking the dynamic properties and makes damage detection possible. In fact, data-driven damage detection is based on monitoring the variation of frequencies over time, assessing environmental effects, and analyzing changes and anomalies via regression and output-only methods (Gentile

DOI: 10.1201/9781003306924-7

and Saisi 2007; Gentile et al. 2016; Ubertini et al. 2016; Azzara et al. 2018) as well as deep learning techniques (Carrara et al. 2022). When coupled with FE numerical modeling, ambient vibration tests become a powerful tool for structural health monitoring. Using the experimental modal properties of a structure makes it possible to calibrate its FE model by adopting model-updating procedures (Ferrari et al. 2019; Girardi et al. 2019; Azzara et al. 2020; Girardi et al. 2021; Rainieri et al. 2023).

This chapter aims to describe the experiments carried out by the authors since 2015 on three masonry towers in the historic center of Lucca. The vibrations of the San Frediano Bell Tower and the Clock Tower were continuously monitored in these past years: high-sensitive seismic stations were installed on the structures to record their response to the dynamic actions of the surrounding environment (wind, traffic, crowd movements, earthquakes). A further monitoring campaign started in 2021 was conducted on the Guinigi Tower, one of the most iconic monuments of Lucca. This chapter is focused on the Guinigi Tower and presents the first results of its dynamic monitoring and FE modeling. For the sake of comparison, the relevant dynamic characteristics of the San Frediano Bell Tower and the Clock Tower are summarized along with the Guinigi Tower's counterparts, pointing out similarities and differences.

The twofold goal of the chapter is to show the effectiveness of the above experiments as a valuable source of information on the structural properties of the towers and sketch the capabilities of experiment-based FE modeling.

7.3 DYNAMIC MONITORING OF MASONRY TOWERS AND FE MODEL UPDATING

The results of long-term dynamic-monitoring campaigns conducted on three towers in Lucca are presented. The San Frediano Bell Tower, the Clock Tower, and the Guinigi Tower are medieval structures located in the restricted vehicular traffic area of the historic center of Lucca (Figure 7.1). The first two towers (Figure 7.2a and 7.2b) were monitored in the period 2015–2018; the monitoring campaign on the Guinigi Tower (Figure 7.2c) took place from June 2021 to October 2022.

The ambient vibrations of the San Frediano Bell Tower were monitored from May to June 2015 and from October 2015 to October 2016, and those of the Clock

FIGURE 7.1 Panoramic view of the main medieval towers in Lucca, Italy.

FIGURE 7.2 (a) The San Frediano Bell Tower, (b) Clock Tower, and (c) Guinigi Tower. (Courtesy of the Municipality of Lucca.)

Tower from November 2017 to March 2018. In both cases, we used seismic stations by SARA Electronic Instruments (Perugia, Italy). Some triaxial velocimeters (SS45 with eigenfrequency 4.5 Hz and SS20 with eigenfrequency 2 Hz), each coupled with a 24-bit digitizer (SL06), were installed on the towers.

Velocities recorded on the towers were processed via the covariance-driven stochastic subspace identification method (SSI/Cov), an OMA technique in the time domain implemented in the MACEC code (Reynders et al. 2016).

The frequencies of the towers were analyzed to highlight possible correlations with air temperature and wind velocities and investigate the influence of vibrations of anthropic origin on their dynamic behavior. Experimental data was then used to calibrate finite element models (FEMs) of the towers and determine some unknown parameters using model-updating techniques. The details of the experimental campaigns and results of such studies are reported in Azzara et al. (2018, 2019, 2020).

The approach followed for the San Frediano Bell Tower and the Clock Tower was also adopted for the Guinigi Tower. Some preliminary outcomes of the experimental investigations on the Guinigi Tower and its numerical modeling are shown in Section 7.3.1. Section 7.3.2 is devoted to comparing the three towers and reports some quantities that describe their dynamic behavior, like velocities and frequencies, including their dependence on temperature.

7.3.1 The Guinigi Tower

The Guinigi Tower, sketched in Figures 7.3 and 7.4, is about 44 m high and dates back to the fourteenth century. It is one of the most famous buildings in Lucca, thanks to the hanging garden with seven holm oaks on its top (Figures 7.2c and 7.3). The tower is managed by the Municipality of Lucca, is open to the public,

FIGURE 7.3 The Guinigi Palace and its tower. View from Via Sant'Andrea. (Courtesy of the Municipality of Lucca.)

and is accessible via a metallic staircase from the level of 23 m to the roof terrace, at a height of about 43 m. The tower is entirely free from an altitude of 23 m to 40 m, without inner diaphragms. The terrace is accessed from a massive masonry vault at about 40 m (Level 5 in Figure 7.4). From the ground floor to the level of 23 m, the tower is incorporated into the surrounding Guinigi Palace from which it is accessed. The thickness of the tower's walls is about 1 m, almost constant along the height.

The seismic stations by SARA Electronic Instruments already employed to monitor the San Frediano Bell Tower and the Clock Tower have been placed on the Guinigi Tower, too. These sensors consist of three-axial electrodynamic velocity transducers, each equipped with a SL06 24-bit digitizer. Three preliminary tests, each lasting about 40 min, were conducted in January 2021, adopting four seismic stations (recording 100 samples per second) and three different sensor layouts.

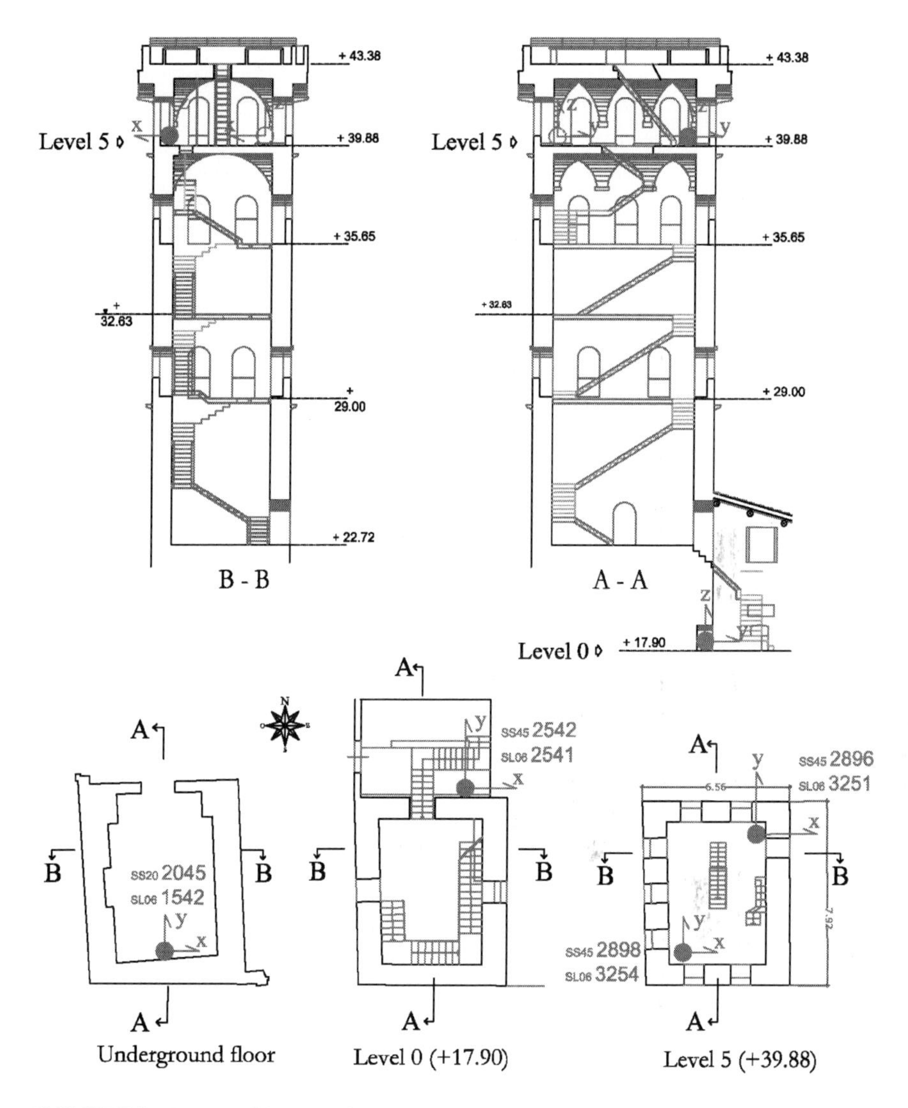

FIGURE 7.4 Layout of the seismic stations on the Guinigi Tower for long-term monitoring.

These tests aimed to assess the tower's natural frequencies and mode shapes and the sensors' performance. In the three layouts, two stations were fixed at Level 5 (Figure 7.4), and the remaining two were placed on the roof terrace (Figure 7.5a, first layout), along the tower's height (second layout) and at the base of the tower near the entrance in the Guinigi Palace (third layout).

Figure 7.4 depicts the position of the four seismic stations installed in June 2021 for long-term monitoring: one station is on the underground floor (SS20 2045), one is at 17.9 m (SS45 2542, Figure 7.5b), and the remaining two stations are at 39.88 m (SS45 2896 and SS45 2898, Figure 7.5c). The monitoring system

FIGURE 7.5 (a) A seismic station on the terrace during a preliminary test. Seismic stations installed on the tower at (b) 17.90 m and (c) 39.88 m during long-term monitoring. (d) The server hosted at the Institute of Information Science and Technologies of the Italian National Research Council (ISTI-CNR).

is completed by two thermo-hygrometers with a sampling time of 60 s. A virtual private network allows sending the data recorded from the instruments to a server hosted at the Institute of Information Science and Technologies of the Italian National Research Council (ISTI-CNR) in Pisa for storage and processing (Figure 7.5d).

Table 7.1 reports some information about the first six natural frequencies of the Guinigi Tower estimated via the SSI/Cov algorithm during the monitoring period. The frequency values are estimated over signal sequences lasting 1 h. Furthermore, the time history of each of the six frequencies measured from August 2021 to April 2022 is analyzed to calculate the minimum value Min, 1st percentile Min_1, average Avg, 99th percentile Max_{99}, maximum value Max, relative difference $\Delta = (Max_{99} - Min_1) / Min_1$, and standard deviation σ.

TABLE 7.1
The Guinigi Tower's Natural Frequencies Calculated via the SSI/Cov Algorithm during the Monitoring Period: Minimum, 1st Percentile, Average, 99th Percentile, Maximum Values, Relative Differences $\Delta = (Max_{99} - Min_1) / Min_1$, and Standard Deviations

Mode	Min (Hz)	Min_1 (Hz)	Avg (Hz)	Max_{99} (Hz)	Max (Hz)	Δ (%)	σ (Hz)
1	1.130	1.180	1.211	1.250	1.260	5.93	0.015
2	1.300	1.304	1.347	1.394	1.400	6.90	0.020
3	2.550	2.560	2.640	2.735	2.749	6.84	0.037
4	3.200	3.213	3.345	3.439	3.450	7.03	0.052
5	4.000	4.005	4.155	4.297	4.300	7.29	0.082
6	6.005	6.217	6.344	6.458	6.498	3.88	0.050

The SSI/Cov method allows calculating the standard deviations σ_{SSI} of the estimated modal properties. These quantities represent the statistical uncertainty due to the output covariances and the propagation of estimation errors and are plotted in the histogram of Figure 7.6 for the first six natural frequencies of the tower calculated over the monitoring period. A comparison with the data in Table 7.1 clearly shows that the variations over time of the fourth, fifth, and sixth frequencies are in the same order of σ_{SSI}, while for the first three frequencies, the estimation inaccuracies are one order of magnitude below the environmental variations. Similar results have been found for the San Frediano Bell Tower (Azzara et al. 2018) and the Clock Tower.

Figure 7.7 shows the behavior of the tower's first two frequencies measured from August 2021 to April 2022. Plots of frequencies versus temperature are

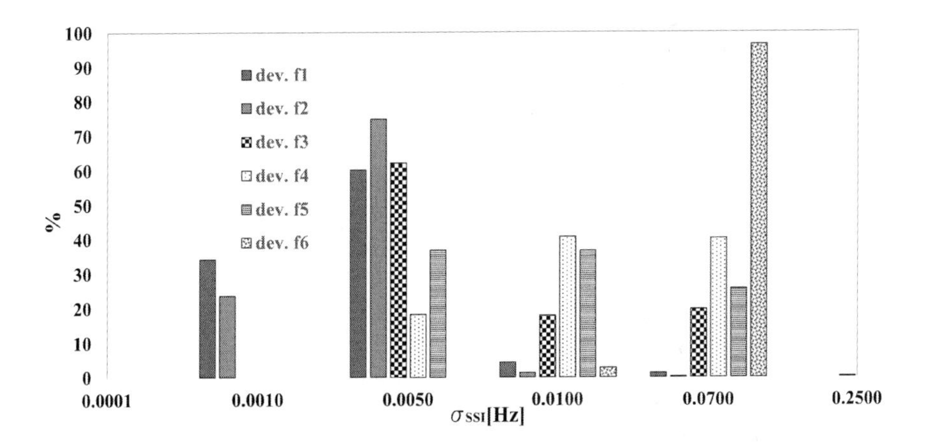

FIGURE 7.6 Histogram of the estimated standard deviations of the Guinigi Tower's first six frequencies.

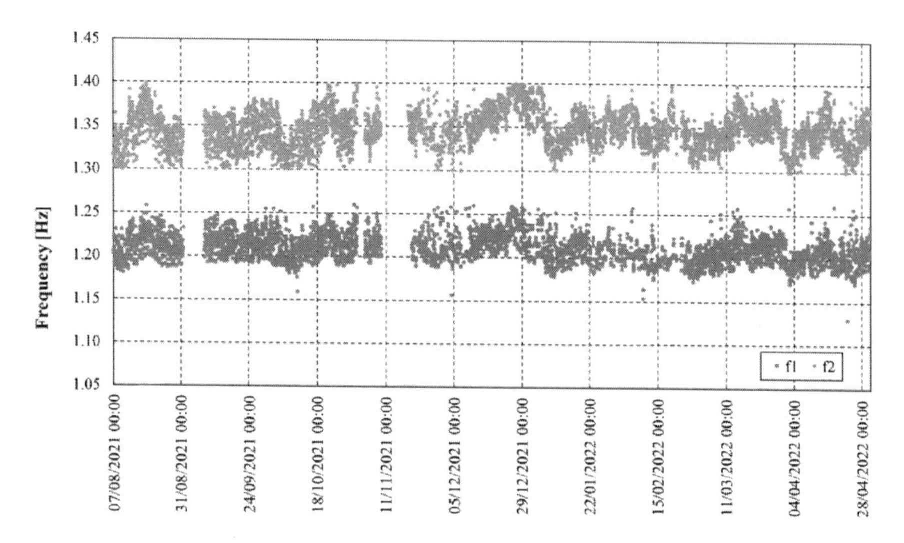

FIGURE 7.7 The Guinigi Tower's first two natural frequencies measured during the monitoring period August 2021–April 2022.

provided in Figure 7.8 for the monitoring period. Finally, Figure 7.9 shows a plot of the tower's first six natural frequencies and the temperature values (plot with solid black line) measured in October 2021.

As reported in Table 7.1, the relative differences Δ of the frequencies (evaluated on the 1st and 99th percentiles) over the monitoring period are at most in the order of 7%. These variations are in agreement with the results presented in Saisi et al. (2015) and Azzara et al. (2018). As far as the frequencies' daily trend is concerned, the maximum daily variations are in the order of 3–4%.

Several long-term vibration-monitoring campaigns on ancient masonry towers (Ramos et al. 2010; Gentile et al. 2016; Ubertini et al. 2016; Azzara et al. 2018) have pointed out that the frequencies tend to increase with temperature as a result of the closing of the cracks due to the thermal expansion. The Guinigi Tower does not exhibit such a trend. Figure 7.8 shows a very low correlation in the monitoring period between temperature and frequencies, which, in turn, are relatively sparse. In the absence of a specific explanation of this experimental result, it is worth noting that this scarce correlation also characterizes the Asinelli Tower in Bologna, which, like the Guinigi Tower, is open to the public and visited annually by many people (Baraccani et al. 2020).

Figures 7.10 and 7.11 show the tower's first six frequencies plotted versus humidity in the whole monitoring period and the frequencies and temperatures (black line) measured in October 2021. The correlation between frequencies and humidity shown by the figures seems quite low.

Regarding the parameters that could influence the velocity levels on the tower, Figure 7.12 plots the maximum velocities per hour recorded in the *x*, *y*,

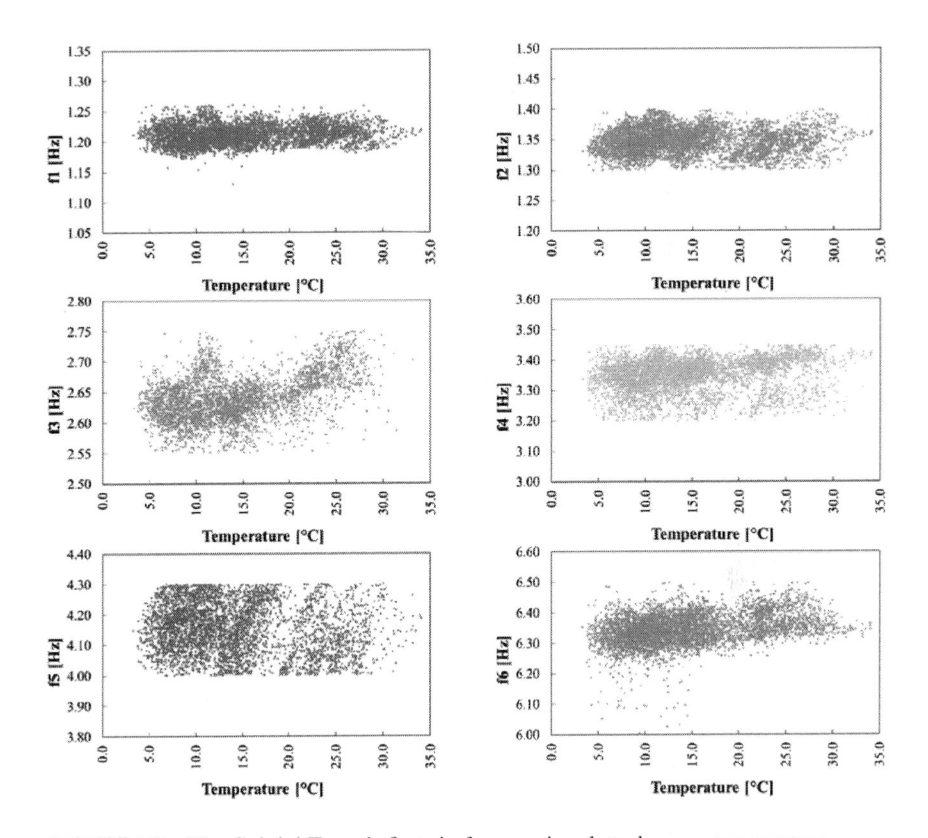

FIGURE 7.8 The Guinigi Tower's first six frequencies plotted versus temperature.

and z directions from 5 to 16 April 2022. The trend of the maximum velocities is clearly related to the presence of the visitors in the tower: the tower's entrance is in fact scheduled from 9:30 a.m. to 7:30 p.m. (weekends included). The effects of the visitors moving inside the tower are highlighted also in Figure 7.13, where the spectral amplitudes of the first two frequencies are plotted versus time and compared to the number of tickets sold during the day. The figure shows this comparison from 2 to 14 July 2021. The tower's opening to visitors produces a sharp and sudden increase in the two spectral amplitudes, which rapidly vanishes at the end of the visiting hours.

Regarding the velocity peaks, Figure 7.14 reports the maximum absolute values per hour of the velocities recorded by velocimeter 2896 set at 39.88 m in the x and y directions, from 5 November to 9 December 2021. Velocity values greater than 2 mm/s occur in the windiest hours of the days in the dashed circles, when the wind velocity exceeds 30 km/h. Data on the maximum wind speeds was recorded at Pieve di Compito, about 10 km from the Lucca historic center, and is available at www.sir.toscana.it.

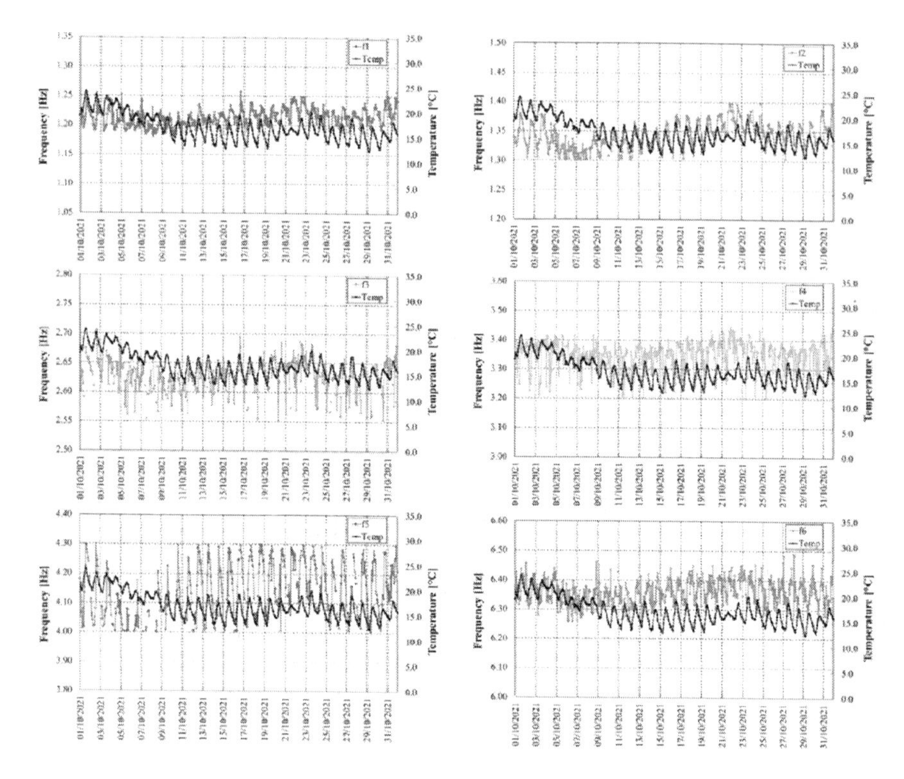

FIGURE 7.9 The Guinigi Tower's first six natural frequencies and the temperature values (black line) measured in October 2021.

The availability of the experimental frequencies makes it possible to calibrate a FE model of the Guinigi Tower via model-updating procedures. The goal is to estimate some unknown parameters (the Young's moduli and the mass densities of the materials constituting the structure) and thus develop a reliable model to carry out FE analyses and simulate the tower's structural response. A FE model of the complex constituted by the tower and the palace was created with NOSA-ITACA (www.nosaitaca.it/software), a software developed by ISTI-CNR for the analysis and calibration of masonry structures. The mesh of the structure, assumed to be perfectly clamped at the base, is shown in Figure 7.15 and consists of 23,161 isoparametric thick-shell and beam elements (elements no. 10 and no. 9 of the NOSA-ITACA library) with 22,665 nodes, for a total of 135,990 degrees of freedom.

The numerical model of the structure was calibrated via the global optimization algorithm implemented in the NOSA-ITACA code and described in Girardi et al. (2021). The algorithm aims at minimizing an objective function, defined as the distance between the structure's natural frequencies evaluated experimentally and their numerical counterparts evaluated by modal analysis, in a feasible set. Such an objective function depends nonlinearly on the unknown parameters

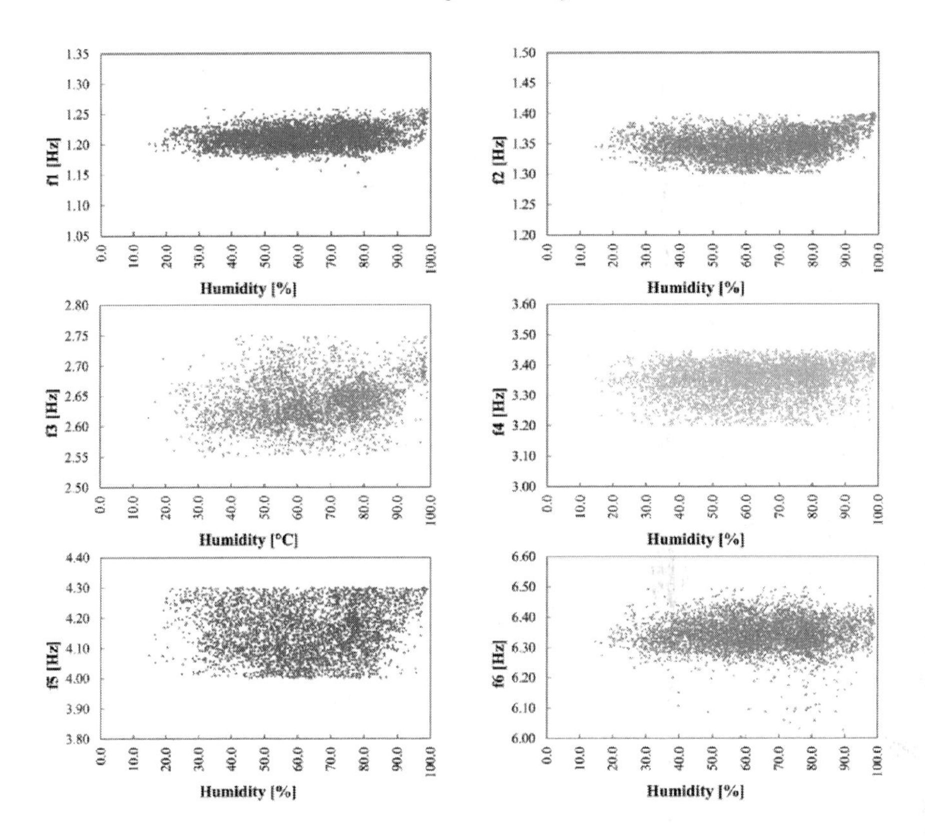

FIGURE 7.10 The Guinigi Tower's first six frequencies plotted versus humidity.

and may have multiple local minimum points. The proposed algorithm consists of a recursive procedure based on the construction of local parametric reduced-order models embedded in a trust-region scheme and is integrated into the NOSA-ITACA code. Given a feasible set into which the parameters are allowed to range, the procedure provides a set of local minimum points, including the global one. Along with the optimal value of each parameter, the procedure also determines two quantities involving the Jacobian of the numerical frequencies that measure how trustworthy the single parameter is. This additional information helps the user to assess the reliability of the parameters without resorting to preliminary sensitivity analyses (Noacco et al. 2019). Such analyses are usually performed before calibrating the numerical model to choose the number of updating parameters and to exclude some uncertain parameters from the model-updating process. The computational cost of sensitivity analysis is very high (on the order of hundreds of modal analyses runs) with respect to the cost of the minimization procedure implemented in NOSA-ITACA, which provides the global minimum point and an assessment of its reliability in few iterations (Girardi et al. 2021).

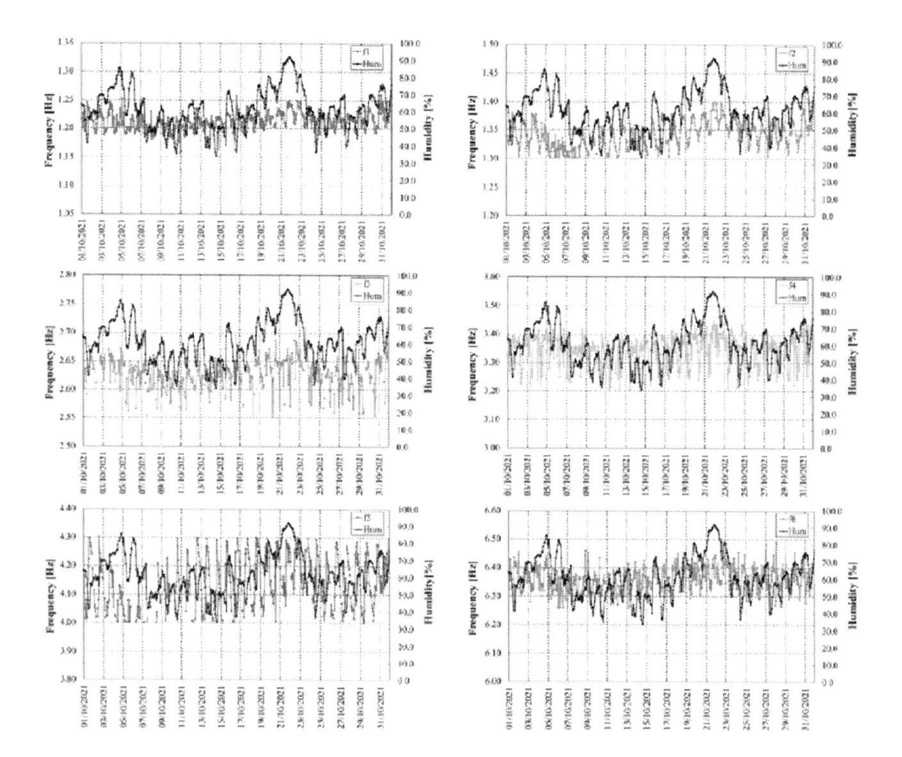

FIGURE 7.11 The Guinigi Tower's first six natural frequencies and the humidity values (black line) measured in October 2021.

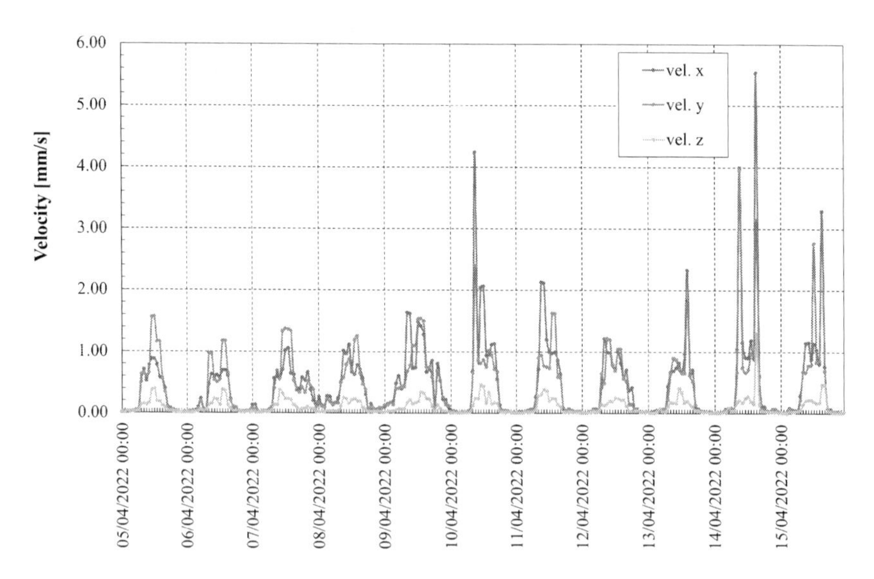

FIGURE 7.12 Maximum absolute values per hour of the velocities recorded by seismic station 2896 (+39.88 m) in the *x*, *y*, and *z* directions from 5 April to 16 April 2022.

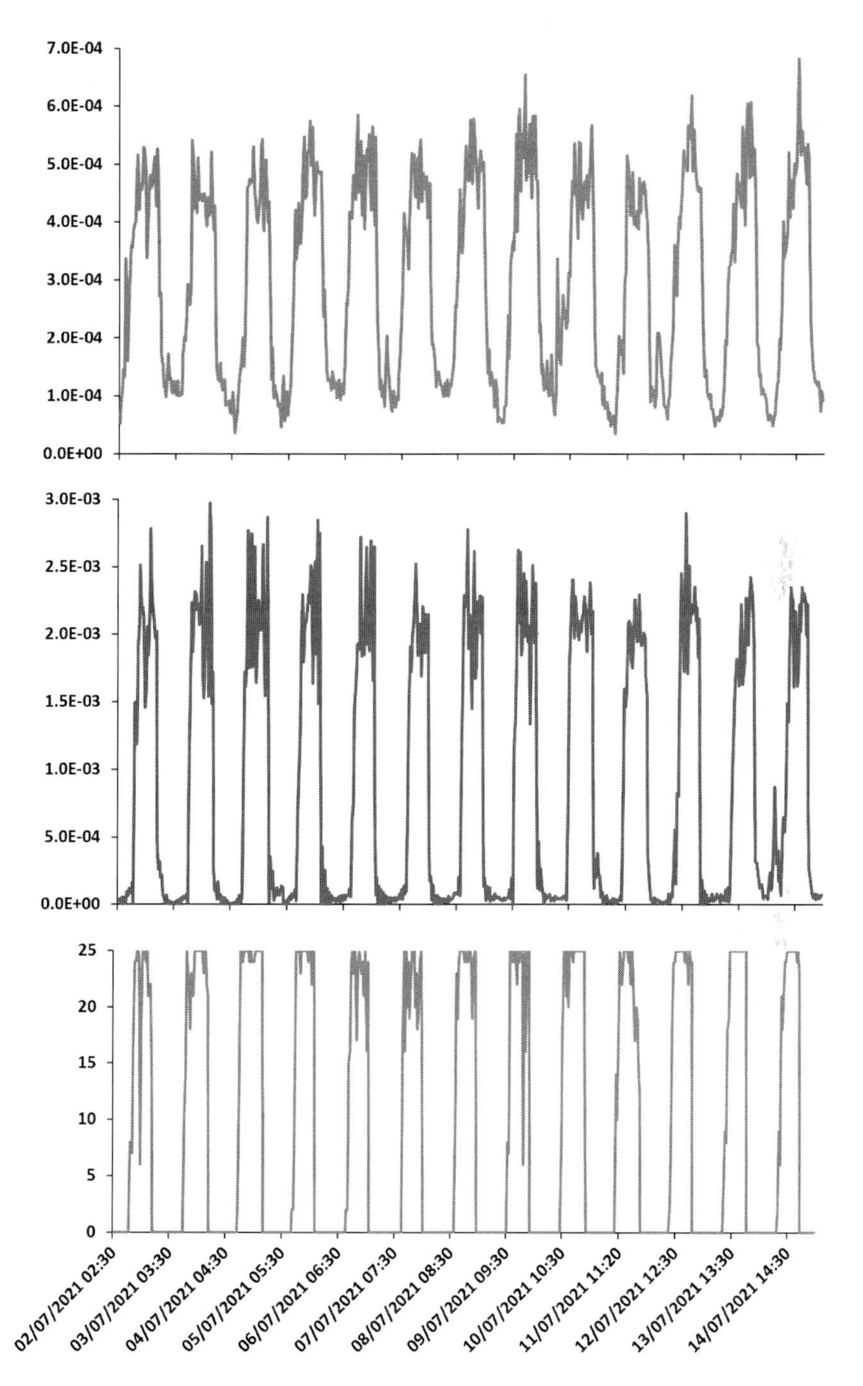

FIGURE 7.13 Daily trend of the first two frequencies' spectral amplitude (m/s/Hz; blue and green lines) and the number of presences (orange line) inside the Guinigi Tower from 2 to 14 July 2021.

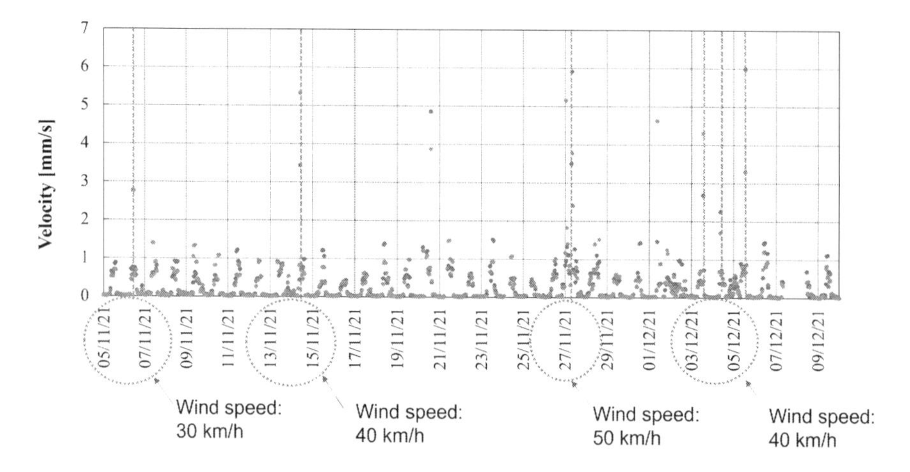

FIGURE 7.14 Maximum absolute values per hour of the velocities recorded by seismic station 2896 (+39.88 m) in the x and y directions from 5 November to 9 December 2021.

The model updating of the structure in Figure 7.15 is carried out by considering the first five experimental frequencies determined in the preliminary tests conducted in January 2021 and reported in the first column of Table 7.2. The sixth frequency corresponding to the axial mode shape is neglected because it is characterized by a high degree of uncertainty.

For calibration, two models of the structure are considered (Figure 7.15), keeping in mind that the number of parameters to be optimized is expected to be no greater than the number of frequencies to match. In model A, the unknown parameters are the Young's modulus E of the palace, the Young's modulus E_1, and the mass density ρ_1 of the tower. In model B, the parameters to be calculated are the Young's modulus E of the palace, the Young's moduli E_1 and E_2, and the mass densities ρ_1 and ρ_2 of the portions of the tower outside the palace and contained in it, respectively. The elastic moduli E_1 and E_2 are

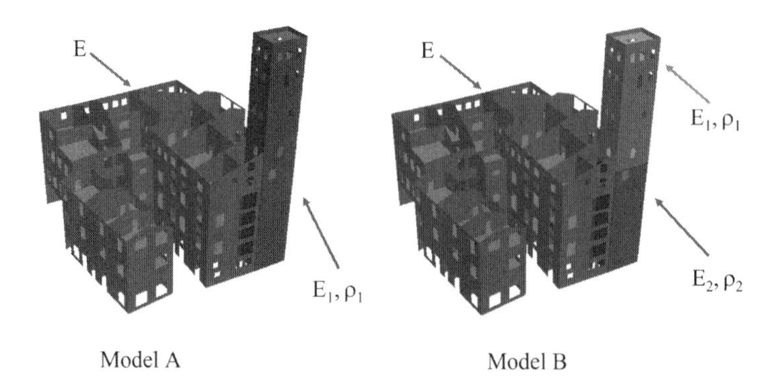

FIGURE 7.15 Finite element discretization of the Guinigi Palace.

TABLE 7.2
Experimental and Numerical Frequencies Calculated for the Optimal Values of the Parameters Calculated by NOSA-ITACA

f_{exp}(Hz)	f_{num}(A)(Hz)	Δf(A)(%)	f_{num}(B)(Hz)	Δf(B)(%)
1.23	1.22	0.81	1.22	0.81
1.34	1.34	0.00	1.35	0.75
2.62	2.58	1.53	2.62	0.00
3.30	3.29	0.30	3.30	0.00
4.15	4.26	2.65	4.16	0.24

allowed to vary within the interval [1.0, 6.0] GPa, while E ranges in [0.5, 6.0] GPa. The mass densities ρ_1 and ρ_2 vary in the interval [1200, 2000] kg/m^3, the mass density of the palace is 1800 kg/m^3, and the Poisson's ratio of all materials is 0.2. Table 7.2 summarizes the numerical frequencies (f_{num1}, f_{num2}, f_{num3}, f_{num4}, f_{num5}) of the structure corresponding to the optimal value of the parameter vector (E, E_1, ρ_1) for model A and (E, E_1, ρ_1, E_2, ρ_2) for model B, and their relative error $|\Delta f|$ with respect to the experimental counterparts. The maximum value of the relative error for the first frequency is 0.81% (for both models A and B). Table 7.3 reports the optimal values of the parameters recovered by NOSA-ITACA for models A and B. The table shows, for each optimal value, the scalars ξ^{-1} and η^{-1} calculated by NOSA-ITACA using the Jacobian of the numerical frequencies (f_{num1}, f_{num2}, f_{num3}, f_{num4}, f_{num5}) with respect to (E, E_1, ρ_1) for model A and (E, E_1, ρ_1, E_2, ρ_2) for model B, calculated at the minimum point (Girardi et al. 2021). Quantities ξ^{-1} and η^{-1} estimate the minimum and maximum percentage errors in assessing the parameters' optimal values under the hypothesis of a 1% error in identifying the experimental frequencies. Table 7.3 shows that, in the worst-case scenario, the Young's modulus E of the palace will be affected, at most, by a 2.14% error. The Young's moduli of the tower will be affected, at most, by a 15.05% error for model A and 10.60% for model B.

TABLE 7.3
Optimal Values of the Parameters Calculated by NOSA-ITACA

Parameter	Optimal Value (A)	ξ^{-1} (A)	η^{-1} (A)	Optimal Value (B)	ξ^{-1} (B)	η^{-1} (B)
E palace (GPa)	0.910	1.73	2.05	0.935	2.08	2.14
E_1 tower (GPa)	2.083	2.97	15.04	1.273	3.61	4.63
ρ_1 tower (kg/m^3)	1657.5	1.80	3.72	1700	1.67	2.10
E_2 tower (GPa)	—	—	—	3.445	5.12	10.60
ρ_2 tower (kg/m^3)	—	—	—	2000	20.00	175.44

The mass density ρ_1 will be affected by a 3.72% error for model A and a 2.10% error for model B; the mass density ρ_2 of the material constituting the portion of the tower contained in the palace (model B) exhibits a very high percentage error (175.44%), likely because ρ_2 has a low influence on the frequencies and cannot be reliably estimated by the model-updating process.

The MAC values calculated between the numerical mode shapes and the experimental counterparts are greater than 0.9.

Model B appears to provide an acceptable trade-off between the relative errors for the frequencies and the reliability of the parameters.

The FE model of the structure suitably calibrated can be used to perform dynamic analyses by assigning to the model the signals of seismic events recorded at the base, thus comparing experimental and numerical velocities. These simulations will be the subject of future work.

7.3.2 Comparative Analysis of the Towers

This subsection summarizes some results of the experimental campaigns conducted on the three towers. A comparison of the dynamic properties of the structures is provided in Tables 7.4 and 7.5, where their main characteristics are recalled, such as height, accessibility, the presence of bells, and the monitoring

TABLE 7.4

Mean Values f_i of the Natural Frequencies during the Monitoring Period and Their Variation $\Delta_i = (f_i^1 - f_i^{99})/f_i^1$, where f_i^1 and f_i^{99} Represent the 1st and 99th Percentiles of the Dataset

Experimental Values	San Frediano Bell Tower h = 52 m Closed to visitors Swinging bells Oct 2015–Oct 2016	Clock Tower h = 48.36 m Closed to visitors in the monitoring period Fixed bells Nov 2017–Mar 2018	Guinigi Tower h = 44.25 m Open to visitors No bells Aug 2021–Apr 2022
f_1 (Hz)	1.099	1.0281	1.211
Δ_1 (%)	5.42	3.65	5.93
f_2 (Hz)	1.382	1.2813	1.347
Δ_2 (%)	6.50	3.40	6.90
f_3 (Hz)	3.447	4.0524	2.640
Δ_3 (%)	5.44	8.67	6.84
f_4 (Hz)	4.614	4.4858	3.345
Δ_4 (%)	5.85	10.68	7.03
f_5 (Hz)	—	—	4.155
Δ_5 (%)	—	—	7.29
f_6 (Hz)	—	—	6.344
Δ_6 (%)	—	—	3.88

TABLE 7.5
***x*, *y*, and *z* Components of the Velocities Recorded on the Top Floor during the Monitoring Period: Maximum, Minimum, and Average Values and 99th Percentile**

Velocities (mm/s)	San Frediano Bell Tower h = 52 m Closed to visitors Swinging bells Oct 2015–Oct 2016	Clock Tower h = 48.36 m Closed to visitors in the monitoring period Fixed bells Nov 2017–Mar 2018	Guinigi Tower h = 44.25 m Open to visitors No bells Aug 2021–Apr 2022
v_x^{max}	4.99	5.37	4.36
v_x^{min}	4.2×10^{-3}	5.57×10^{-3}	5.4×10^{-3}
v_x^{av}	0.10	0.26	0.33
v_x^{99}	2.11	5.36	1.65
v_y^{max}	1.18	5.34	5.14
v_y^{min}	3.2×10^{-3}	5.25×10^{-3}	5.1×10^{-3}
v_y^{av}	0.05	0.26	0.31
v_y^{99}	1.01	5.25	1.67
v_z^{max}	0.32	5.46	10.55
v_z^{min}	2.2×10^{-3}	1.74×10^{-3}	2.5×10^{-3}
v_z^{av}	0.02	0.19	0.13
v_z^{99}	0.17	5.31	0.43

period. Table 7.4 reports, for each tower, the first frequencies and their variations in the relative monitoring period. Figure 7.16 shows the first two frequencies of the towers versus the temperature measured in the Botanic Garden in Lucca. Table 7.5 summarizes the most relevant values of velocity recorded on the top of the towers. The minimum, maximum, and average values of the velocities recorded along the three directions x, y, and z are listed along with the 99th percentile of the dataset.

The conclusions drawn from this comparison are as follows:

1. The average velocities measured on the towers, located in a restricted-traffic area, are very low. The velocity peaks are in the order of 5 mm/s.
2. The most important vibration sources recognized in the towers are given by the wind, the bells swinging (for the San Frediano Bell Tower), and the tourists accessing the tower's structure (for the Guinigi Tower).
3. The main frequencies of the towers exhibit a variation in the order of 5–6% during a one-year monitoring period.
4. The correlation between frequencies and temperature is positive for the San Frediano Bell Tower and the Clock Tower. For the Guinigi Tower, the experimental data recorded in the monitoring campaign exhibits a rather low correlation.

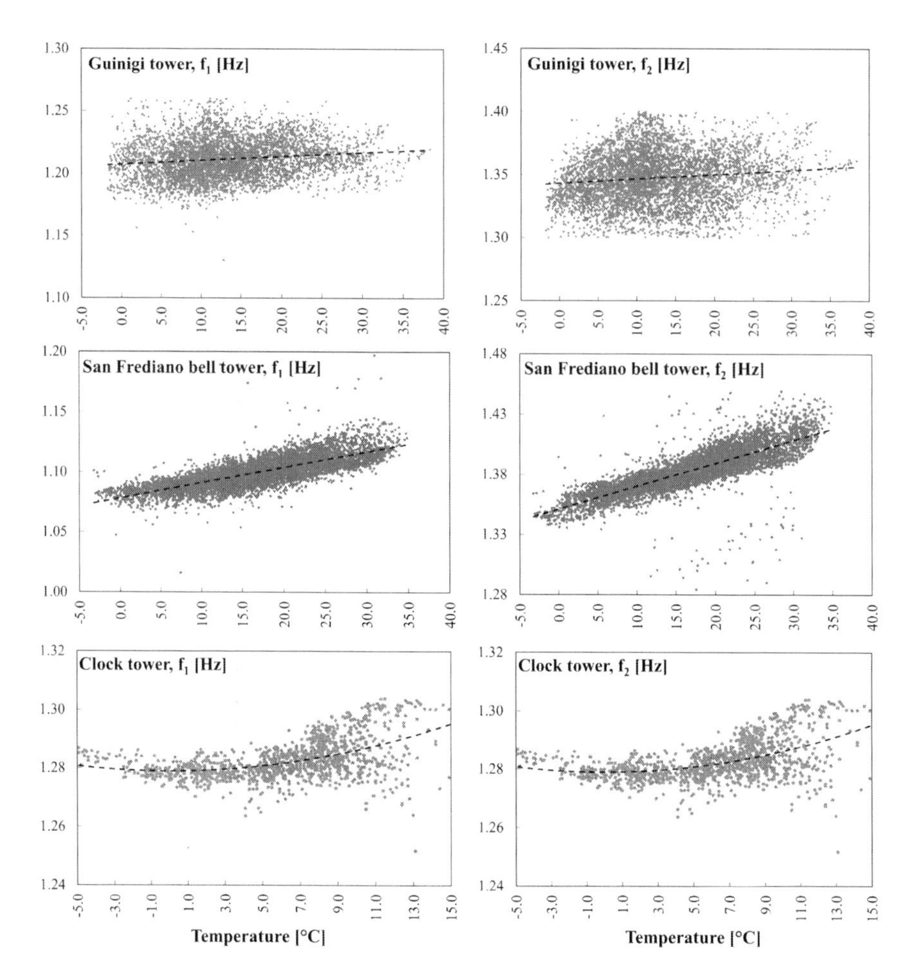

FIGURE 7.16 First two frequencies of the Clock Tower (blue dots), San Frediano Bell Tower (red dots), and Guinigi Tower (green dots) versus temperature.

5. The only one of the three structures accessed by visitors during the monitoring period was the Guinigi Tower; the presence of the tourists strongly influences the dynamics of the building and seems not to allow evaluating the effects of the environmental parameters on the tower's modal properties.

7.4 CONCLUSIONS

This chapter presents the results of long-term vibration-monitoring campaigns conducted on three medieval towers located in the historic center of Lucca, Italy. Results for the San Frediano Bell Tower and the Clock Tower have already been published and are summarized in the chapter, which focuses on the dynamic monitoring of the Guinigi Tower, the most iconic tower in Lucca. A FEM of the tower is presented and fine-tuned through a model-updating procedure implemented in the FE code NOSA-ITACA developed by the authors. This procedure allows calibrating the unknown parameters with good accuracy and evaluating the reliability of the optimal solution without resorting to expensive sensitivity analyses.

Finally, a comparison among the main dynamical features of the three towers is also presented in terms of natural frequencies, velocities, frequency variations in the monitoring period, and correlation with temperature.

The example applications described in the chapter show the effectiveness of long-term continuous-monitoring protocols in providing information on the vibration sources and vibration levels acting on ancient monuments, even in restricted-traffic areas. The datasets recorded exploit only ambient vibrations and allow tracking of the structure's dynamical properties over time. Integrating experimental investigations and numerical modeling via model-updating techniques makes it possible to estimate the structure's mechanical properties and its safety conditions.

ACKNOWLEDGMENTS

This research has been partially supported by the Fondazione Cassa di Risparmio di Lucca (SOUL project, 2019–2022) and the Italian National Research Council (REVOLUTION Project, Progetti di Ricerca @CNR, 2022–2024). This support is gratefully acknowledged. The authors wish to thank the Municipality of Lucca for the technical support in the experimental campaign and the information provided on the Guinigi Tower.

REFERENCES

Azzara, R.M., De Roeck, G., Girardi, M., Padovani, C., Pellegrini, D., Reynders, E. 2018. The influence of environmental parameters on the dynamic behaviour of the San Frediano bell tower in Lucca. *Eng Struct.* 156: 175–187.

Azzara, R.M., Girardi, M., Iafolla, V., Lucchesi, D., Padovani, C., Pellegrini, D. 2019. Ambient vibrations of age-old masonry towers: Results of long-term dynamic monitoring in the historic centre of Lucca. *Int J Archit Heritage.* doi: 10.1080/15583058.2019.1695155.

Azzara, R.M., Girardi, M., Iafolla, V., Padovani, C., Pellegrini, D. 2020. Long-term dynamic monitoring of medieval masonry towers. *Front Built Environ.* 6(9): doi: 10.3389/fbuil.2020.00009.

Baraccani, S., Azzara, R.M., Palermo, M., Gasparini, G., Trombetti, T. 2020. Long-term seismometric monitoring of the two towers of Bologna (Italy): Modal frequencies identification and effects due to traffic induced vibrations. *Front Built Environ.* 6: 85.

Brincker, R., Ventura, C. 2015. *Introduction to Operational Modal Analysis.* New York: John Wiley and Sons.

Carrara, F., Falchi, F., Girardi, M., Messina, N., Padovani, C., Pellegrini, D. 2022. Deep learning for structural health monitoring: An application to heritage structures. XXV *AIMETA* Conference, 4–8 September 2022, Palermo.

Ferrari, R., Froio, D., Rizzi, E., Gentile, C., Chatzi, E.N. 2019. Model updating of a historic concrete bridge by sensitivity-and global optimization-based Latin Hypercube Sampling. *Eng Struct.* 179: 139–160.

Gentile, C., Guidobaldi, M., Saisi, A. 2016. One-year dynamic monitoring of a historic tower: Damage detection under changing environment. *Meccanica.* 51(11): 2873–2889.

Gentile, C., Saisi, A. 2007. Ambient vibration testing of historic masonry towers for structural identification and damage assessment. *Constr Build Mater.* 21(6): 1311–1321.

Girardi, M., Padovani, C., Pellegrini, D., Robol, L. 2019. Model updating procedure to enhance structural analysis in FE Code NOSA-ITACA. *J Perform Constr Facil.* 33(4): 04019041.

Girardi, M., Padovani, C., Pellegrini, D., Robol, L. 2021. A finite element model updating method based on global optimization. *Mech Syst Signal Process.* 152: 107372.

Noacco, V., Sarrazin, F., Pianosi, F., Wagener, T. 2019. Matlab/R workflows to assess critical choices in Global Sensitivity Analysis using the SAFE toolbox. *MethodsX.* 6: 2258–2280.

Rainieri, C., Rosati, I., Cieri, L., Fabbrocino, G. 2023. Development of the Digital Twin of a Historical Structure for SHM Purposes. In *European Workshop on Structural Health Monitoring*, pp. 639–646. Springer: Cham.

Ramos, L.F., Marques, L., Lourenço, P.B., De Roeck, G., Campos-Costa, A., Roque, J. 2010. Monitoring historical masonry structures with operational modal analysis: Two case studies. *Mech Syst Signal Process.* 24: 1291–1305.

Reynders, E., Maes, K., Lombaert, G., De Roeck, G. 2016. Uncertainty quantification in operational modal analysis with stochastic subspace identification: Validation and applications. *Mech Syst Signal Process.* 66: 13–30.

Saisi, A., Gentile, C., Guidobaldi, M. 2015. Post-earthquake continuous dynamic monitoring of the Gabbia Tower in Mantua, Italy. *Constr Build Mater.* 81: 101–112.

Ubertini, F., Comanducci, G., Cavalagli, N. 2016. Vibration-based structural health monitoring of a historic bell tower using output-only measurements and multivariate statistical analysis. *Struct Health Monit.* 15(4): 438–457.

8 Development of an Adaptive Linear Quadratic Gaussian (LQG) Controller for Structural Control Using Particle Swarm Optimization

Gaurav Kumar, Wei Zhao, M. Noori, and Roshan Kumar

8.1 OVERVIEW

In this work, particle swarm optimization and the maximum dominant period approach are used to improve on the conventional linear quadratic Gaussian (LQG) controller. To obtain an appropriate command signal, the LQG controller's control weighting matrix is varied. This command signal is sent to the magnetorheological (MR) damper, which generates the appropriate amount of counter-control force for seismic vibration mitigation. The proposed controller is evaluated using a benchmark three-story structure with an MR damper at the base. On this structure, the proposed controller is tested under three different earthquake time histories, different soil conditions, and a situation wherein power is lost at the peak of the earthquake. According to the outcomes and discussion, the proposed control strategy outperforms the conventional LQG controller.

8.2 INTRODUCTION

The aim of using a semi-active control system is to improve the structure's response by mitigating the effects of dynamic loadings, including earthquakes. Because of the nonlinear actuator dynamics, resonance conditions, dynamic coupling, uncertainties, and measurement limitations, this is a challenging task. To protect structures from seismic and wind loadings, a diverse set of control algorithms has been used. These control algorithms work well for some structures but

DOI: 10.1201/9781003306924-8

not so well for others. To dispense the appropriate electrical control signal in a semi-active control scheme, a controller that is realizable, simple, fault-tolerant, optimal, and robust must be developed [1–6].

In their research, Dyke et al. covered on-off controllers. Although these controllers are simple and easily implemented, the problem here is that there is only one level of voltage regardless of whether the structure is moving away from or toward the center [7]. Agrawal and Cha proposed two new methods: simple passive control (SPC) and decentralized output feedback polynomial control (DOPFC). It was discovered that the DOPFC controller outperforms the passive on-off controller [8]. A quasi-bang-bang controller was also proposed, in which the voltage to the magnetorheological (MR) damper is determined by two different rules. These rules are determined by the structure's reference position [9–11]. In this case, intermediate voltage levels between the center and the extreme have been ignored. As a solution, a modified quasi-bang-bang controller approach was proposed [10]. The variable weights in this control law are like those of the fuzzy logic controller (FLC). This controller's weights are constant and determined by trial and error, which is a disadvantage.

Many researchers have used proportional-integral-derivative (PID) family controllers for structure control and compared their performance to that of more advanced controllers such as sliding mode controllers (SMCs) or FLCs [12–15]. According to studies, PID controllers alone are not as useful as they are in hybrid form (e.g., PID SMCs or FLCs) [15]. To reduce the vibration of a four-degree-of-freedom structure, R. Guclu designed a PID controller and an SMC [12]. The PID controller's efficiency was compared to that of the SMC. The author compared the results using time histories for the first and third floors, specifically the 1999 Marmara earthquake in Turkey.

Two proportional-derivative (PD) controllers were used by R. Guclu et al. to control two actuators installed on the first and top floors of a 15-story structure model. The performance of the PID family controllers, on the other hand, was found to be unsatisfactory because proper tuning of the PID gain was extremely difficult to attain [16].

Furthermore, the H-infinity control method is a popular linear robust controller. The H-infinity represents Hardy space as defined by the infinity norm [17, 18]. To solve the optimization problem, the authors implemented a numerically efficient solution algorithm in this work. The authors also discovered that the H-infinity control algorithm performs admirably in multiple-input multiple-output (MIMO) systems. In their work, Liu et al. considered various time delays. The H-infinity controller is created by employing a matrix inequality and parameter adjustment method [19].

The SMC is regarded as a nonlinear control algorithm. It is an excellent choice for structural control because it is resistant to model uncertainties and extremely robust [20]. Yakut et al. demonstrated SMC using a neural network. It has advantageous characteristics, such as SMC robustness and neural network flexibility. The neural network is used to eliminate SMC chattering [21]. To avoid the chattering effect in SMC, fuzzy logic was used. The resulting controller is known as the fuzzy sliding mode controller (FSMC). The authors created FSMC to reduce the chattering effect of the SMC while maintaining the SMC's robustness and insensitivity to parameter changes [22]. The authors observed that the SMC controller's efficiency is satisfactory, but the chattering effect caused by imperfections in the sliding surface due to high-frequency switching may cause damage to mechanical components, such as actuators. The chattering effect is a significant issue in the SMC algorithm and should be avoided. According to the reported literature, this can be accomplished in two ways: by appropriately smoothing the control force and when sliding mode control is achieved by employing a continuous SMC algorithm [23–26].

Furthermore, the FLC is a model-free approach to structural control. FLC design entails the intelligent selection of input and output variables, data manipulation methods, membership functions, and rule-based design [16, 27–30]. To control seismic vibrations, Ramaswamy et al. created an FLC for active tuned-mass dampers [31, 32]. K. M. Choi et al. presented a study that used a semi-active fuzzy control technique on the ground floor of a three-story structure, resulting in a decrease in seismic response. In addition, some researchers worked on a hybrid approach to designing the controller with fuzzy logic controller (FLC) [39–41]. Choi et al. proposed an FLC based on modern time-domain control theory for structural seismic vibration mitigation. The authors investigated the observer capability of the Kalman filter for state estimation [33, 34]. In addition, the authors used a low-pass filter to eliminate the spillover issue. The FLC algorithm was proposed by Das et al. for the semi-active control scheme. The authors fuzzified the MR damper characteristics, removing the need for mechanical modeling of the MR damper [35]. However, the FLC has some drawbacks in parameter determination, such as membership functions, control rules, and insufficient stability analysis. Also, the FLC and NN controllers do not consider feedback from the actuator (MR damper), and these controllers rely solely on structural response measurements, which are difficult to obtain accurately during a seismic event.

Furthermore, the linear quadratic regulator (LQR) controller is a well-known and well-studied controller [36–39]. In optimal control theory, a cost function must be minimized to achieve the desired or optimum results. This cost function is determined by the controller and system parameters [40, 41]. To determine the control signal (voltage/current) to the MR damper, Dyke et al. proposed the LQR controller in conjunction with the on-off switch-based controller. Due to the highly nonlinear nature of the MR damper, which is difficult to model precisely, it is difficult to maintain the relationship between the input voltage and output force of the MR damper [2, 24, 42, 43]. As a result, most of the proposed control strategies modify the voltage through on-off rules rather than using a model. This controller became well-known in structural control and is commonly referred to as the clipped optimal controller. Using the damper's force in feedback, the authors created a command signal (voltage) for the MR damper using the LQR controller. The command signal (voltage) was set using the clipped control law by comparing the desired force to the force of the available damper [7].

In the presence of this noise, the system states are unknown for the application of the control action. As a result, an observer known as a Kalman filter is used to estimate the system's states. *Linear quadratic Gaussian* (LQG) refers to the combination of the Kalman observer and the LQR controller [18]. Jansen et al. compared the performance of recently proposed controllers such as the Lyapunov controller, decentralized bang-bang controller, and moderated homogeneous friction procedure to the performance of the clipped optimal LQG controller used in a semi-active control scheme. In this study, the clipped optimal LQG/LQR controllers were found to be the most effective [44]. However, determining the optimal weighting matrix for optimal performance remained a research topic. Panariello et al. proposed an algorithm for carrying up-to-date weighting matrices for the gain of the LQR controller from a database of documented earthquake excitations in this direction. The limitation of the preceding studies is the lack of an offline repository of known earthquakes [45].

Alavinasab et al. presented an energy-based approach for determining the LQR controller's gain matrices. The authors worked to eliminate the need for the trial-and-error method of determining suitable gain matrices [39, 46]. Basu et al. developed the modified time variable LQR (TVLQR) method, which involves updating weighting matrices with a constant multiplier using discrete wavelet transform (DWT) analysis [47]. Although the weighting matrices in this method vary at resonance conditions, the constant multiplier was determined offline. As a result, offline data was still required. Amini et al. presented a novel technique for determining the best control forces for the active tuned-mass damper to solve this problem. This technique employed three distinct procedures: DWT analysis, particle swarm optimization (PSO), and the LQR [48].

The aim of this work is to create an optimized LQG controller by combining a maximum dominant period (τ_p^{max}) approach with PSO in three steps. First, adaptive LQG is developed by modifying their parameters in real time using PSO-τ_p^{max} approaches. Second, the proposed controllers' performance is investigated using numerical simulations on a scaled three-story building with an MR damper between the ground and first floors. Concurrently, the structural responses produced by the widely used LQR/LQG-based clipped optimal controller are compared to the responses produced by the proposed controllers.

8.3 DEVELOPMENT OF A MODIFIED LQG CONTROLLER USING PSO-τ_p^{max} APPROACHES

A basic mathematical understanding of the LQG controller is presented before the modification of LQG. This modified controller is advisable for use in circumstances where uncertainties and noise are present, as the states of the systems may not be available all the time as assumed in the LQR controller.

Assume a linear time-invariant (LTI) system represented by Eqs. (8.1–8.2):

$$\dot{z} = \boldsymbol{A}z + \boldsymbol{B}f + \boldsymbol{E}w \tag{8.1}$$

$$y = \boldsymbol{C} z + \boldsymbol{D} f + v \tag{8.2}$$

where $\boldsymbol{w}$ and $\boldsymbol{v}$ are the disturbance input and measurement error, respectively. Both are assumed as uncorrelated and white Gaussian random processes with zero means.

For this system, the cost function is defined as given in Eq. (8.3):

$$J_i(z, u) = \frac{\text{Lim}}{t \to \infty} \left[\int_0^t \left(z^T \boldsymbol{Q}_i \; z(t) + u^T \boldsymbol{R}_i \; u(t) \right) d(t) \right] \tag{8.3}$$

where $\boldsymbol{Q}$ is the state weighting matrix, and it is semidefinite; whereas $\boldsymbol{R}$ is the control weighting matrix, and it is a positive definite matrix. The LQG controller that solves the LQG control problem is formulated by Eqs. (8.4–8.5):

$$\dot{\hat{z}} = \boldsymbol{A} \hat{z} + \boldsymbol{B} u + \boldsymbol{L}_{\mathbf{Kal}} \left(\boldsymbol{y} - \boldsymbol{C} \hat{z} - \boldsymbol{D} \boldsymbol{u} \right) \tag{8.4}$$

$$\boldsymbol{u} = -\boldsymbol{K}_{\mathbf{LQR}} \hat{z} \tag{8.5}$$

where $\hat{z}$ is the observed state or the next state. As described earlier, the LQG controller is the combination of the LQR controller and the Kalman filter, and the calculations of the Kalman filter gain $\boldsymbol{L}_{\mathbf{Kal}}$ and the LQR controller gain $\boldsymbol{K}_{\mathbf{LQR}}$ are calculated separately using the algebraic Ricatti equation. These gains are given independently by Eqs. (8.6–8.7):

$$\boldsymbol{L}_{\mathbf{Kal}} = \boldsymbol{R}^{-1} \boldsymbol{B}^T \boldsymbol{P}_{\mathbf{LQR}} \tag{8.6}$$

$$\boldsymbol{K}_{\mathbf{LQR}} = \boldsymbol{P}_{\mathbf{Kal}} \boldsymbol{V}^{-1} \boldsymbol{C}^T \tag{8.7}$$

The Kalman filter is used to design an observer by measuring the available data. This observer minimizes the spread of the estimated error probability density in the process. The block diagram for the LQG controller is shown in Figure 8.1.

Furthermore, coming to the methodology to modify the LQG using PSO and the $\tau_p^{\max}$ approach, it is understood that the structure's response reflects similar properties of the earthquake excitation. So, the entire duration of the response $(0, t_i)$ is divided further into smaller time windows, with the ith window being (t_{i-1}, t_i). The maximum predominant period $\tau_p^{\max}$ is used to find the dominant frequency for each time window. This keeps the system always in the time domain, and thus the system becomes inherently fast. Originally, the idea of the maximum predominant period τ_p was first introduced by Nakamura [49] to classify large and small earthquakes based on frequency content present in the earthquake signal. The parameter τ_p can be calculated from the acceleration time

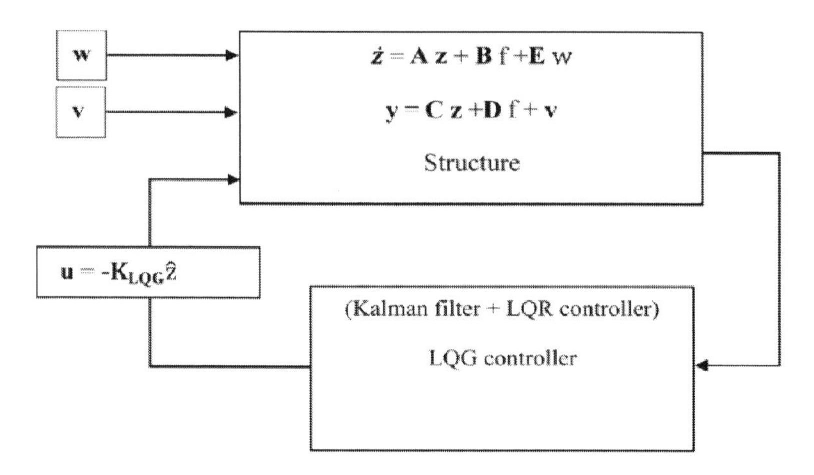

FIGURE 8.1 The block diagram of the linear quadratic Gaussian (LQG) controller.

series for each time step in real time, according to the following relations given in Eqs. (8.8–8.10).

$$\tau_{p,\,i} = 2\pi\sqrt{\frac{V_i}{A_i}} \tag{8.8}$$

$$V_i = aV_{i-1} + v_i^2 \tag{8.9}$$

$$A_i = aA_{i-1} + \left(\frac{dv}{dt}\right)_i^2 \tag{8.10}$$

where v_I is the recorded ground velocity; I is the smoothed ground velocity squared; I_i is the smoothed acceleration squared; and the smoothing parameter a has a value between 0 and 1. The maximum predominant period $\tau_p^{\max}$ is the maximum value of τ_p in the selected time window. Thus, Eq. (8.11) can obtain the maximum dominant frequency of a selected time window.

$$f_d = \frac{1}{\tau_p^{\max}} \tag{8.11}$$

This dominant frequency determines the quasi-resonance stances where the value of $\boldsymbol{R}$ is to be modified. Here, the PSO algorithm is used to find the optimal value of $\boldsymbol{R}$ that gives the optimal structural response with lesser control effort. The PSO algorithm helps to find weighting matrices $\boldsymbol{R}$ on the quasi-resonant bands. The benefit of this specific local optimal solution is that it can change the estimation of matrix $\boldsymbol{R}$ on an odd frequency at which quasi-resonance occurs, unlike the clipped optimal LQR, which has a global value of $\boldsymbol{R}$ during an earthquake. The cost function to be minimized for this modified LQR problem is formulated by having state weighting matrix $\boldsymbol{Q}_i$ and control weighting matrix $\boldsymbol{R}_i$ for the ith window, and it is given in Eq. (8.12):

$$J_i(x, u) = \int_0^t \left(x^T Q_i x(t) + u^T R_i u(t)\right) dt \tag{8.12}$$

The result of this modified optimal control problem with cost function J_i leads to a control law given in Eq. (8.13):

$$u = -\boldsymbol{G}_i\, x \tag{8.13}$$

The solution of the Ricatti matrix differential equation for every windowed interval gives the gain matrix $\boldsymbol{G}_i$, and the anticipated control force required to counter the effect of quasi-resonance can be found by applying this gain of the ith window. The flowchart of the development of the adaptive LQR controller is shown in Figure 8.2, and the flowchart for updating the matrix $\boldsymbol{R}$ using the PSO algorithm is shown in Figure 8.3.

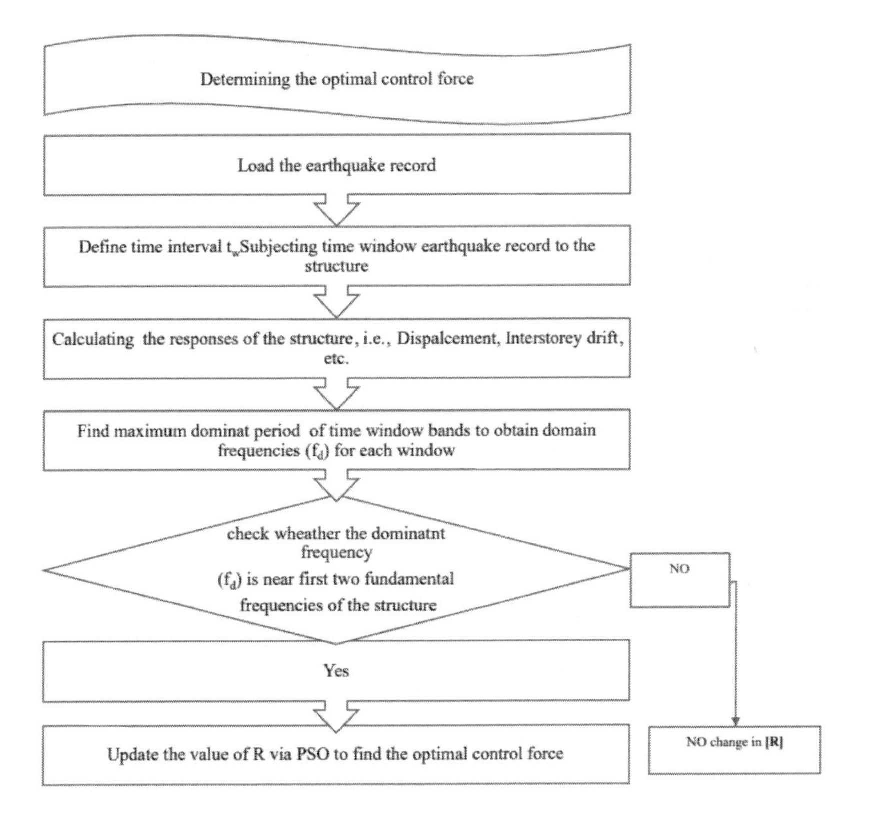

FIGURE 8.2 The flowchart of the development of an adaptive LQR controller using the PSO-$\tau_p^{\max}$ approach.

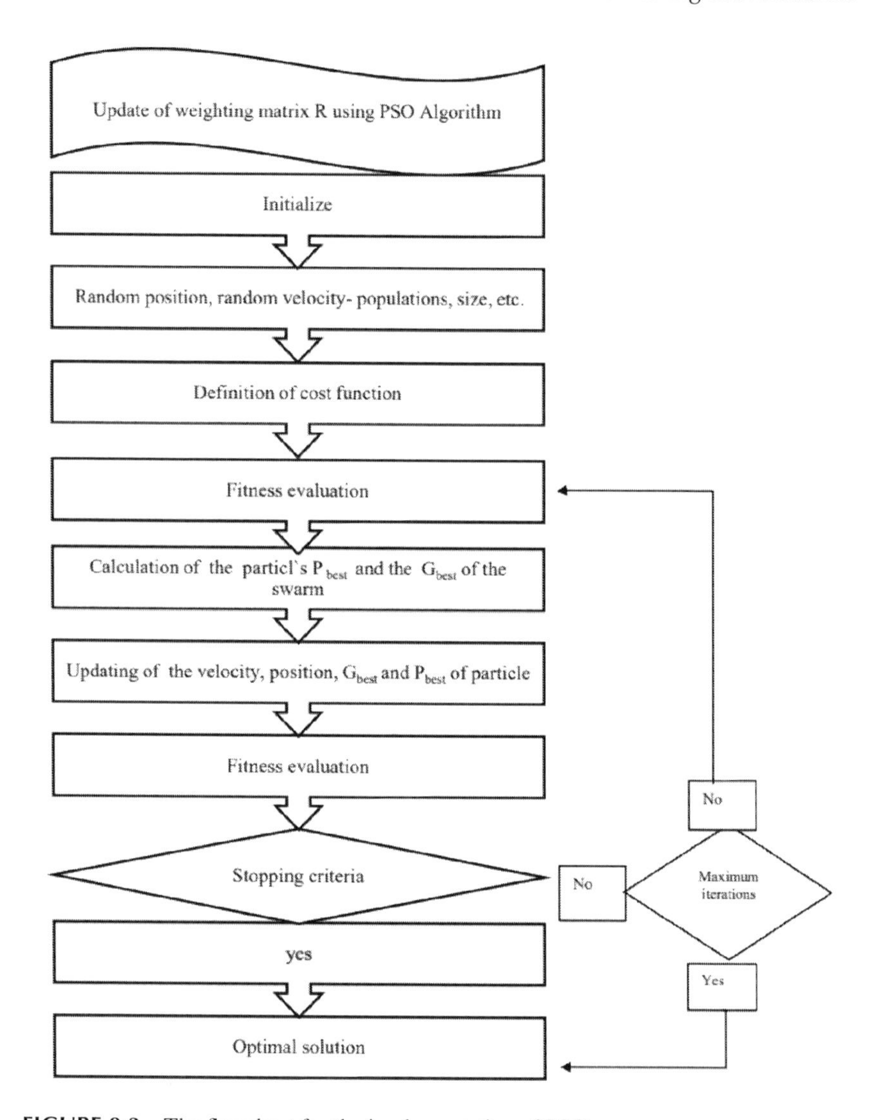

FIGURE 8.3 The flowchart for the implementation of PSO.

8.4 RESULTS AND DISCUSSION

A comprehensive performance analysis of the proposed controllers is carried out by comparing the structural responses of the three-story test structure obtained using the proposed controller and the conventional LQG controller under the following conditions.

1. Using different earthquake time histories
2. Using an earthquake recorded in different soil conditions
3. Considering a situation in which power is lost at the peak of the earthquake

There exist two main regulatory control objectives in the structural control, namely, acceleration mitigation and displacement mitigation. Acceleration mitigation is a serviceability criterion, whereas displacement mitigation deals with structural integrity. The acceleration criterion allows higher robustness for the control algorithm due to the lesser concern of structural integrity. Although acceleration mitigation is important, displacement mitigation is a prevalent concern during earthquake excitations because structural integrity is at stake. For structural integrity, it is essential to minimize stresses and strains in structural members. Therefore, in the present discussion, emphasis is given to displacement mitigation. Furthermore, a new parameter of the performance analysis, cumulative energy, confined in the displacement signal of the top floor, is introduced. This parameter gives the magnitude of the disruptive energy content of the displacement signal. The cumulative energy (W) for any continuous-time signal $x(t)$ is given by the following Eq. (8.14):

$$W = \int_{0}^{t} |x(t)|^2 \, dt \tag{8.14}$$

8.4.1 Using Different Earthquake Time Histories

For analysis under this condition, the following three earthquake time histories are used.

1. 1940 El Centro earthquake (also called the 1940 Imperial Valley earthquake), California, USA
2. 1999 Chi-Chi earthquake, Nantou County, Taiwan
3. 1999 Gebze earthquake, Turkey

The displacement response of the uncontrolled structure is shown in Figure 8.4a, whereas a comparison between the uncontrolled displacement response and the response due to the CO-LQG and PSO-τ_p^{max}-modified-LQG controller is shown in Figure 8.4b. The visual inspection of Figure 8.4b suggests that the displacement response is reduced. As shown in Table 8.1, this reduction in the peak values is 78% using CO-LQG and 82% using PSO-τ_p^{max}-modified-LQG. Furthermore, in Figure 8.4c, a comparison of the displacement due to CO-LQG and PSO-τ_p^{max}-modified-LQG is presented. Figure 8.4c illustrates that the displacement is reduced using the proposed controller throughout the time history presented. In Table 8.1, the reduction in the peak values of relative displacement using the proposed controller is 15% for the first floor, 21% for the second floor, and 22% for the third floor as compared with CO-LQG.

Furthermore, the adaptive variation in $\boldsymbol{R}$ is shown in Figure 8.4e. The value of $\boldsymbol{R}$ for the CO-LQG remains the same through the seismic event, whereas in the proposed algorithm, there are variations in the values of $\boldsymbol{R}$ for each time window (t_w) according to the quasi-resonance between the domain frequency and the first

TABLE 8.1
Peak Responses of the Structure Due to the Conventional CO-LQG Controller and PSO-τ_p^{max}-Modified-LQG Controller

	El Centro Earthquake			Chi-Chi Earthquake			Gebze Earthquake		
Control Algorithm	Uncontrolled	CO-LQG	PSO-τ_p^{max}-modified-LQG	Uncontrolled	CO-LQG	PSO-τ_p^{max}-modified-LQG	Uncontrolled	CO-LQG	PSO-τ_p^{max}-modified-LQG
Displacement (cm)	0.55	0.12	0.10	0.14	0.02	0.015	0.074	0.0180	0.017
	0.83	0.19	0.15	0.22	0.04	0.031	0.117	0.0353	0.029
	0.97	0.22	0.17	0.27	0.05	0.040	0.138	0.0513	0.040
Inter-Story Drift (i_d) (cm)	0.55	0.12	0.10	0.14	0.02	0.015	0.074	0.018	0.017
	0.29	0.07	0.05	0.08	0.02	0.016	0.042	0.017	0.012
	0.14	0.03	0.02	0.05	0.01	0.009	0.022	0.016	0.011
Acceleration (cm/s^2)	870	733	526	181	98	48	126	91	47
	1070	755	410	268	81	60	150	74	61
	1400	723	525	317	97	84	185	113	101
Force (N)	0	971	736	0	1178	1098	0	1278	1167

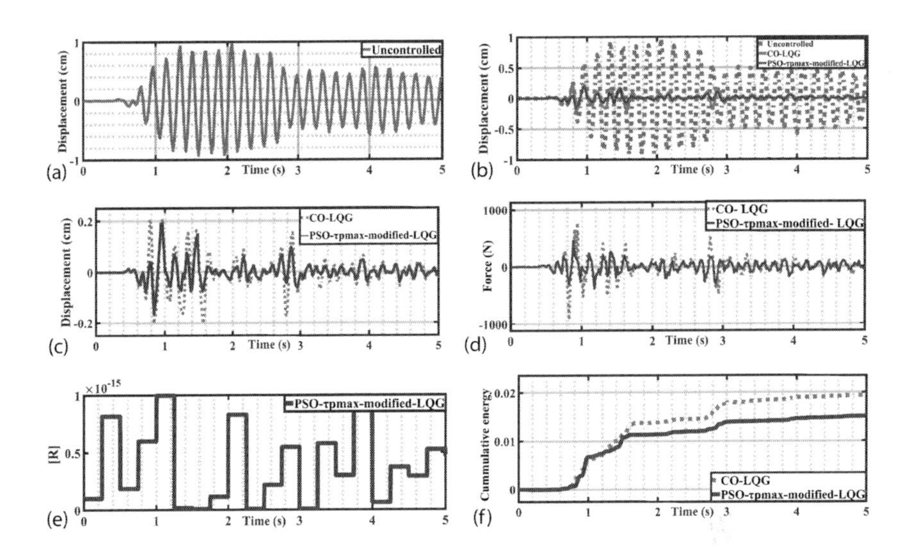

FIGURE 8.4 Structural responses of the structure subjected to the 1940 El Centro earthquake: (a) Displacement response of the third floor of an uncontrolled structure, (b) comparison of controlled responses of the third floor using CO-LQG and PSO-$\tau_p^{\max}$-modified-LQG algorithms, (c) control forces for the CO-LQG and the PSO-$\tau_p^{\max}$-modified-LQG, (d) variation of $\boldsymbol{R}$ with time, and (e, f) comparison of cumulative energies of the third floor's displacement by applying CO-LQG and PSO-$\tau_p^{\max}$-modified-LQG.

two fundamental frequencies of the structure. With the reduction in the peak values of the inter-story drift, the proposed control algorithm achieved a reduction of 29% between the first and second floors and 33% between the second and third floors as compared with the CO-LQG. The reduction in peak values of the absolute acceleration is shown in Table 8.1. Observations from Table 8.1 reveal that the proposed algorithm can reduce the peak values of absolute acceleration by 28%, 46%, and 27% for the first, second, and third floors as compared with the CO-LQG. It is necessary to point out here that all these reductions in the structural responses are achieved using lesser control force, as shown in Figure 8.4d.

In Table 8.1, the proposed controller utilized 24% lesser force (peak value) to achieve the above-mentioned results as compared with the CO-LQG. The energy confined in the controlled signal of the relative displacement of the third floor due to CO-LQG and the PSO-$\tau_p^{\max}$-modified-LQG is shown in Figure 8.4f.

Similar discussions are presented for the Chi-Chi and Gebze earthquakes. For the Chi-Chi earthquake (see Figure 8.5), the displacement response of the third floor of the uncontrolled structure is shown in Figure 8.5a, whereas a comparison of uncontrolled displacement response and response due to the CO-LQG and PSO-$\tau_p^{\max}$-modified-LQG controller is shown in Figure 8.5b. A visual inspection of Figure 8.5b suggests that the displacement response is reduced. A comparison of the displacement response due to CO-LQG and PSO-$\tau_p^{\max}$-modified-LQG is presented in Figure 8.5c for the Chi-Chi earthquake, whereas the same is

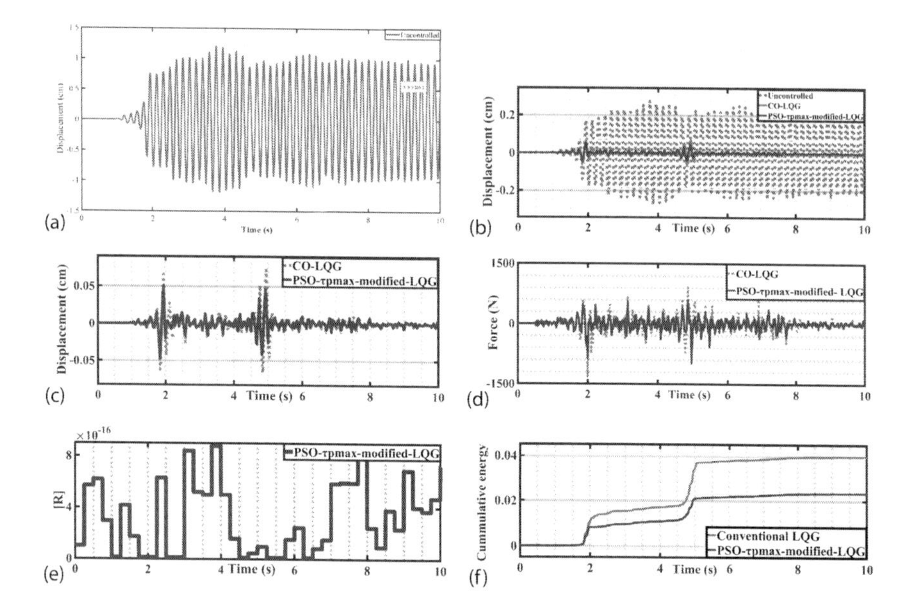

FIGURE 8.5 Structural responses for the structure subjected to the 1999 Chi-Chi earthquake: (a) displacement response of the third floor of an uncontrolled structure, (b) comparison of controlled responses of the third floor using CO-LQG and PSO-τ_p^{max}-modified-LQG algorithms, (c) control forces for the CO-LQG and the PSO-τ_p^{max}-modified-LQG, (d) variation of $\boldsymbol{R}$ with time, and (e, f) comparison of cumulative energies of the third floor's displacement by applying CO-LQG and PSO-τ_p^{max}-modified-LQG.

presented in Figure 8.6c for the Gebze earthquake. Observations conclude that the displacement is reduced using the proposed controller throughout the time history.

In Table 8.1, the reduction in the peak values of relative displacement using the proposed controller as compared with the CO-LQG is 26% for the first floor, 23% for the second floor, and 20% for the third floor in the Chi-Chi earthquake, and 6% for the first floor, 18% for the second floor, and 22% for the third floor for the Gebze earthquake. Variations of the control weighting matrix $\boldsymbol{R}$ are shown in Figure 8.5e for the Chi-Chi earthquake, whereas for the Gebze earthquake it is shown in Figure 8.6e. Regarding the inter-story drift for the Chi-Chi earthquake, in Table 8.1 the proposed control algorithm achieved a reduction of 20% between the first and second floors and 10% between the second and third floors. For the Chi-Chi earthquake, it is 31% for both (i.e., between the first and second floors and the second and third floors).

Similarly, for the Gebze earthquake, in Table 8.1, the proposed controller achieved a reduction in the inter-story drift of 31% between the first and second floors and between the second and third floors, each as compared to the CO-LQG. The absolute accelerations of all floors of the structure subjected to the Chi-Chi earthquake and Gebze earthquake are also presented in Table 8.1.

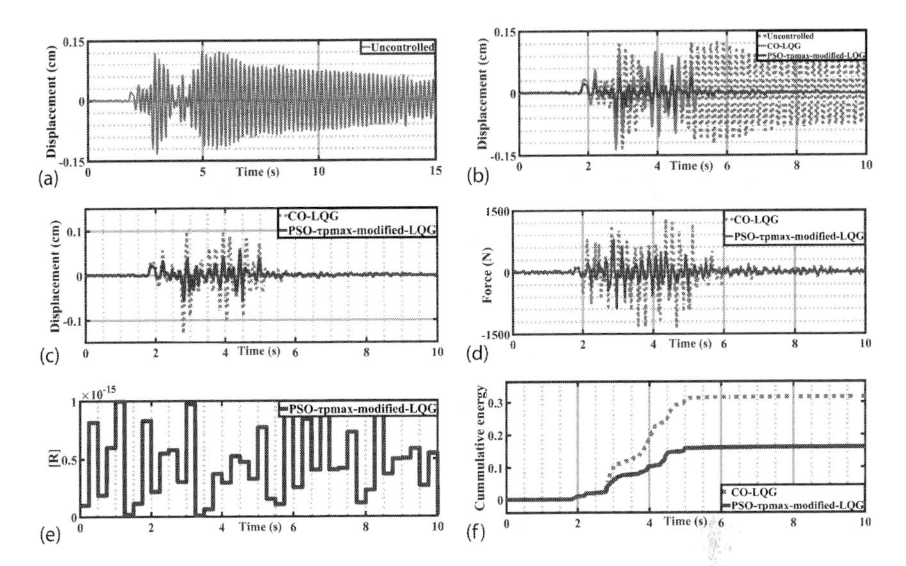

FIGURE 8.6 Structural responses for the structure subjected to the 1999 Gebze earthquake: (a) Displacement response of the third floor of an uncontrolled structure, (b) comparison of controlled responses of the third floor using CO-LQG and PSO-$\tau_p^{\max}$-modified-LQG algorithms, (c) control forces for the CO-LQG and the PSO-$\tau_p^{\max}$-modified-LQG, (d) variation of $\boldsymbol{R}$ with time, and (e, f) comparison of cumulative energies of the third floor's displacement by applying CO-LQG and PSO-$\tau_p^{\max}$-modified-LQG.

The observations from Table 8.1 reveal that the proposed algorithm can reduce the peak values of the acceleration by 51% for the first floor, 25% for the second floor, and 13% for the third floor as compared with the CO-LQG, whereas for the Gebze earthquake, reductions of 48% for the first floor, 17% for the second floor, and 10% for the third floor are achieved by the proposed control algorithm.

Comparisons of the time histories of the control force due to CO-LQG and the proposed controller are shown in Figure 8.5d for the Chi-Chi earthquake and in Figure 8.6d for the Gebze earthquake. From Table 8.1, the proposed controller utilized lesser force to achieve the above-described results. A comparison of the cumulative energies of the third floor's displacement by applying CO-LQG and the PSO-$\tau_p^{\max}$-modified-LQG is shown in Figure 8.5f for the Chi-Chi earthquake and Figure 8.6f for the Gebze earthquake.

8.4.2 Using an Earthquake Recorded in Different Soil Conditions

This earthquake record is taken from the Kyoshin Network (K-NET). The soil type is determined according to the Federal Emergency Management Agency's (FEMA) *FEMA-356* based on the shear wave velocity (v_s). A performance analysis of the proposed controller is carried out for the structure subjected to hard, medium, and soft soil. For the performance assessment in hard soil, the uncontrolled displacement response of the third floor of the structure is considered in Figure 8.7a.

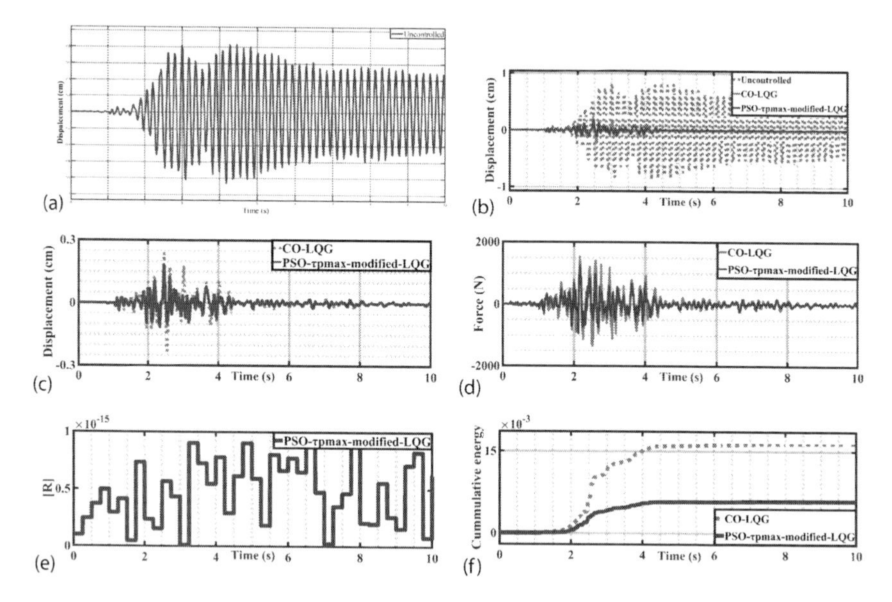

FIGURE 8.7 Structural responses for the structure subjected to a hard-soil earthquake: (a) Displacement response of the third floor of an uncontrolled structure, (b) comparison of controlled responses of the third floor using CO-LQG and PSO-$\tau_p^{\max}$-modified-LQG algorithms, (c) control forces for the CO-LQG and the PSO-$\tau_p^{\max}$-modified-LQG, (d) variation of $\boldsymbol{R}$ with time, and (e, f) comparison of cumulative energies of the third floor's displacement by applying CO-LQG and PSO-$\tau_p^{\max}$-modified-LQG.

A comparison of the third floor's displacement responses of the uncontrolled structure and semi-actively controlled structure using the CO-LQG and PSO-$\tau_p^{\max}$-modified-LQG control algorithm is shown in Figure 8.7b. Figure 8.7b shows the mitigation of the displacement response through the seismic event. It can be seen from Table 8.2 that the reduction in the peak value of the displacement response is 72% using CO-LQG and 80% using the proposed PSO-$\tau_p^{\max}$-modified-LQG as compared with the uncontrolled structure.

A comparison of the displacement response due to CO-LQG and PSO-$\tau_p^{\max}$-modified-LQG is shown in Figure 8.7c. This figure shows that the proposed controller is effectively reducing the relative displacement response of the structure. It can also be observed from Table 8.2 that the proposed controller achieved 46%, 33%, and 28% more reduction for the first, second, and third floors in peak values of the relative displacement as compared with the CO-LQG controller.

It can be seen from Table 8.2 that the inter-story drift is reduced by 12% for the first and second floors and 25% for the second and third floors when applying the PSO-$\tau_p^{\max}$-modified-LQG controller instead of CO-LQG in the semi-active control scheme. The reduction in the absolute acceleration using the proposed controller is moderate as compared with the CO-LQG. The absolute accelerations for the first, second, and third floors are reduced by 4%, 2%, and 14%, respectively, by applying the proposed controller. Furthermore, the proposed controller attains

TABLE 8.2
Peak Responses Due to CO-LQG and PSO-τ_p^{max}-Modified-LQG for Structure Subjected to Earthquakes Recorded in Different Soil Conditions

	Earthquake (Hard Rock)			Earthquake (Medium Soil)			Earthquake (Soft Soil)		
Control Algorithm	**Uncontrolled**	**CO-LQG**	**PSO-τ_p^{max}-modified-LQG**	**Uncontrolled**	**CO-LQG**	**PSO-τ_p^{max}-modified-LQG**	**Uncontrolled**	**CO-LQG**	**PSO-τ_p^{max}-modified-LQG**
Displacement (cm)	0.66	0.13	0.07	0.60	0.38	0.18	0.99	0.22	0.15
	0.80	0.22	0.15	0.93	0.45	0.24	1.26	0.55	0.39
	0.89	0.25	0.18	1.20	0.51	0.29	1.53	0.76	0.49
Inter-Story Drift (i_d) (cm)	0.66	0.13	0.07	0.60	0.38	0.18	0.99	0.22	0.15
	0.14	0.09	0.08	0.33	0.07	0.06	0.27	0.33	0.24
	0.09	0.04	0.03	0.27	0.06	0.05	0.27	0.21	0.10
Acceleration (cm/s^2)	1167	512	493	830	615	520	852	580	540
	1287	570	561	1018	702	602	1217	770	660
	1356	854	735	1157	970	840	1299	904	814
Force (N)	—	1533	1224	—	1642	1129	—	1690	1367

these reductions in structural responses using 20% less force than the CO-LQG controller. Figure 8.7d shows the comparison of the time histories of the forces used by the proposed controller and the CO-LQG. The variations of the control weighting matrix $\boldsymbol{R}$ are shown in Figure 8.7e, and a comparison of the cumulative energy of the displacement is shown in Figure 8.7f. This comparison of the cumulative energy concludes that the energy content in the displacement signal of the third floor of the structure is less. Hence, the structural integrity is protected.

For earthquakes recorded in the medium soil, a comparison of the displacement time histories of an uncontrolled and controlled structure employing CO-LQG and the proposed control algorithm is shown in Figure 8.8b. The preeminence of the proposed control over the CO-LQG can be observed in Figure 8.8c, in which a comparison of the time histories of the displacement responses due to CO-LQG and the proposed controller is shown. Observing Table 8.2, the proposed controller achieves a reduction in the displacement by 53%, 47%, and 43% for the first, second, and third floors, respectively, as compared with the CO-LQG. The inter-story drift between the first and second floors and between the second and third floors is reduced by 14% and 17% more, respectively, using the proposed controller in place of CO-LQG. Likewise, the proposed controller reduced the absolute accelerations for the first, second, and third floors by 15%, 14%, and 13%, respectively, as compared with the CO-LQG. The proposed

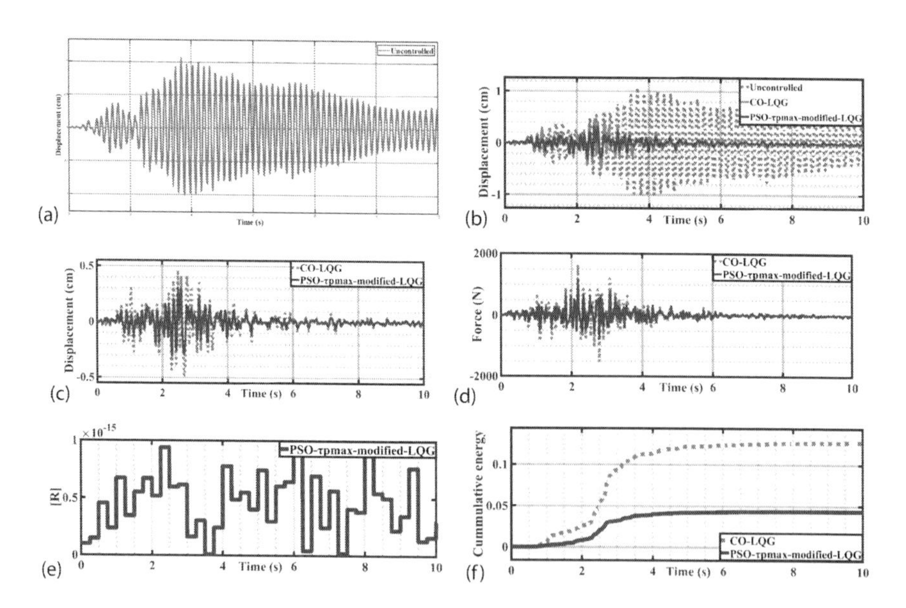

FIGURE 8.8 Structural responses for the structure subjected to a medium-soil earthquake: (a) Displacement response of the third floor of an uncontrolled structure, (b) comparison of controlled responses of the third floor using CO-LQG and PSO-$\tau_p^{\max}$-modified-LQG algorithms, (c) control forces for the CO-LQG and the PSO-$\tau_p^{\max}$-modified-LQG, (d) variation of $\boldsymbol{R}$ with time, and (e, f) comparison of cumulative energies of the third floor's displacement by applying CO-LQG and PSO-$\tau_p^{\max}$-modified-LQG.

controller attains these reductions in structural responses using 31% less force than the CO-LQG, as can be seen in Figure 8.8d. Variation of the values of the control weighting matrix $\boldsymbol{R}$ is shown in Figure 8.8e. A comparison of the cumulative energies of the displacement of the third floor is shown in Figure 8.8f, which shows that the displacement signal obtained using the proposed controller has less energy for destruction.

For earthquakes recorded in soft soil, the displacement response of the third floor of the uncontrolled structure is shown in Figure 8.9a, which is to be reduced by employing the semi-active control scheme. A comparison of the time histories of the displacement response of the uncontrolled structure obtained by employing CO-LQG and the proposed controller is shown in Figure 8.9b, which shows the effectiveness of the semi-active control scheme as compared to the uncontrolled structure. Also, it can be seen from Table 8.2 that the proposed controller achieved a reduction in the displacement by 32% for the first floor, 29% for the second floor, and 36% for the third floor as compared with the CO-LQG.

The interstory drifts between the first and second floors and between the second and third floors are reduced by 27% and 52%, respectively, as compared to CO-LQG. Similarly, the proposed controller reduced the accelerations for the first, second, and third floors by 7%, 14%, and 10%, respectively, as compared

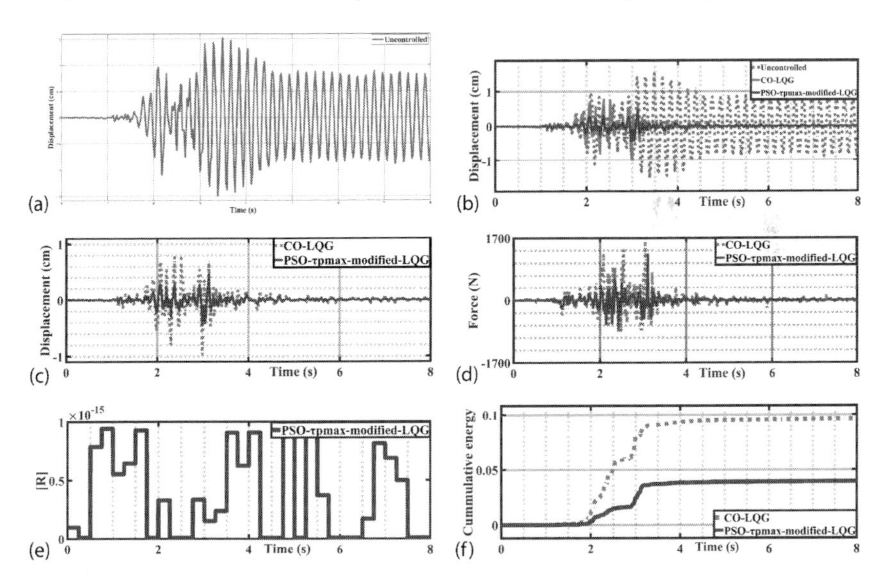

FIGURE 8.9 Structural responses for the structure subjected to a soft-soil earthquake: (a) Displacement response of the third floor of an uncontrolled structure, (b) comparison of controlled responses of the third floor using CO-LQG and PSO-τ_p^{max}-modified-LQG algorithms, (c) control forces for the CO-LQG and the PSO-τ_p^{max}-modified-LQG, (d) variation of $\boldsymbol{R}$ with time, and (e) comparison of cumulative energies of the third floor's displacement by applying CO-LQG and PSO-τ_p^{max}-modified-LQG, and (f) comparison of cumulative energy contained in third floor of the structure using CO-LQG and PSO τ_p^{max}-modified-CO-LQG controllers.

with the CO-LQG. These reductions in structural responses were achieved by the proposed controller using 19% less force than the CO-LQG. To validate these results, a comparison of the time histories of the CO-LQG and PSO-τ_p^{max}-modified-LQG is demonstrated in Figure 8.9d. Furthermore, variations of the values of the control weighting matrix $\boldsymbol{R}$ are shown in Figure 8.9e. These variations are obtained based on the quasi-resonance that occurred between the natural frequency of the prototype three-story structure and the dominant frequencies of each time window for the earthquake. In the case of quasi-resonance, the larger force is required to control the increased vibrations; hence, an appropriate lower value of weighting matrix $\boldsymbol{R}$ is determined by the PSO algorithm, and vice versa. Figure 8.9f shows the cumulative energy content of the displacement signal of the third floor of the structure obtained using the proposed controller.

8.4.3 Effect of a Power Cutoff at the Peak of an Earthquake

Here, analysis is done for a fictitious circumstance in which power disappears at the peak of an earthquake event. For simulation, the El Centro earthquake has been considered. When considering the power loss at 0.9 s of the El Centro time history, a comparison of the third floor's relative displacement between the uncontrolled structure and the PSO-τ_p^{max}-modified-LQG controlled structure is shown. When electricity is provided continuously and disappears during the earthquake's peak, Figure 8.10a compares the displacement of an uncontrolled building's third floor to that of a controlled structure using the PSO-τ_p^{max}-modified-LQG (Figure 8.10a). Although the displacement response works best when power is always present, it is better seen using a controlled system when the power goes out at the peak of earthquake activity. This is because after losing power, the controller turns into a passive off controller, which is preferable to an uncontrolled structure.

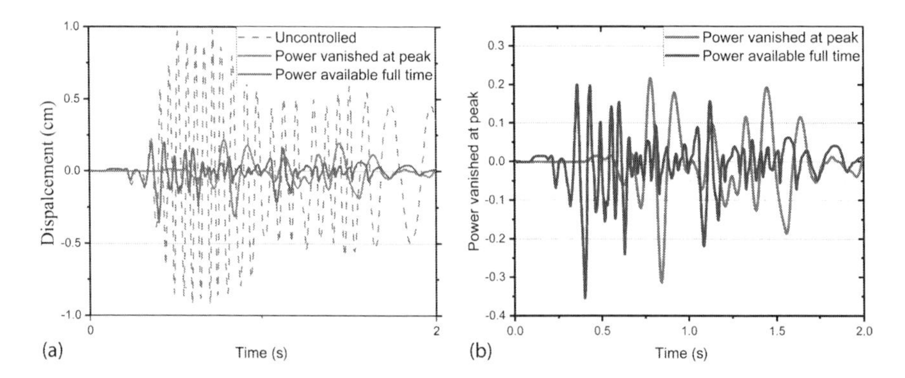

FIGURE 8.10 (a) Performance analysis of PSO-τ_p^{max}-modified-LQG for the 1940 El Centro earthquake's time history, considering a fictitious situation in which power goes off during the peak of the earthquake. (b) Comparison of the third floor's displacement of an uncontrolled structure with the PSO-τ_p^{max}-modified-LQG controlled displacement when power is available full-time and when it is cut off during the peak of an earthquake.

The enlarged version of the activity at the instant when the peak of the earthquake occurs is shown in Figure 8.10b. Around 0.9 s, it can be observed that the displacement response is lesser compared to the uncontrolled system when the power vanishes at the peak of an earthquake.

8.5 CONCLUSIONS

This study discusses the development process for the PSO-τ_p^{max}-modified-LQG controller. By adjusting the control weighting matrix $\boldsymbol{R}$ over each small-time window-based PSO-τ_p^{max} algorithm when the quasi-resonance occurred, the traditional LQG controller was changed to create this controller. During the earthquake, these updated values of $\boldsymbol{R}$ adaptively give the necessary optimal control force.

For the PSO-τ_p^{max} algorithm to estimate the quasi-resonance between the first two fundamental frequencies and the earthquake, the signal is evaluated in the time domain rather than the frequency domain or time–frequency domain, as in fast Fourier transform (FFT), short-time Fourier transform (STFT), DWT, and so on. It is an advance over earlier research where determining the dominant frequency for each window required analysis of the signal in the frequency or time–frequency domain. The benefits of the suggested approach are that it always operates in the time domain and that the PSO algorithm determines the gain matrices in an adaptable manner. In contrast to typical LQG controllers where the gain matrix is left unchanged, the gain matrices for each time window are determined adaptively using the PSO method.

To obtain the regulated response of the prototype three-story structure with a single MR damper, the PSO-τ_p^{max}-modified-LQG controller is assessed. The findings show that the suggested controllers outperform the traditional LQG controller by a significant margin. Additionally, the developed controller is numerically tested under a variety of scenarios, including putting the structure through three separate seismic events that happen in various soil conditions, and assuming a scenario where power goes out during the peak of the seismic activity. This controller is a popular option for vibration control because of its inherent adaptability in the design of the proposed PSO-τ_p^{max}-modified-LQG controllers to account for the quasi-resonance by the change of $\boldsymbol{R}$.

REFERENCES

1. T. K. Datta, "A state-of-the-art review on active control of structures," *ISET J Earthq Technol*, vol. 40, no. 1, pp. 1–17, 2003.
2. V. Bhaiya, M. K. Shrimali, S. D. Bharti, and T. K. Datta, "Modified semiactive control with MR dampers for partially observed systems," *Eng Struct*, vol. 191, pp. 129–147, 2019, doi: 10.1016/j.engstruct.2019.04.063.
3. T. E. Alqado, G. Nikolakopoulos, and L. Dritsas, "Semi-active control of flexible structures using closed-loop input shaping techniques," *Struct Control Health Monit*, vol. 24, no. 5, p. e1913, 2017, doi: 10.1002/stc.1913.

4. M. D. Symans and M. C. Constantinou, "Semi-active control systems for seismic protection of structures: A state-of-the-art review," *Eng Struct*, vol. 21, no. 6, pp. 469–487, 1999, doi: 10.1016/S0141-0296(97)00225-3.
5. R. Ahamed, S. B. Choi, and M. M. Ferdaus, "A state of art on magneto-rheological materials and their potential applications," *J Intell Mater Syst Struct*, vol. 29, no. 10, pp. 2051–2095, 2018, doi: 10.1177/1045389X18754350.
6. T. E. Saaed, G. Nikolakopoulos, J. E. Jonasson, and H. Hedlund, "A state-of-the-art review of structural control systems," *J Vib Control*, vol. 21, no. 5, pp. 919–937, 2015, doi: 10.1177/1077546313478294.
7. L. M. Jansen and S. J. Dyke, "Semiactive control strategies for MR dampers: Comparative study," *J Eng Mech*, vol. 126, no. 8, pp. 795–803, 2002, doi: 10.1061/(asce)0733-9399(2000)126:8(795).
8. Y.-J. Cha and A. K. Agrawal, "Decentralized output feedback polynomial control of seismically excited structures using genetic algorithm," *Struct Control Health Monit*, vol. 20, no. 3, pp. 241–258, 2013, doi: 10.1002/stc.486.
9. G. Kumar and A. Kumar, "Fourier transform and particle swarm optimization based modified LQR algorithm for mitigation of vibrations using magnetorheological dampers," *Smart Mater Struct*, vol. 26, no. 11, p. 115013, 2017, [Online]. Available: http://stacks.iop.org/0964-1726/26/i=11/a=115013
10. G. Kumar, A. Kumar, and R. S. Jakka, "The particle swarm modified quasi bang-bang controller for seismic vibration control," *Ocean Engineering*, vol. 166, pp. 105–116, 2018, doi:10.1016/j.oceaneng.2018.08.002.
11. G. Kumar, A. Kumar, and R. S. Jakka, "An adaptive LQR controller based on PSO and maximum predominant frequency approach for semi-active control scheme using MR damper," *Mech Ind*, vol. 19, no. 1, p. 109, 2018, doi: 10.1051/meca/2018018.
12. R. Guclu, "Sliding mode and PID control of a structural system against earthquake," *Math Comput Model*, vol. 44, no. 1, pp. 210–217, 2006, doi: 10.1016/j.mcm.2006.01.014.
13. S. Y. Zhang and X. M. Wang, "Study of Fuzzy-PID control in MATLAB for two-phase hybrid stepping motor," in Energy Research and Power Engineering, 2013, vol. 341, pp. 664–667. doi: 10.4028/www.scientific.net/AMM.341-342.664.
14. X. Xiang, C. Liu, H. Su, and Q. Zhang, "On decentralized adaptive full-order sliding mode control of multiple UAVs," *ISA Trans*, vol. 71, pp. 196–205, 2017, doi: 10.1016/j.isatra.2017.09.008.
15. W. Yu and S. Thenozhi, *Active Structural Control with Stable Fuzzy PID Techniques*. Springer International Publishing, 2016.
16. R. Guclu and H. Yazici, "Vibration control of a structure with ATMD against earthquake using fuzzy logic controllers," *J Sound Vib*, vol. 318, no. 1, pp. 36–49, 2008, doi: 10.1016/j.jsv.2008.03.058.
17. K. Liu, L. Chen, and G. Cai, "H∞ control of a building structure with time-varying delay," *Adv Struct Eng*, vol. 18, no. 5, pp. 643–657, 2015, doi: 10.1260/1369-4332.18.5.643.
18. G. Kumar, R. Kumar, A. Kumar, and B. Mohan Singh, "Development of modified LQG controller for mitigation of seismic vibrations using swarm intelligence," 2022. doi: 10.1504/IJAAC.2023.10049079.
19. A. M. A. Soliman and M. M. S. Kaldas, "Semi-active suspension systems from research to mass-market – A review," *J Low Freq Noise Vib Act Control*, vol. 40, no. 2, pp. 1005–1023, 2021, doi: 10.1177/1461348419876392.
20. V. Utkin, "Variable structure systems with sliding modes," *IEEE Trans Automat Contr*, vol. 22, no. 2, pp. 212–222, 1977, doi: 10.1109/TAC.1977.1101446.

21. O. Yakut and H. Alli, "Neural based sliding-mode control with moving sliding surface for the seismic isolation of structures," *J Vib Control*, vol. 17, no. 14, pp. 2103–2116, 2011, doi: 10.1177/1077546310395964.
22. H. Alli and O. Yakut, "Fuzzy sliding-mode control of structures," *Eng Struct*, vol. 27, no. 2, pp. 277–284, 2005, doi: 10.1016/j.engstruct.2004.10.007.
23. S. K. Gorade, S. R. Kurode, and P. S. Gandhi, "Modeling and sliding mode control of flexible structure," in *International Conference on Control, Automation and Systems*, 2014. doi: 10.1109/ICCAS.2014.6987886.
24. Q. P. Ha, M. T. Nguyen, J. Li, and N. M. Kwok, "Smart structures with current-driven MR dampers: modeling and second-order sliding mode control," *IEEE ASME Trans Mechatron*, vol. 18, no. 6, pp. 1702–1712, 2013, doi: 10.1109/TMECH.2013.2280282.
25. A. Yesmin and M. K. Bera, "Design of event-triggered sliding mode controller based on reaching law with time varying event generation approach," *Eur J Control*, vol. 48, pp. 30–41, 2019, doi: 10.1016/j.ejcon.2018.12.003.
26. A. T. Azar and Q. Zhu, "Advances and applications in sliding mode control systems," *Stud Comput Intell*, 2015, doi: 10.1007/978-3-319-11173-5.
27. A. Ahlawat and A. Ramaswamy, "Multi-objective optimal structural vibration control using fuzzy logic control system," *J Struct Eng-Asce*, vol. 127, no. 11, pp. 1330–1337, 2001, doi: 10.1061/(ASCE)0733-9445(2001)127:11(1330).
28. Sk. F. Ali and A. Ramaswamy, "Optimal fuzzy logic control for MDOF structural systems using evolutionary algorithms," *Eng Appl Artif Intell*, vol. 22, no. 3, pp. 407–419, 2009, doi: 10.1016/j.engappai.2008.09.004.
29. A. Rama Mohan Rao and K. Sivasubramanian, "Multi-objective optimal design of fuzzy logic controller using a self-configurable swarm intelligence algorithm," *Comput Struct*, vol. 86, no. 23–24, pp. 2141–2154, 2008, doi: 10.1016/j.compstruc.2008.06.005.
30. K. Takin, R. Doroudi, and S. Doroudi, "Vibration control of structure by optimising the placement of semi-active dampers and fuzzy logic controllers," *Aust J Struct Eng*, vol. 22, no. 3, pp. 222–235, 2021, doi: 10.1080/13287982.2021.1957198.
31. Sk. F. Ali and A. Ramaswamy, "Testing and modeling of MR damper and its application to SDOF systems using integral backstepping technique," *J Dyn Syst Meas Control*, vol. 131, no. 2, pp. 21009–21011, 2009, [Online]. Available: http://dx.doi.org/10.1115/1.3072154
32. Sk. F. Ali and A. Ramaswamy, "GA-optimized FLC-driven semi-active control for phase-II smart nonlinear base-isolated benchmark building," *Struct Control Health Monit*, vol. 15, no. 5, pp. 797–820, 2008, doi: 10.1002/stc.272.
33. K. M. Choi, S. W. Cho, D. O. Kim, and I. W. Lee, "Active control for seismic response reduction using modal-fuzzy approach," *Int J Solids Struct*, vol. 42, no. 16–17, pp. 4779–4794, 2005, doi: 10.1016/j.ijsolstr.2005.01.018.
34. K. M. Choi, S. W. Cho, H. J. Jung, and I. W. Lee, "Semi-active fuzzy control for seismic response reduction using magnetorheological dampers," *Earthq Eng Struct Dyn*, vol. 33, no. 6, pp. 723–736, 2004, doi: 10.1002/eqe.372.
35. D. Das, T. K. Datta, and A. Madan, "Semiactive fuzzy control of the seismic response of building frames with MR dampers," *Earthq Eng Struct Dyn*, vol. 41, no. 1, pp. 99–118, 2012, doi: 10.1002/eqe.1120.
36. S. J. Dyke and B. F. Spencer, "A comparison of semi-active control strategies for the MR damper," in *Proceedings Intelligent Information Systems. IIS'97*, 1997, pp. 580–584. doi: 10.1109/IIS.1997.645424.
37. B. Spencer, S. Dyke, M. Sain, and M. Carlson, "Phenomenological model for magnetorheological dampers," *J Eng Mech*, vol. 123, no. 3, pp. 230–238, 1997, doi: 10.1061/(ASCE)0733-9399(1997)123:3(230)

38. I. Halperin, G. Agranovich, and Y. Ribakov, "Optimal LQR Control of Structures using Linear Modal Model," *Ariel.Ac.Il*, 2005.
39. A. Alavinasab, H. Moharrami, and A. Khajepour, "Active control of structures using energy-based LQR method," *Comput-Aided Civ Infrastruct Eng*, vol. 21, no. 8, pp. 605–611, 2006, doi: 10.1111/j.1467-8667.2006.00460.x.
40. Ü. Önen, A. Çakan, and İ. İlhan, "Particle swarm optimization based LQR control of an inverted pendulum," *Eng Technol J*, 2017, doi: 10.18535/etj/v2i5.01.
41. K. Miyamoto, J. She, D. Sato, and N. Yasuo, "Automatic determination of LQR weighting matrices for active structural control," *Eng Struct*, vol. 174, pp. 308–321, 2018, doi: 10.1016/j.engstruct.2018.07.009.
42. L. M. Jansen and S. J. Dyke, "Semiactive control strategies for MR dampers: Comparative study," *J Eng Mech*, vol. 126, no. 8, pp. 795–803, 2000, doi: 10.1061/(ASCE)0733-9399(2000)126:8(795).
43. A. Dominguez, R. Sedaghati, and I. Stiharu, "Modeling and application of MR dampers in semi-adaptive structures," *Comput Struct*, vol. 86, no. 3–5, pp. 407–415, 2008, doi: 10.1016/j.compstruc.2007.02.010.
44. E. G. Collins and M. F. Selekwa, "A fuzzy logic approach to LQG design with variance constraints," *IEEE Trans Control Syst Technol*, vol. 10, no. 1, pp. 32–42, 2002, doi: 10.1109/87.974336.
45. G. F. Panariello, R. Betti, and R. W. Longman, "Optimal structural control via training on ensemble of earthquakes," *J Eng Mech*, vol. 123, no. 11, pp. 1170–1179, 1997, doi: 10.1061/(ASCE)0733-9399(1997)123:11(1170).
46. B. Karimpour, A. Keyhani, and J. Alamatian, "New active control method based on using multiactuators and sensors considering uncertainty of parameters," *Adv Civ Eng*, vol. 2014, p. 10, 2014, doi: 10.1155/2014/180673.
47. B. Basu and S. Nagarajaiah, "A wavelet-based time-varying adaptive LQR algorithm for structural control," *Eng Struct*, vol. 30, pp. 2470–2477, 2008, doi: 10.1016/j.engstruct.2008.01.011.
48. F. Amini, N. K. Hazaveh, and A. A. Rad, "Wavelet PSO-based LQR algorithm for optimal structural control using active tuned mass dampers," *Comput-Aided Civ Infrastruct Eng*, vol. 28, no. 7, pp. 542–557, 2013, doi: 10.1111/mice.12017.
49. Y. Nakamura and J. Saita, "UrEDAS, the earthquake warning system: Today and tomorrow," in *Earthquake Early Warning Systems*, Berlin, Heidelberg: Springer, 2007, pp. 249–281. doi: 10.1007/978-3-540-72241-0_13.
50. Prestandard and commentary for the seismic rehabilitation of buildings. A technical report published by American society of civil engineers Reston, Virginia Prepared for Federal emergency Management Agency. Washington, D.C., November 2000.

9 Application of AI Tools in Creating Datasets from a Real Data Component for Structural Health Monitoring

Minh Q. Tran, Hélder S. Sousa, and José C. Matos

9.1 OVERVIEW

Structural health monitoring (SHM) based on dynamic methods is becoming a widely applied method, mainly due to its high accuracy combined with the fact that it is not necessary to limit service activities during monitoring. In order to accurately identify the vibration characteristics of a complex structure, such as frequency, mode shape, and damping ratio, it is necessary to arrange a dense network of acceleration sensors, which might be a challenge due to onsite conditions. Most sensors are fixed and can only be used for a single building during their lifetime. This may result in a waste of resources if there are not enough sensors for a specific structure or even if their layout is inefficient, thus resulting in insufficient or missing data. To overcome this, a new approach based on the application of artificial intelligence (AI) may be considered. Specifically, an artificial neural network (ANN) may be used to generate missing data to determined areas from the position of one or more fixed measurement points. For that case, an ANN model should be trained and tested on a number of projects to ensure accuracy during operation. The results show that the data generated is accurate and the data storage capacity is optimized. A major benefit is that a large number of sensors can be removed from the building to serve other purposes, optimizing the costs for SHM. This chapter presents the approach and a case study that was carried out on a cable-stayed bridge in Vietnam. The obtained results show the potential of applying AI in creating virtual data, serving larger goals in SHM.

DOI: 10.1201/9781003306924-9

9.1.1 Background

Maintenance costs and the life cycle assessment of large structures are complex challenges for managers and policymakers. During a structure's service period, damage can emerge anywhere and with different levels of severity. This damage can lead to a significant overall downgrade of service or shorten the structure's lifespan. Early detection of damage and timely maintenance are therefore essential to improve the service quality of these constructions. Structural health monitoring (SHM), an interdisciplinary engineering field, has become an effective approach for assessment and timely failure detection. The tasks of a SHM system are to monitor the structure for early detection of damage based on measurement data and to analyze and evaluate the severity of these damages before making decisions concerning intervention, such as repair or replacement.

An effective SHM strategy must incorporate global and local monitoring techniques. Local methods focus on a part of the structure, such as ultrasonic or acoustic methods and X-ray inspection, among others. These methods provide different alternatives for detecting damage that can be combined for better accuracy. However, since these methods are local approaches, they require prior information about the location of the damaged area. In many cases, it can be difficult to determine with certainty those damaged areas due to the inaccessibility of their location. On the other hand, vibration-based method is a global technique that can assess the safety and integrity of the monitored structures. Vibration-based damage detection appeared during the late 1970s, being applied in the aerospace and offshore oil industries [1].

The potential of vibration-based methods applied to existing constructions is significantly promising and has gotten the attention of many researchers [2–10]. The presence of damage shifts the structural parameters (e.g., stiffness, mass, flexibility, and energy dissipation) of the considered structures. Subsequently, it changes frequency response functions and modal parameters (e.g., natural frequencies, mode shapes, and damping ratios). Among these characteristics, natural frequencies are easier to identify with just one or a few sensors. Nevertheless, natural frequencies only determine the occurrence of damage. The location and severity of the damage are not assessed [11, 12]. Moreover, the effect of environmental factors (e.g., temperature variation) can be larger than that of a damage in the structure. In contrast, being more sensitive to local damage and less sensitive to temperature changes is a typical feature of displacement mode shapes. Xia et al. [13], Ndambi et al. [14], Maia et al. [15], and Ismail et al. [16] used the mode shape as a key parameter to determine the defects present on different structures. These studies show that using mode shape can detect damage, thus monitoring structural health with a higher level of accuracy [17, 18]. However, measuring the mode shape is more laborious and time-consuming than measuring the natural frequency. At the same time, this requires using a sensor system with multiple measuring points placed on optimal positions of the structure [19]. The appearance of singularities in the mode shape due to the existence of defects is an important factor in failure recognition [20].

9.1.2 Problems Encountered during Data Collection

As shown above, defining the mode shapes is highly effective in SHM. Determining the mode shape of a structure (especially large structures such as cable-stayed bridges and cantilever bridges, among others) requires a large number of sensors spread over the structure's surface. The sensor system in this case usually consists of many sensors, which are installed to cover the structure. Managing and ensuring the good working condition of all sensors are complicated tasks. In fact, errors during SHM by the sensor grid often occur and are detected only when processing data [21]. This significantly increases the cost of SHM, not to mention that active sensors can fail, leading to data interruption (Figure 9.1).

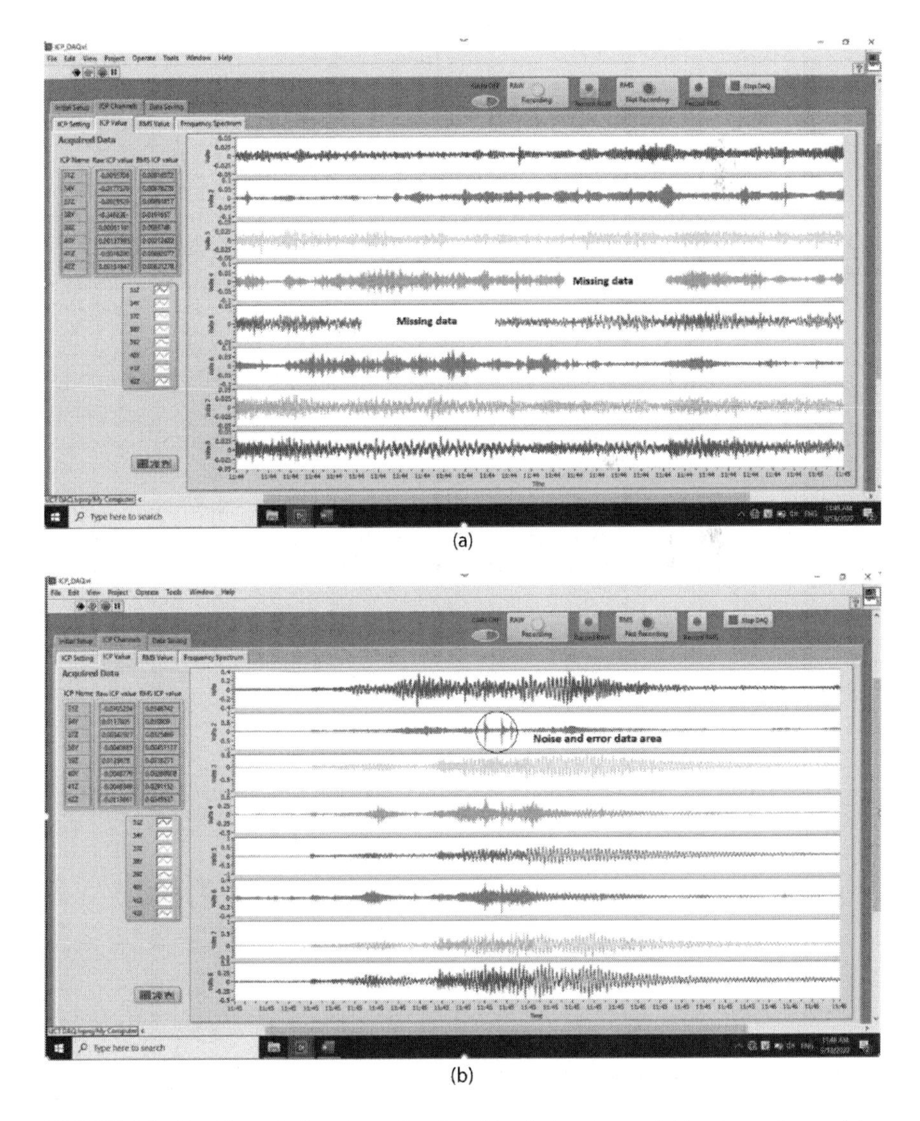

FIGURE 9.1 Problems during data collection: (a) missing data and (b) noisy and error data.

The sensors are densely placed on the structure, and although this is good for data collection, it is difficult to control the quality of the data obtained. The structure is always subjected to the effects of loads such as wind, temperature, and live loads, and the sensors are extremely sensitive to external influences. With just a small impact, the data obtained can be incorrect. In such cases, it is extremely difficult to recover the data or the subsequent data is not accurate enough. This work will present a solution to overcome these limitations. An ANN model is trained based on reliable data collected on a structure. Then, based on this trained network, data from optimal sensor locations can be used to predict and generate virtual data.

9.2 RESEARCH METHODS AND APPROACH

Time series data is one of the main data types in SHM. Usually, time series data that is unstable or unstructured when generated by machine learning or AI networks will not give accurate results [22]. When the data is unstructured, it is very difficult or impossible to separate the characteristics of the data. Then, the data has absolutely no value in machine learning. The newly generated data is not accurate. However, it is possible to create virtual vibrations data on a structure. A structure consists of a combination of masses derived from each one of its elements. These masses, because they are on the same body, will move relative to each other and are related to each other. Following this concept, if one point is used as a reference, data at other points can be generated through machine learning algorithms.

An ANN is a distributed network inspired by the human nervous system. One of the outstanding features of the ANN is its ability to learn from experience to improve its performance by creating different paths to connect the neurons of the network. Applications of ANN include classification, pattern recognition, control systems, and image processing. An ANN consists of three main components, namely, the input layer, the hidden layer, and the output layer. Each layer consists of sets of neurons that are interconnected by training parameters (weight and bias). Each neuron consists of a processing element with synaptic input connections based on the number of process neurons in the previous layer.

Figure 9.2 shows an example of the signal transmission between neurons in the different layers. The input layer receives input samples and then transmits the signal to the hidden layer. The hidden layer consists of a certain number of neurons that play an essential role as a bridge between the input layer and the output layer. Each neuron in the preceding layers is fully connected to the next layers, and the connections are based on the training. The signal transmission is based on two functions: summation function and activation functions.

With the application of the theory of ANNs, Section 9.3 will present a case study where virtual data was created from real monitoring data.

In this work, the architecture of the ANN includes three layers (input layer, hidden layer, and output layer). Input data includes data collected from points that may in the future become permanent installation points for sensors. In this case, the input data is the data from the reference point (red point in the measuring grid; see Figure 9.5). The output data is, in turn, the data at the other

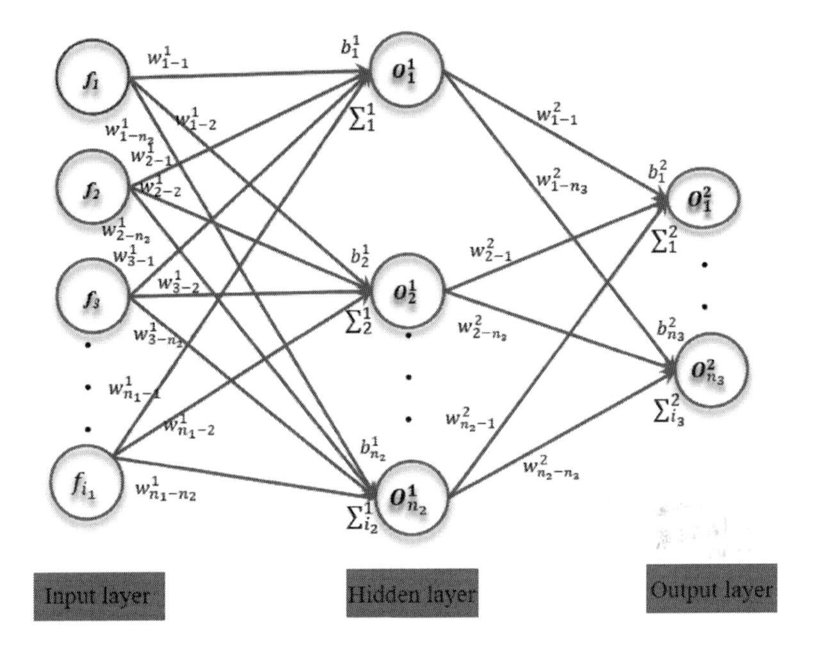

FIGURE 9.2 Example of an artificial neural network (ANN) architecture.

measurement points. The training of the network is repeated with different measurement points. Figure 9.3 shows the architecture of the ANN used in this study.

The number of neurons in the hidden layer heavily influences the results of the process of training the network. It is assumed that the choice of how many neurons depends on the specific problems that need to be solved. If the ANN has too few hidden layers, the ANN is too simple, which makes it difficult to deal with

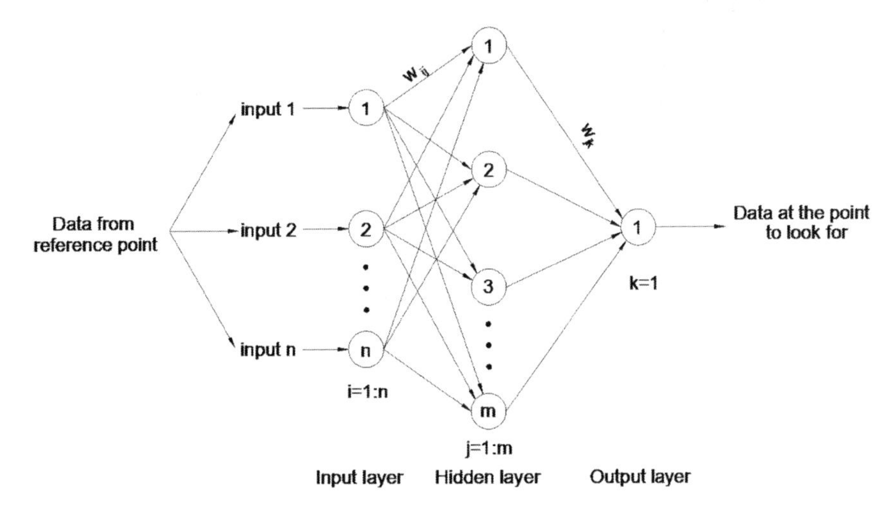

FIGURE 9.3 Architecture of the ANN used in this research.

the problems to be solved. In contrast, if the ANN has too many hidden layers, the network is too complex, which leads to overconsumption of computer resources, which in turn can easily lead to overfitting. A loop process is performed to select the most suitable number of hidden layers. The number of hidden layers for the loop process will range from 1 to 50. An ANN employs the Levenberg–Marquardt backpropagation algorithm to train the network. Data split in the training process 70–15–15% is used for damage identification. The maximum number of epochs is set to 1000. These are default parameters and are suitable for large datasets [23]. In the case of epochs greater than 1000 but still not reaching the best value, it is necessary to implement network optimization solutions.

9.3 CASE STUDY

9.3.1 Collecting Data for Network Training

9.3.1.1 Introduction to Kien Bridge, Vietnam

Kien Bridge (Figure 9.4) is a large-span cable-stayed bridge spanning the Cam River. This bridge, located on National Highway 10, connects the Thuynguyen and Anduong districts in Haiphong City, Vietnam. The bridge has two lanes

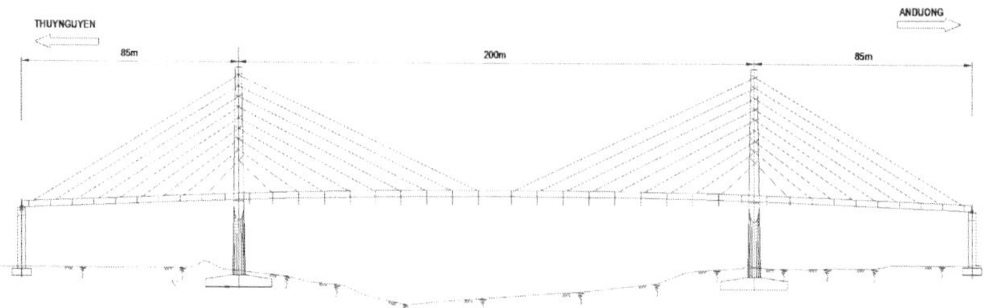

FIGURE 9.4 Kien Bridge, Haiphong, Vietnam.

and sidewalks on both sides for pedestrians. The main structure is a continuous pre-stressed concrete box girder cable-stayed bridge with three spans (85 + 200 + 85 m). The approach bridge consists of 12 simply supported pre-stressed concrete spans, each of which is 34 m long. The main cable-stayed bridge was built using the balanced cantilever method. The overall width of the bridge is 16.7 m. The cross section of the main bridge is a three-cell box girder at a constant height of 2.2 m; meanwhile, the approach bridge consists of six pre-stressed concrete I-beams.

The bridge consists of two cable planes that are 17.6 m apart. The pillar has a height of 85 m and a solid cross section. Seventy-two cables are strung down diagonally to the sides of the pylons and anchored to the beams. The arrangement of stay cables at Kien Bridge is in a fan diagram. The angles of inclination of the cables range from 30.44° to 50.37°.

9.3.1.2 Collecting Vibration Data from Kien Bridge

In the first data collection, a comprehensive vibration measurement campaign was carried out on the entire main bridge section of Kien Bridge. The excitation sources during vibration measurement were the water flow and wind (or ambient load). A full-bridge measurement grid was designed to obtain the mode shape of the entire bridge and enough input data to train the network (Figure 9.5). The red points are the reference points, and the remaining points are the moving points. Points numbered as 103,104,301,303,306,308,501, and 502 and S1 5,9, T1 12,16,21,25 and T2 29 and 32 will later be considered for data to generate data for other points.

For the determination of the coordinates, a Cartesian coordinate system was used. The z-axis was vertical with a positive direction upward, the y-axis was in the transversal direction of the bridge, and the x-axis was in the longitudinal direction. At each bearing, two sensors with an x-axis and y-axis were installed. At other points, the sensors were placed in the y-axis or z-axis or in the y- and z-axes. With an optimized design done, at least 57 sensors were found to be needed. A laptop was used to control the measurement process and collect

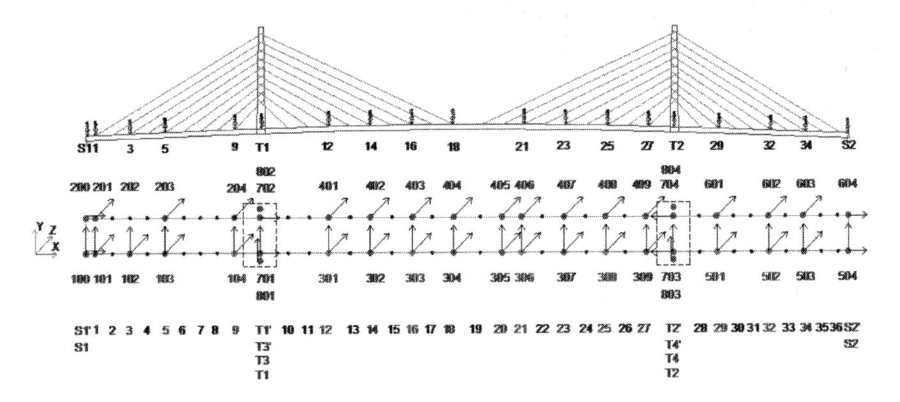

FIGURE 9.5 Arrangement of the measurement grid on Kien Bridge.

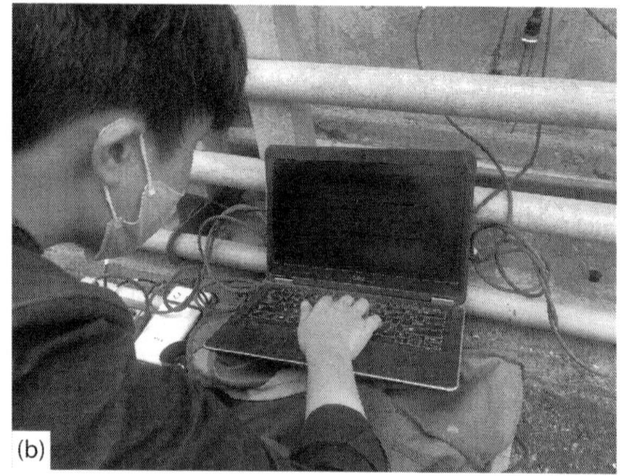

FIGURE 9.6 Collecting vibration data from Kien Bridge. (a) Installing sensors on the bridge; (b) collecting and storing data on a computer.

dynamic responses (Figure 9.6). Rapid deployment, on-site accessibility, and wire length are important factors for planning measurement setups and ensuring the optimization of resources.

Vibration data for the whole bridge was collected in 20 minutes. The sampling frequency was 1651 Hz. In the process of data collection, it was necessary to pay attention to the quality of the data as well as to record the conditions when measuring so that noise could be eliminated during the training of the network. Figure 9.7 shows data over time for one specific measurement point and the combination of several measurement points.

9.3.2 Training the ANN

We performed network training as described in Section 9.2. The performance of the ANN model after training is shown in Figures 9.8–9.10, for a representative point.

Figure 9.8 shows the regression values of all training cases that are higher than 0.99. The training, evaluation, and test datasets are located along the target line (45° line). This shows that the predicted value and the actual value are similar. With linear regression models, the regression values (R – is coefficient of regression) range from 0 to 1. The calculated and target results are equivalent if R is close to the upper bound (1). Figure 9.9 presents the histogram of the errors of the calculated and the desired outputs. The zero-error line locates the position where calculated and target results are analogous. The difference between the target and the output is very small. Figure 9.10 shows the training performance in the datasets. The best validation performance value of 0.00354 is achieved at epoch 109. That graph also shows that the values in the datasets are quite convergent and no overfitting phenomenon occurs. The performance graph of the R values, MSE (mean squared error, or tolerance), and error histogram shows that the model has been trained successfully.

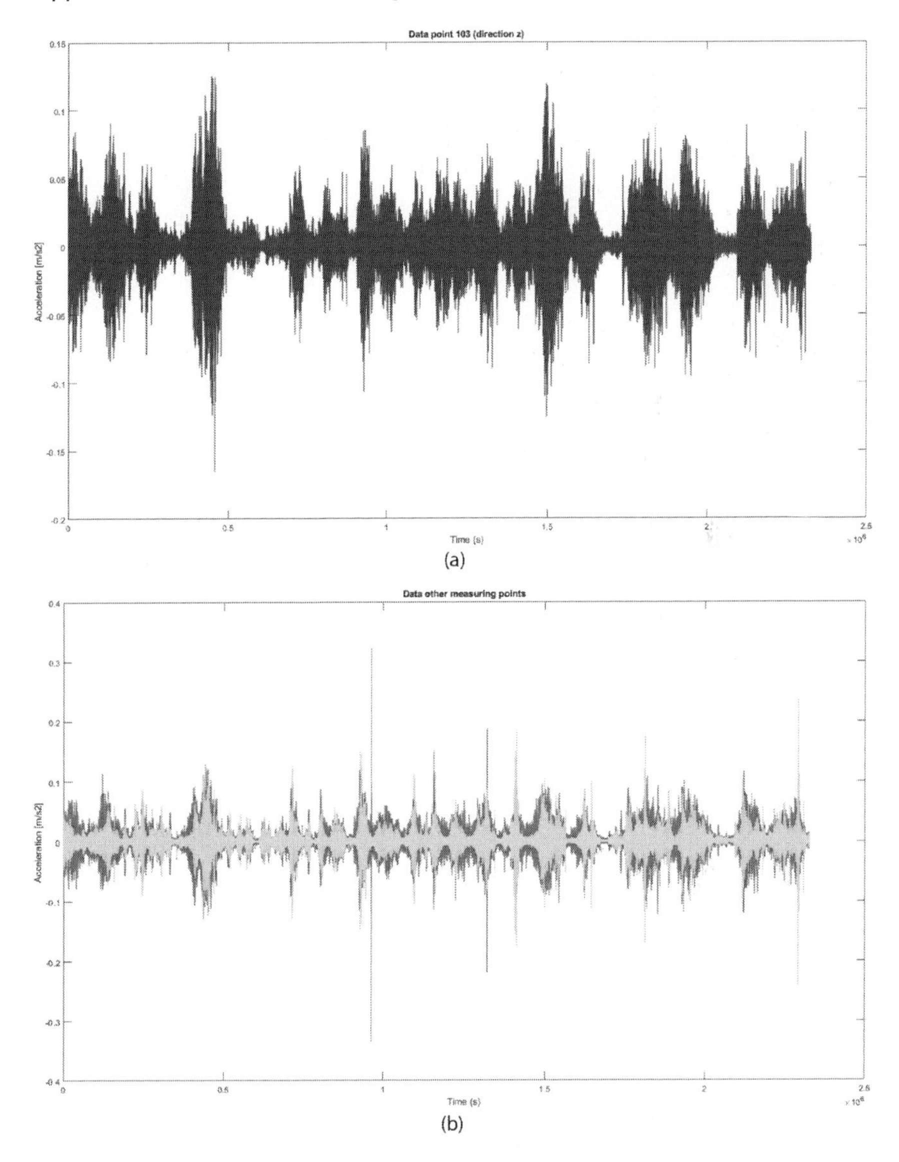

FIGURE 9.7 Vibration data collection: (a) Point 103 (direction z), and (b) other measuring points.

9.3.3 Creating Virtual Data and Evaluating the Results

In this section, data from Kien Bridge was collected for the second time. The second data collection was performed in the same way as the first time. In this implementation, the data was fully measured and analyzed. The final result was the mode shape of the bridge. Then, some data of the points would be hidden and recreated using the network trained in Section 9.3.2. The input is the data at the reference point. The trained network returns the data of the hidden point. Virtual

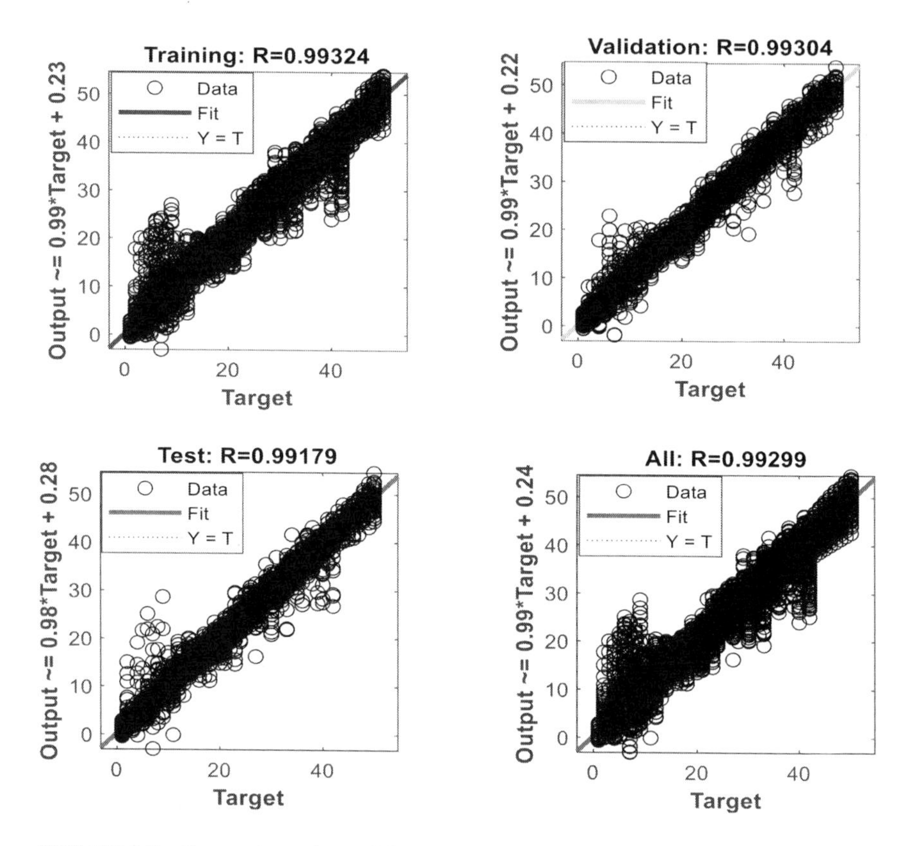

FIGURE 9.8 Regression values of the ANN.

data is generated and used for analysis similar to real data. The analysis results of two real and virtual datasets are used to evaluate the performance of the method. Figure 9.11 shows the data at measuring point 103 of the second measurement. Figure 9.12 shows the virtual data at point 102 (representative point) generated from the trained network.

With the two datasets obtained, processing was carried out using the MACEC tool [24, 25]. The procedure of data processing is described in Figure 9.13.

The stochastic subspace identification (SSI) algorithm is employed to perform system identification for the output-only (also operational) modal analysis (OMA) of structures. The dynamic characteristics are estimated by a stabilization diagram. Figure 9.14 presents a stabilization diagram of the two datasets under consideration.

After processing vibration measurement data in the time domain by parametric method through the program to identify vibration characteristics by the MACEC tool, the summary table of natural frequency results is listed in Table 9.1.

In both cases of real datasets and virtual data, the standard deviations of natural frequencies (*std*.f) were calculated to evaluate the quality of the identified modes.

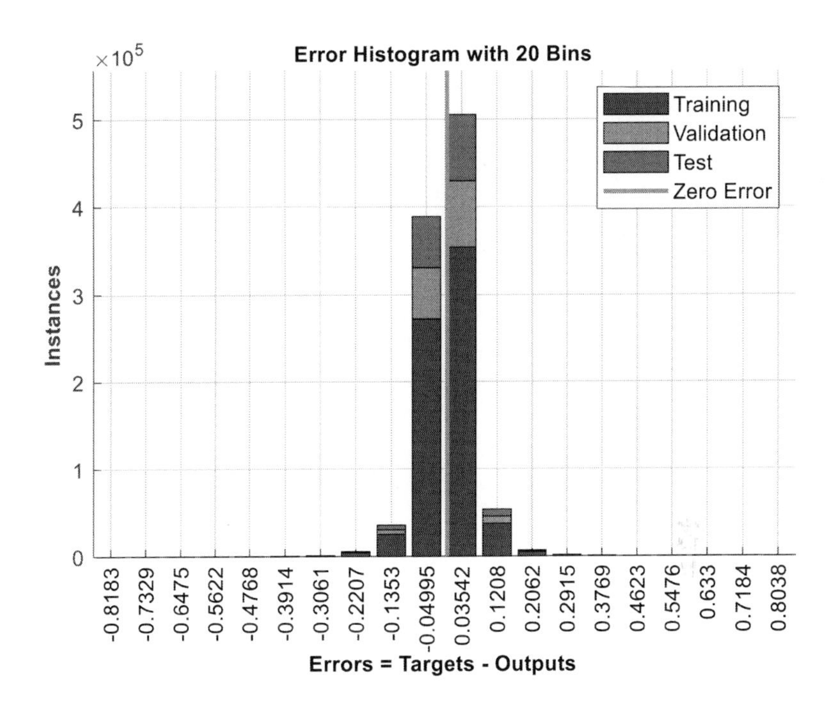

FIGURE 9.9 Error histogram of the ANN.

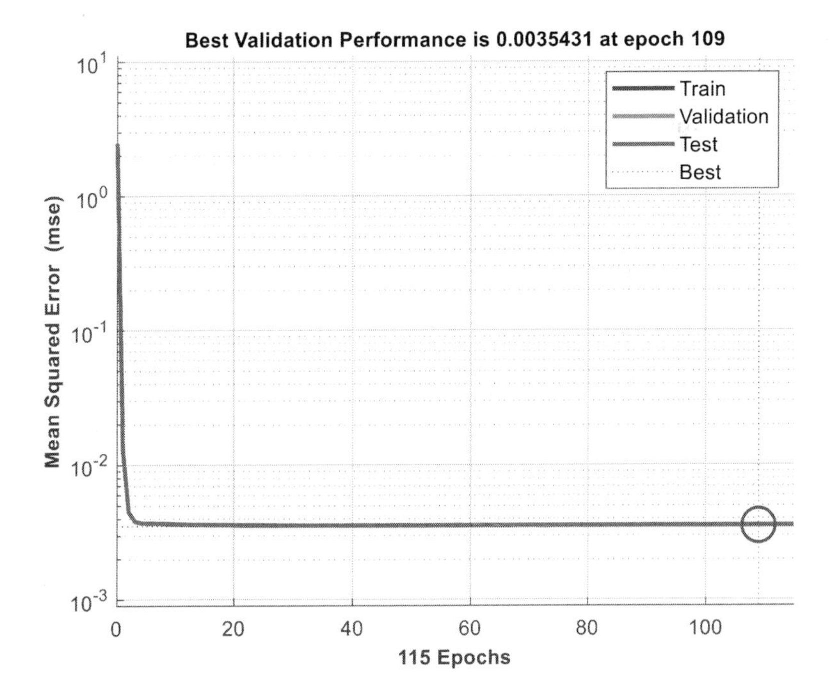

FIGURE 9.10 Tolerance of the network.

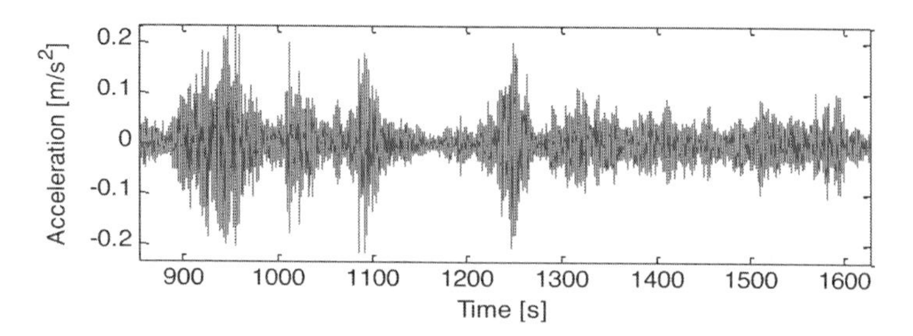

FIGURE 9.11 Data collected from a representative point (point 103).

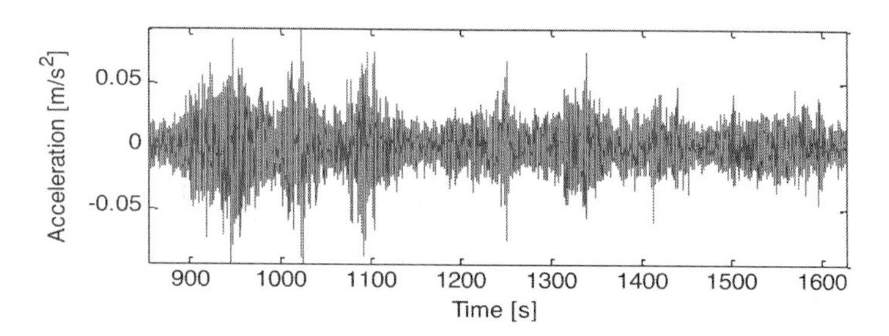

FIGURE 9.12 Virtual data at point 102 generated from the trained network.

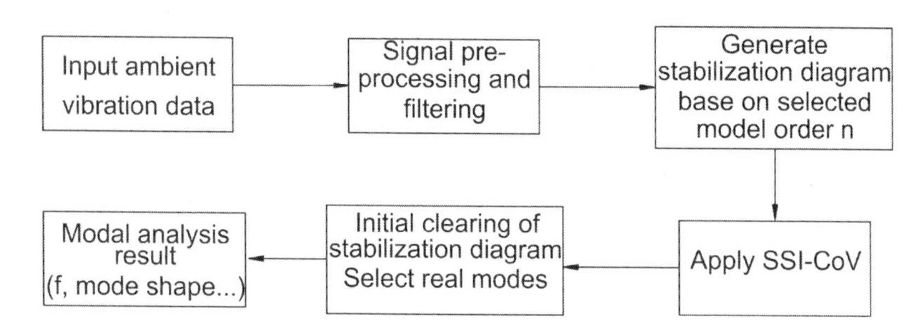

FIGURE 9.13 Process of system identification.

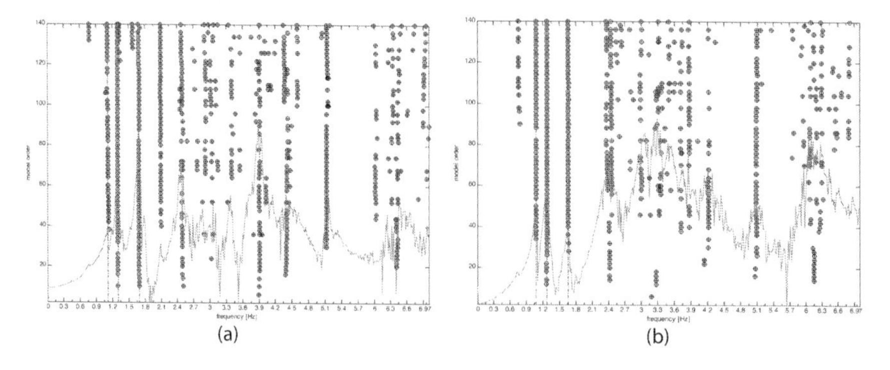

FIGURE 9.14 Stabilization diagram from 0 to 7 Hz: (a) real data and (b) virtual data.

TABLE 9.1
Summarization of Natural Frequency Values of the Five Identified Modes

Mode	Real Data	Virtual Data (Hz)	Error (%)	Mode Shape
1	0.452	0.477	5.24	Vertical bending 1
2	0.766	0.806	4.96	Vertical bending 2
3	1.197	1.252	4.39	Vertical bending 3
4	1.293	1.373	5.82	Vertical bending 4
5	1.691	1.589	6.41	Vertical bending 5

Each setup's system identification quality is good because the values of *std.f* are negligible. Modal phase collinearity (MPC) represents the deviation from a real valued mode shape, with MPC = 1 corresponding to a pure real mode. All MPC values of real datasets and virtual data are greater than 0.998. A high MPC value is generally a good indication of an accurately identified mode. The mode shapes from the two datasets are analyzed and shown in Figures 9.15–9.19.

The moderate assurance criterion (MAC) is a statistical indicator that is most sensitive to large differences and relatively sensitive to small differences in oscillator shapes. This gives a good statistical indicator and a degree of consistency

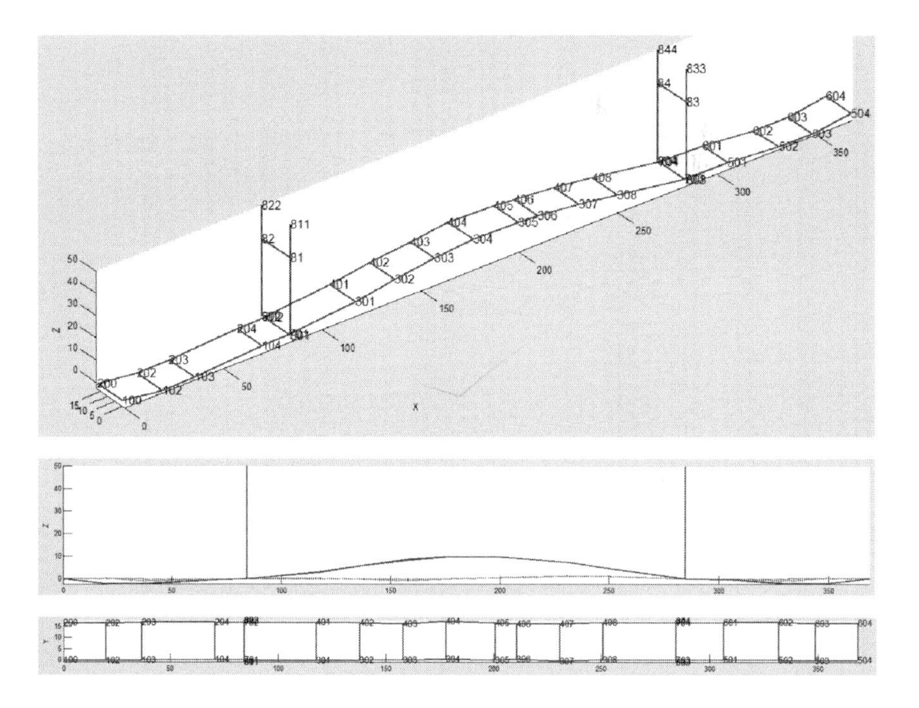

FIGURE 9.15 Mode shape 1: Vertical bending 1.

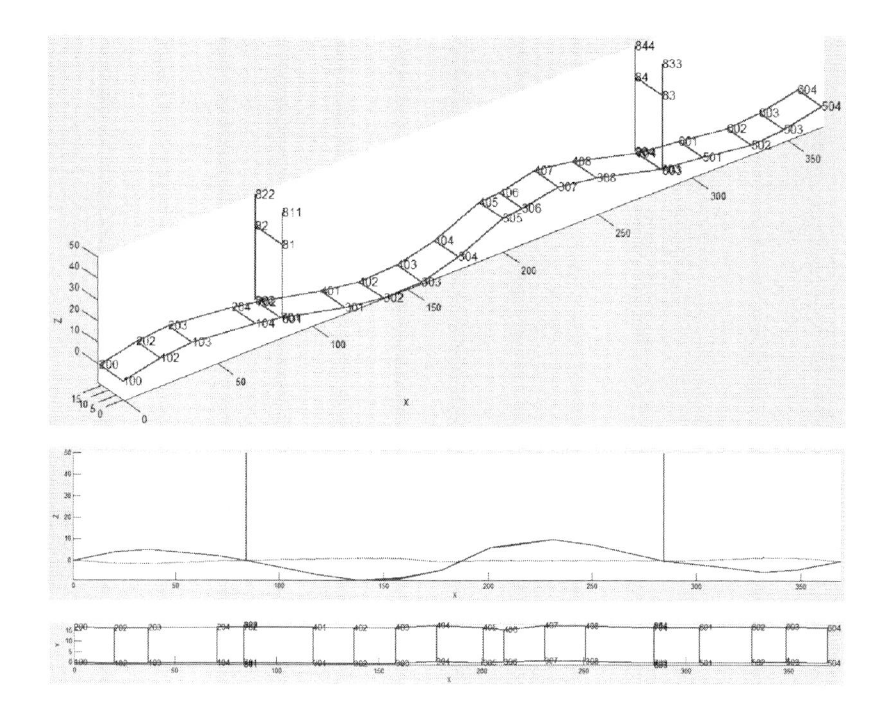

FIGURE 9.16 Mode shape 2: Vertical bending 2.

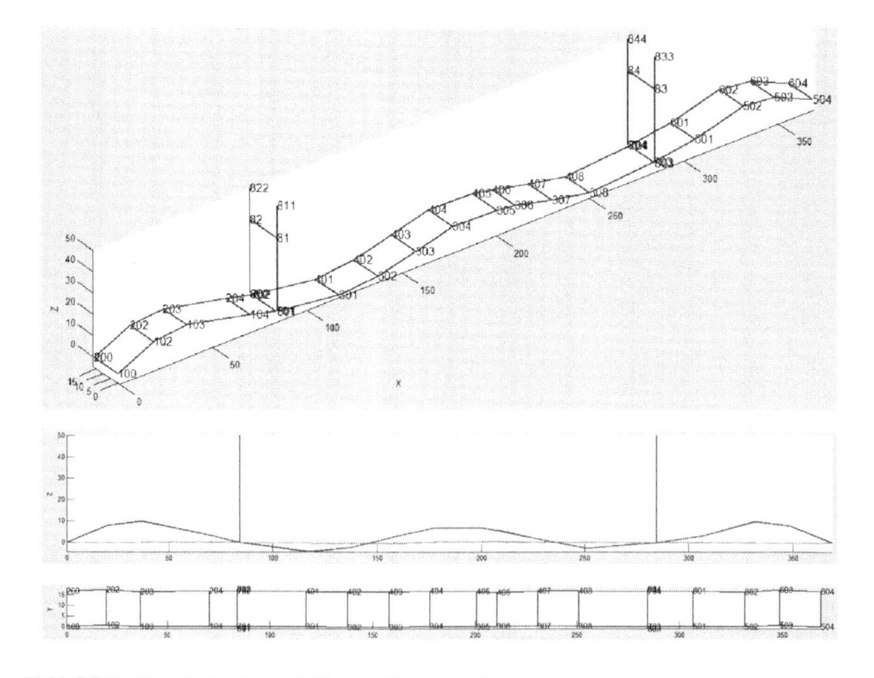

FIGURE 9.17 Mode shape 3: Vertical bending 3.

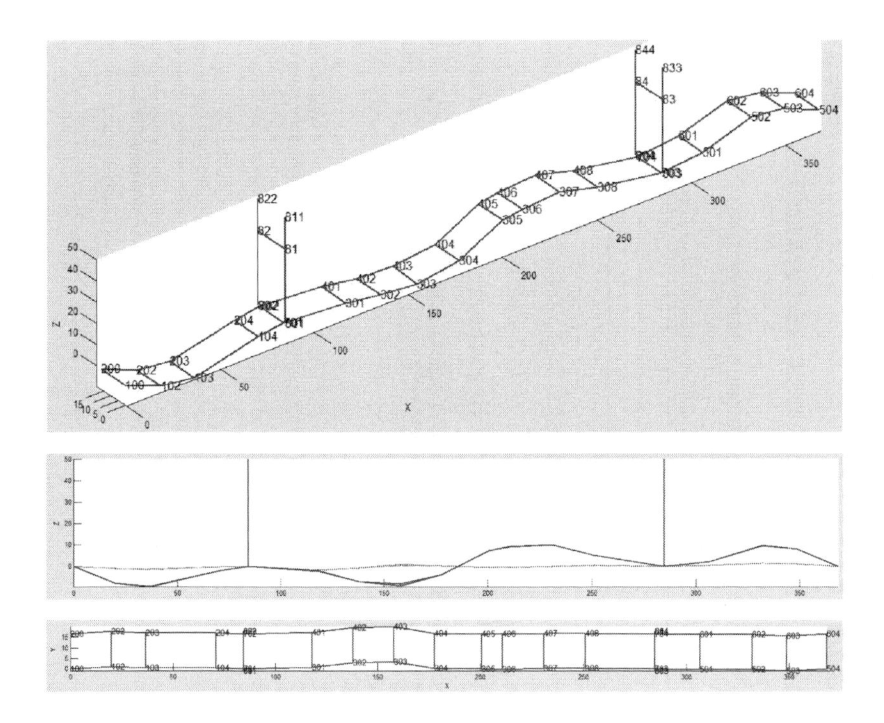

FIGURE 9.18 Mode shape 4: Vertical bending 4.

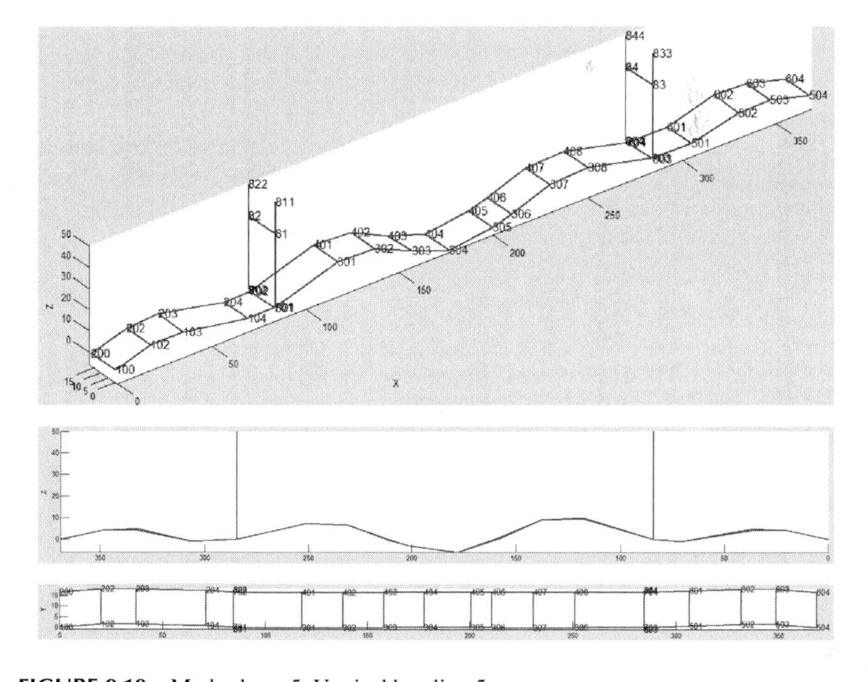

FIGURE 9.19 Mode shape 5: Vertical bending 5.

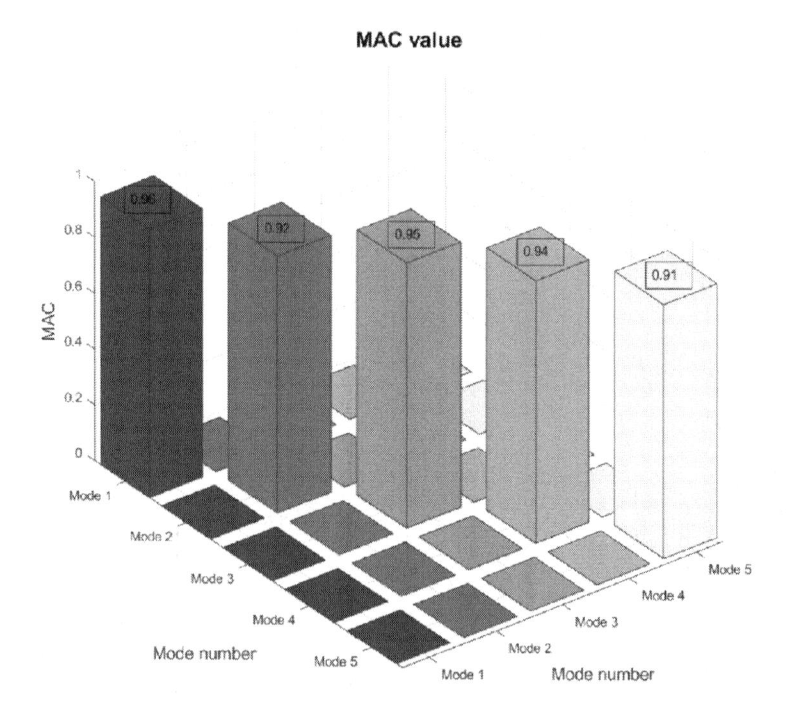

FIGURE 9.20 The moderate assurance criterion (MAC) value between a real dataset and a virtual dataset.

between mode shapes. In this study, the MAC value is used to evaluate two oscillation shapes of two real and virtual datasets (Figure 9.20).

The results of vibration analysis show that two real datasets and virtual data have negligible errors (the largest error at natural frequency is 6.41%). The mode shapes obtained from the two datasets are the same and have a high degree of similarity (shown by the MAC value and the analyzed mode shape).

9.4 CONCLUSIONS

Through an AI-based approach, this study presents a solution to generate data where it is needed. Through testing and practical application at Kien Bridge, the method achieved positive results. The main conclusions are:

- ANNs can be used to create virtual data for structural health monitoring using dynamic methods.
- Although performing network training using measured data is relatively time-consuming, the generated dataset is quite similar to the real dataset. Training time with the amount of data like the Kien Bridge is up to 40 minutes for one data point (to train for all moving points, it takes up to 27 hours). The generated virtual dataset has high accuracy; the degree

of coincidence with the actual dataset is high. This is shown by the MAC value when comparing the results of real and virtual data (no MAC value lower than 0.9).

- Since they do not contain as much information as real datasets (real datasets also include system-related data), the generated virtual datasets take up less storage space (mainly in the form of matrix numbers). Virtual datasets take up only about 50% of the storage space of real data.
- The method can be applied to minimize the number of sensors in the case of continuous monitoring of structures, minimizing the cost of structural health monitoring for large structures.

ACKNOWLEDGMENTS

This research was funded by FCT/MCTES through national funds (PIDDAC) from the R&D Unit Institute for Sustainability and Innovation in Structural Engineering (ISISE), under the reference UIDB/04029/2020, and from the Associate Laboratory Advanced Production and Intelligent Systems ARISE, under the reference LA/P/0112/2020. This research was supported by a doctoral grant (reference no. PRT/BD/154268/2022) financed by the Portuguese Foundation for Science and Technology (FCT), under the MIT Portugal Program (2022 MPP2030-FCT). The second author acknowledges the funding by FCT through the Scientific Employment Stimulus—4th Edition.

REFERENCES

1. C. R. Farrar, S. W. Doebling, Damage Detection and Evaluation II. In *Modal Analysis and Testing*. NATO Science Series, vol 363. Springer, Dordrecht. Doi: 10.1007/978-94-011-4503-9_17.
2. E. P. Carden, P. Fanning, "Vibration based condition monitoring: A review," *Structural Health Monitoring*, doi: https://doi.org/10.1177/1475921704047500.
3. F. N. Catbas, D. L. Brown, A. E. Aktan, "Use of modal flexibility for damage detection and condition assessment: Case studies and demonstrations on large structures," *American Society of Civil Engineers*, Volume 132, Issues 11, 2006, Pages 1699, doi: https://doi.org/10.1061/(ASCE)0733-9445.
4. A. Alvandi, C. Cremon, "Assessment of vibration-based damage identification techniques," *Journal of Sound and Vibration*, Volume 292, Issues 1–2, 25 April 2006, Pages 179–202, doi: https://doi.org/10.1016/j.jsv.2005.07.036.
5. O. Avci et al., "A review of vibration-based damage detection in civil structures: From traditional methods to Machine Learning and Deep Learning applications," *Mechanical Systems and Signal Processing*, Volume 147, 15 January 2021, Pages 107077, doi: https://doi.org/10.1016/j.ymssp.2020.107077.
6. H.- P. Chen, Y.- Q. Ni, *Structural Health Monitoring of Large Civil Engineering Structures*. John Wiley & Sons Ltd. [Online]. doi: https://doi.org/10.1002/9781119166641.
7. J. P. Amezquita-Sanchez, H. Adeli, "Signal processing techniques for vibration-based health monitoring of smart structures," *Archives of Computational Methods in Engineering*. 2016, doi: https://doi.org/10.1007/s11831-014-9135.

8. R. P. Bandara, T. H. T. Chan, D. P. Thambiratnam, "Frequency response function based damage identification using principal component analysis and pattern recognition technique," *Engineering Structures*. 2014, doi: https://doi.org/10.1016/j.engstruct.2014.01.044.
9. B. Kostic, M. Gül, "Vibration-based damage detection of bridges under varying temperature effects using time-series analysis and artificial neural networks," *Journal of Bridge Engineering*. 2017, doi: https://doi.org/10.1061/(ASCE)BE.1943-5592.0001085.
10. N. T. C. Nhung, T. Q. Minh, J. C. Matos, H. S. Sousa, "Research and application of indirect monitoring methods for transport infrastructures to monitor and evaluate structural health," in *Recent Advances in Structural Health Monitoring and Engineering Structures - Select Proceedings of SHM&ES 2022*.
11. V. M. Karbhari, L. S.- W. Lee, "Vibration-based damage detection techniques for structural health monitoring of civil infrastructure systems," *Structural Health Monitoring of Civil Infrastructure Systems*, Woodhead Publishing Series in Civil and Structural Engineering, doi: https://doi.org/10.1533/9781845696825.1.177.
12. C. R. Farrar, S. W. Doebling D. A. Nix, "Vibration–based structural damage identification," *Royal Society*, Volume 359, Issue 1778, doi: https://doi.org/10.1098/rsta.2000.0717.
13. Y. Xia, H. Hao, J. M. W. Brownjohn, P.-Q. Xia, "Damage identification of structures with uncertain frequency and mode shape data," doi: https://doi.org/10.1002/eqe.137.
14. J.-M Ndambi, J. Vantomme, K Harri, "Damage assessment in reinforced concrete beams using eigenfrequencies and mode shape derivatives," *Engineering Structures*, Volume 24, Issue 4, April 2002, Pages 501–515, doi: https://doi.org/10.1016/S0141-0296(01)00117-1.
15. N. M. M. Maia, J. M. M. Silva, E. A. M. Almas, R. P. C. Sampaio, "Damage detection in structures: From mode shape to frequency response function methods," *Mechanical Systems and Signal Processing*, Volume 17, Issue 3, May 2003, Pages 489–498, doi: https://doi.org/10.1006/mssp.2002.1506.
16. Z. Ismail, H. A. Razak, A. G. A. Rahman, "Determination of damage location in RC beams using mode shape derivatives," *Engineering Structures*, doi: https://doi.org/10.1016/j.engstruct.2006.02.010.
17. M. Fayyadh, H. A. Razak, "Detection of damage location using mode shape deviation: Numerical study," *International Journal of Physical Sciences*, Volume 6, Issue 24, Pages 5688–5698, doi: https://doi.org/10.5897/IJPS11.971.
18. M. Dahak, N. Touat, M. Kharoubi, "Damage detection in beam through change in measured frequency and undamaged curvature mode shape," *Inverse Problems in Science and Engineering*, Volume 27, Issue 1, 2019, doi: https://doi.org/10.1080/17415977.2018.1442834.
19. S. W. Doebling, C. R. Farrar, M. B. Prime, D. W. Shevitz, "Damage identification and health monitoring of structural and mechanical systems from changes in their vibration characteristics: A literature review." [Online]. Available: https://doi.org/10.2172/249299.
20. G. Sha, M. Radzienski, R. Soman, M. Cao, W. Ostachowicz, W. Xua, "Multiple damage detection in laminated composite beams by data fusion of Teager energy operator-wavelet transform mode shapes," *Composite Structures*, doi: https://doi.org/10.1016/j.compstruct.2019.111798.
21. UCT Company, "Report on inspection of Chuong Duong bridge in 2022."
22. V. Flovik, "How (not) to use Machine Learning for time series forecasting: Avoiding the pitfalls." Towards Data Science. [Online]. Available: https://towardsdatascience.com/

23. MathWorks, "Divide data for optimal neural network training." [Online]. Available: https://www.mathworks.com/
24. E. Reynders, M. Schevenels, G. De Roeck, Macec - The MATLAB toolbox for experimental and operational modal analysis, vol. MACEC 3.4. [Online]. Available: https://bwk.kuleuven.be/bwm/macec/macec.pdf
25. E. Reynders, M. Schevenels, G. D. Roeck, MACEC 3.2: A MATLAB toolbox for experimental and operational modal analysis, Department of Civil Engineering, Department of Civil Engineering, KU Leuven.

10 Ambient Vibration Prediction of a Cable-Stayed Bridge by an Artificial Neural Network

Melissa De Iuliis, Cecilia Rinaldi, Francesco Potenza, Vincenzo Gattulli, Thibaud Toullier, and Jean Dumoulin

10.1 OVERVIEW

Large-scale civil infrastructures play a vital role in society, as they ensure smooth transportation and improve the quality of people's daily life. However, they are exposed to several continuous external dynamic actions such as wind loads, vehicular loads, and environmental changes. Interaction assessment between external actions and civil structures is becoming more challenging due to the rapid development of transportation. Data-driven models have lately emerged as a viable alternative to traditional model-based techniques. They provide different advantages: timely damage detection, prediction of structural behaviors, and suggestions for optimal maintenance strategies. This chapter aims to describe the advantages and the characteristics of data-driven techniques to predict the dynamic behavior of civil structures through an artificial neural network (ANN). The applicability and effectiveness of the proposed approach are supported by the results achieved by processing the measurements coming from a monitoring system installed on a cable-stayed bridge (the Éric Tabarly Bridge in Nantes, France). Accelerations recorded by a network of 16 mono-axial accelerometers and Nantes Airport weather data acquired with the observation platform of the METAR (MEteorological Terminal Aviation Routine Weather Report) Station Network have been used as training to predict the structural response and to statistically characterize the behavior through a nonlinear autoregressive (NAR) prediction network. The performance has been evaluated through statistical analysis of the error between the measured and predicted values also related to both environmental conditions and the number of signals. The results show that the forecast network could be useful to detect the trigger of anomalies, hidden in the dynamic response of the bridge, at a low computational cost.

DOI: 10.1201/9781003306924-10

10.2 INTRODUCTION

Engineering structures and infrastructures, such as bridges, gradually suffer from aging because of different loads and environmental conditions. Proper and periodic assessment operations are paramount for structural safety to prevent the effects of environmental hazards, reduce maintenance costs, and improve the level of safety of bridges. Therefore, structural identification has become an increasingly intensive research topic for structural health monitoring (SHM), damage evaluation, and safety assessment of existing engineering structures (Lee et al. 2006; Lee and Shinozuka 2006; Ribeiro et al. 2014; Yu et al. 2017; Domaneschi et al. 2021; Gattulli et al. 2021; Gattulli et al. 2022b). A significant number of studies of SHM systems focusing on damage detection of bridges can be found in the literature (Lee et al. 2007; Magalhães et al. 2010). For instance, Dutta and Talukdar (2004) proposed a modal-based approach in a finite-element framework for damage detection of complex structures like bridges. Magalhães et al. (2010) presented a statistically based damage detection ability through short-term vibration-monitoring datasets. Liu et al. (2009) evaluated the structural performance of existing bridges using a strain-based monitoring system and a condition assessment of structural components. A strain-based monitoring system was integrated into a structural performance assessment of a steel girder bridge by Orcesi and Frangopol (2010), who used the monitoring data to carry out a structural reliability analysis of critical sections. A load rating of a bridge was evaluated through diagnostic load testing based on a strain-based SHM system by Phares et al. (2003).

Although many studies have been conducted on SHM systems to predict the behavior of engineering structures, traditional techniques are mostly based on dynamical models of the structure. The performance of conventional approaches strictly depends on the accuracy of the analytical and numerical dynamical models that are, in most cases, very challenging to derive (Gattulli et al. 2016; Gattulli et al. 2019). Furthermore, the development of such models and the definition of the parameters required to represent the behavior of structures may be time-consuming and computationally intensive. With the recent developments in computer technology for communication, data acquisition, and signal processing and analysis, an alternative solution to structural identification can be data-driven models (Gattulli et al. 2022a). Unlike the traditional physics-based SHM models, data-driven models exploit patterns arising from spatial and temporal correlations in measurements and offer bottom-up solutions that include damage detection, remaining life estimation, and also structural control (Zhao et al. 2019; Di Girolamo et al. 2020). In addition, data-driven techniques require a training dataset comprising measurements that represent the baseline conditions of a bridge (Kromanis and Kripakaran 2017). Consequently, data-driven models have recently drawn considerable attention in the civil engineering community. ANNs are among the most common machine learning techniques employed for structural system identification and model updating (Park et al. 2020). Barai and Pandey (1995) presented an ANN-based approach for damage identification in steel bridges. A model updating scheme with an adaptive neural network was proposed by Chang et al. (2000) for structural health assessment methodologies and control strategies. In their work, the model updating and network training were repeated until achieving a good agreement between the calculated and measured modal responses of the bridge. In Fang et al. (2005), ANNs were employed to study the identification of structural damage in beams. A cantilever beam was divided into a given number of elements, and the damage was associated with loss of stiffness in one or more elements. Bakhary et al. (2007) detected simulated damage in a numerical steel portal frame model through ANNs, accounting also for uncertainties derived from the model and the measured vibration data. A methodological framework for predicting the dynamic responses of the vehicle–bridge interaction system using ANNs was illustrated by Li et al. (2021). Furthermore, de Oliveira Dias Prudente dos Santos et al. (2016) presented a neural network (NN) to simulate and predict the structural behavior of a bridge using the air temperature as an input to the network. Taking into account the changes of the environmental conditions (e.g., climate, traffic load, and degradation mechanisms) is paramount in the SHM process and the development of fault detection methods. Yun-Lai and Wahab (2017) employed an auto-associative NN combined with transmissibility to identify damage location in a 10-floor structure. Furthermore, in their study, the effect of noise on natural frequencies was evaluated. A backpropagation algorithm in an ANN was used to apply the change of static properties (i.e., strains and displacements) for damage detection in a cantilever beam (Maity and Saha 2004). Cascardi et al. (2017) employed ANNs to estimate the strength of fiber-reinforced polymer (FRP)-confined concrete, showing that the proposed model could provide a good agreement with measurements. A novel data-driven measurement approach to predict traffic-induced and thermal responses of bridges was presented by Kromanis and Kripakaran (2017). In their work, prediction error signals are created and then interpreted with anomaly detection techniques. Wang et al. (2022) developed a long short-term memory (LSTM) approach for condition assessment of bridges to provide an early warning protocol through analysis of the time series measured data of deflection and temperature.

Among different types of recurrent NNs, time delay neural networks (TDNNs) (Haykin and Lippmann 1994; Ji and Chee 2011), layer recurrent networks (Haykin and Lippmann 1994), and NAR networks (Chow and Leung 1996; Markham and Rakes 1998) have been widely used in time series forecasting. A TDNN is a straightforward dynamic network consisting of a feedforward network with a tapped delay line at the input layer. A NAR network is a dynamic recurrent network with feedback connections including different layers of the network. In Benmouiza and Cheknane (2016), hybrid approaches on time series data using an autoregressive moving average (ARMA) with a TDNN or an ARMA with a NAR network are compared, and results presented showed that a NAR network tends to perform well in providing more precise results for multi-step-ahead prediction (Benmouiza and Cheknane 2016).

In the framing of the research field presented above, this study is concerned with investigation and implementation of ANNs for dynamic behavior prediction of bridges. A SHM system, installed in a cable-stayed bridge (the Éric Tabarly Bridge in Nantes, France), is used as a case study to demonstrate and verify the method. First, accelerations recorded by 16 sensors and the Nantes Airport's METAR (MEteorological Terminal Aviation Routine Weather Report) weather measurements station data, which provide the weather observer's interpretation of the weather conditions, are used as training to predict the structural response and to statistically characterize the behavior through a NAR prediction network. Statistical analysis of the error between measured and predicted values is performed. It is shown that the proposed network could constitute a useful tool to detect the presence of anomalies in the dynamic response of a bridge at a reasonable computational cost.

The chapter is organized as follows: Section 10.3 briefly introduces ANNs and their underlying computational methodologies. Section 10.4 describes the geometrical and structural features of the cable-stayed bridge used as the case study. Furthermore, the characteristics of a vibration-based SHM system installed on the bridge are illustrated. Section 10.5 discusses the results obtained by the application of the data-driven approach. Finally, conclusions and future developments are drawn.

10.3 ARTIFICIAL NEURAL NETWORKS (ANNs): BASIC PRINCIPLES

ANNs have been developed to model complex relationships where explicit formulae are difficult to carry out (Chang et al. 2000). ANNs are defined as nonlinear nonparametric systems consisting of three main components: architecture, a learning algorithm, and activation functions. Multilayered feedforward NNs are the most widely used network models in structural engineering applications. A typical NN comprises an input layer, an output layer, and one or more hidden layers of neurons (computational units) that are interconnected through weighted connections. Figure 10.1 schematizes the architecture of a NN, along with the layers (i.e., input and output layers) and neuron numbers (Hasançebi and Dumlupınar 2013; Bal and Demir 2021).

Let us consider a neuron j with n_i inputs, as illustrated in Figure 10.2. If neuron j is in the first layer, all of its inputs would be connected to the inputs of the network. If connections across layers are allowed, its inputs can be connected to outputs of other neurons or to network inputs. Node y can have two indices ($y_{j,i}$) as the ith input of neuron j, or it can be implemented as y_j to define the output of neuron j (with one index).

The output node of neuron j is computed as:

$$y_j = f_j\left(net_j\right) \tag{10.1}$$

where f_i is the activation function of neuron j, and net value net_j is the sum of weighted input nodes of neuron j:

$$net_j = \sum_{i=1}^{n} w_{j,i} y_{j,i} + w_{j,0} \tag{10.2}$$

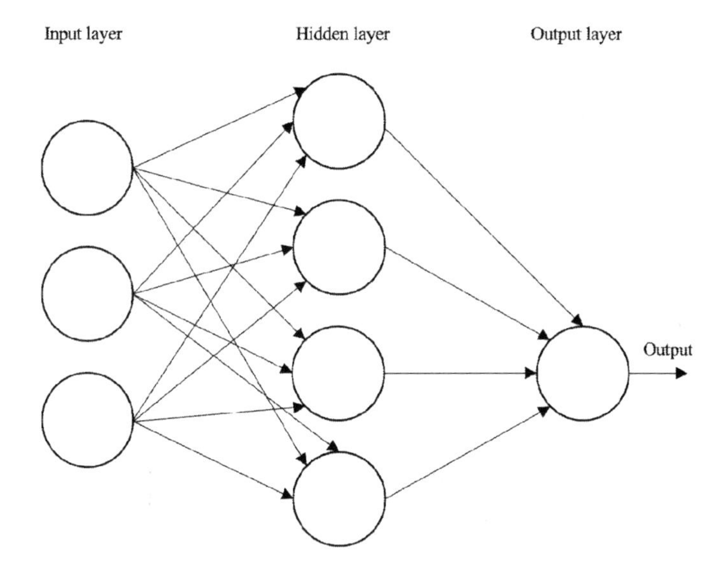

FIGURE 10.1 Artificial neural network architecture. (Adapted from Bal and Demir 2021.)

where $y_{j,i}$ is the ith input node of neuron j, weighted by $w_{j,i}$; and $w_{j,0}$ is the bias weight of neuron j.

Activation functions are linear and nonlinear components of NNs. They are responsible for mapping between input and target values. The S-shaped sigmoidal functions, such as tangent-sigmoid and logistic-sigmoid functions, are widely used as activation functions due to their nonlinear mapping ability within the hidden layer of the network. Usually, the output layer contains linear activation functions such as step and purelin functions (Bal and Demir 2021). The activation

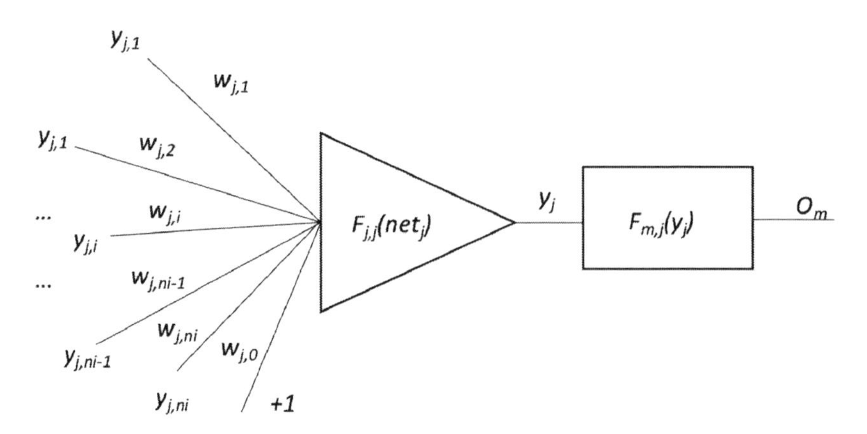

FIGURE 10.2 Connection of a neuron j with the rest of the network. (Extracted from Yu and Wilamowski 2011.)

function in NNs takes the input $y_{j,i}$ multiplied by its weight $w_{j,i}$. Bias allows shifting of the activation function by adding a constant to the input, and it has the effect of shifting the activation function by a constant amount ($w_{j,0}$). A complex nonlinear relationship between the output node y_j of a hidden neuron j and the network output o_m can be defined:

$$o_m = F_{m,j}(y_j) \tag{10.3}$$

where o_m is the mth output of the network. The complexity of the mentioned nonlinear function $F_{m,j}(y_j)$ depends on the number of neurons between neuron j and network output m.

10.3.1 Training Algorithms

The training of a network is performed to adjust the weight connections between the considered neurons. During the training process, the input data is given to the network, the estimated output of the network is computed by using the current value set of weight coefficients, and it is compared with the desired output. Then, the error between the estimated and desired output of the network is calculated, and it is used to adjust the weight coefficients (initially set based on the Nguyen–Widrow initialization method; Nguyen and Widrow 1990; Pavelka and Procházka 2004). In NNs, the learning algorithm is the component of the training process that makes NNs learn from data lags. Learning algorithms of NNs are basically backpropagation algorithms that use error functions derivatives as gradients. To evaluate the training process for all training patterns and network outputs, the sum square error (SSE) is implemented as:

$$E(x,w) = \frac{1}{2}\sum_{p=1}^{P}\sum_{m=1}^{M} e_{p,m}^2 \tag{10.4}$$

where p is the index of patterns ranging from 1 to P (i.e., the number of patterns), m is the index of outputs ranging from 1 to M (i.e., the number of outputs), x is the input vector, w is the weight vector, and $e_{p,m}$ is the training error at output m when applying pattern p, and it is defined as:

$$e_{p,m} = d_{p,m} - o_{p,m} \tag{10.5}$$

where d is the desired output vector, and o is the output vector from the model.

A suitable algorithm for training small- and medium-sized problems in the ANN field is Levenberg–Marquardt backpropagation, which provides a numerical solution to the problem of minimizing a nonlinear function, which, in this case, is the error function E. This function uses the Jacobian for calculations, which assumes that performance is a mean or sum of squared errors. Therefore, networks trained with this function must use either the mean squared error (MSE) or SSE performance function.

The Levenberg–Marquardt algorithm combines two algorithms for the training process: around the area with complex curvature, it shifts to the steepest descent algorithm; and, when the local curvature becomes suitable for a quadratic approximation, it converts to the Gauss–Newton algorithm, which can significantly accelerate convergence (Marquardt 1963; Yu and Wilamowski 2011).

The prediction performance of a network strictly depends on the network parameters, training process, and quality of the training dataset. Generally, extensive parametric studies performed on different networks are necessary to get the best performance. Further details on ANNs as well as their implementation steps are provided in Hagan et al. (1997), Maier and Dandy (2000), Yun and Bahng (2000), Rafiq et al. (2001), Tang et al. (2003), and Feng et al. (2006).

10.3.2 Neural Network Time Series Prediction

Dynamic NNs are generally employed to forecast future values of a time series $y(t)$ only from past values of the actual time series. Time series prediction using the NN approach is nonparametric; therefore, it is not necessary to know any information about the process that generates the signal (Denton 1995; Markham and Rakes 1998; Zhang 2003). This form of prediction is known as the NAR model. The NAR network is a feedforward NN with three layers: input, hidden, and output layers. The NAR NN is trained in a series-parallel configuration. In the training phase, the true output is provided, and it is used as the input to the network. In the testing phase, the computer output is fed back to the network to estimate the next value of the output in a parallel configuration.

The NAR network consists of feedback connections, or layers, and it is determined by the following equation:

$$\hat{y}(t) = f(y(t-1) + y(t-2) + \cdots + y(t-d)) \tag{10.6}$$

where f is a nonlinear function, where the future values depend only on regressed d earlier values of the output signal. Since the network performs only a one-step-ahead prediction after the training in the NAR network, it is necessary to use the closed-loop network to perform a multi-step-ahead prediction. Subsequent time steps can be predicted in a sequence through closed-loop forecasting by using the previous predictions as input. For the aim of this study, a one-step-ahead prediction also was used in the testing phase. The flowchart in Figure 10.3 shows the implementation of a NN algorithm to time series data and the sketch of the procedure proposed for SHM application. The performance of the trained network is evaluated by comparing real and simulated acceleration signals through the evaluation of the following indicators: standard deviation (StD) difference, root mean square error (RMSE), and coefficient of determination (R^2). For structural monitoring purposes, the obtained NN model, which includes the number of inputs (d, time delay), the number of neurons and hidden layers, and the values of the weights determined through a training process, can be applied to new acceleration data coming from a monitoring system, related to different ambient

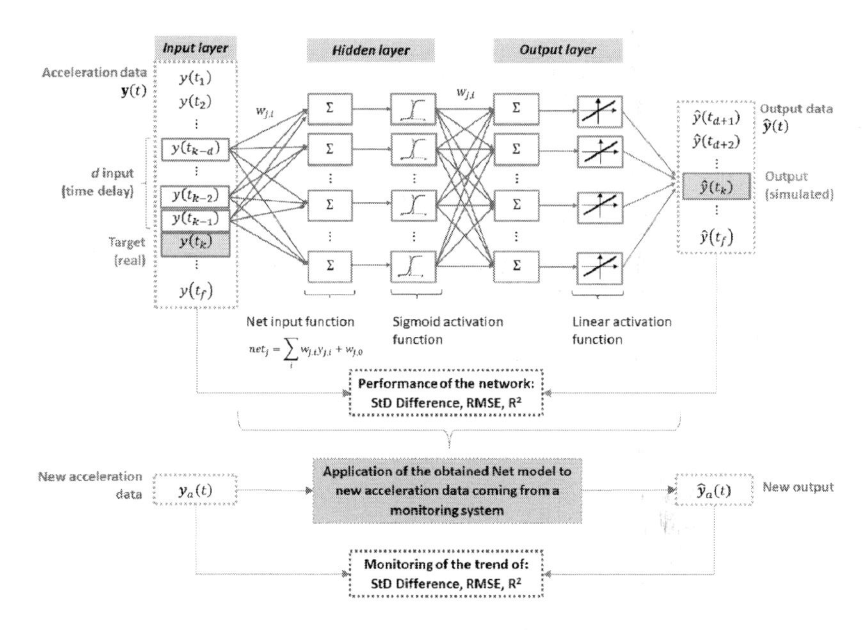

FIGURE 10.3 Neural network time series prediction: implementation of the one-step-ahead prediction in the testing phase, evaluation of the network performance, and application of the obtained model for structural monitoring purposes.

conditions (e.g., temperature or traffic load). The variation of the indicators of the network performance according to the environmental conditions can be considered as an alert for anomalies detection in the structural behavior.

10.4 MONITORING SYSTEM OF A CABLE-STAYED BRIDGE

The Éric Tabarly Bridge in Nantes, France, is a 210 m long cable-stayed road bridge crossing the Loire River. It is composed of a 27 m wide steel deck divided into two spans by a 57 m high steel pylon, and the main span is 143 m long (Figure 10.4).

In 2016, the instrumentation was installed for monitoring the vibrational behavior of the bridge. That is, 16 single-axis accelerometers were installed onto the bridge with two different acquisition zones: eight accelerometers in the deck and eight accelerometers in the pylon. Data is acquired through a PEGASE generation 2 acquisition card marketed by the A3IP company under an IFSTTAR (now Université Gustave Eiffel) license for each of the two zones. Each acquisition card exports voltage measurements at a sampling frequency of 100 Hz. Silicon Designs 2210 accelerometers, which can collect acceleration from –10G to 10G and are usable with a 700 Hz acquisition frequency, were used. Figure 10.5 depicts the locations of the accelerometers (red dots) and details of the deck and pylon. In the deck, each slot contains two single-axis accelerometers, with one along the Y-direction (vertical) and one along the Z-direction

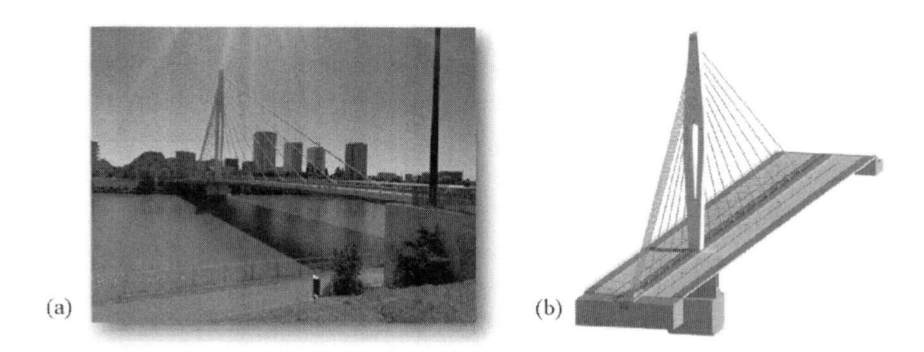

FIGURE 10.4 (a) Photo and (b) building information modeling (BIM) model of the Éric Tabarly Bridge in Nantes, France.

(horizontal-transversal). Two accelerometers along the Y-direction are placed in X3 (see Figure 10.5a). In the pylon (see Figure 10.5b), three of the eight accelerometers follow the Z-direction, while the others are along the X-direction. The directions of the different accelerometers along with the sensitivities of the sensors are listed in Tables 10.1 and 10.2.

A global schematic view of the monitoring system and its connection to a cloud database solution is presented in Figure 10.5. As shown, the full system is connected to a cloud database through a 3G/UMTS modem. Due to a few bandwidth limitations induced by some communication components of the first monitoring system solution, only data issued from accelerometers were feeding the database. All accelerometers were integrated in a designed cube accessory connected to the structure.

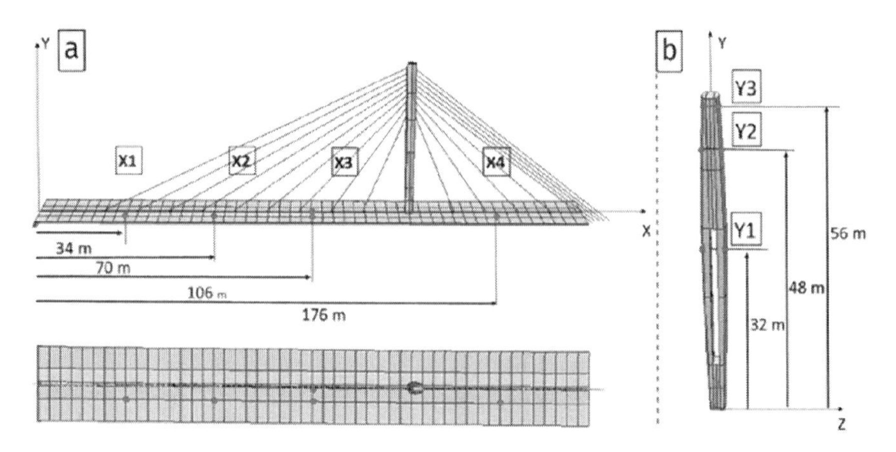

FIGURE 10.5 Layout of the accelerometric sensors placement for the (a) deck and (b) pylon.

TABLE 10.1
Direction and Sensitivities of the Deck Accelerometers

Deck				
Location	**Channel**	**Serial Number**	**Direction**	**Sensitivity (mV/g)**
X1	1	26188	$\vec{Y}$	2009
	2	26192	$\vec{Z}$	2009
X2	3	26187	$-\vec{Y}$	2013
	4	26193	$\vec{Z}$	2012
X3	5	26182	$-\vec{Y}$ (west)	2004
	6	26184	$-\vec{Y}$ (east)	2012
X4	7	26197	$-\vec{Y}$	2010
	8	26190	$\vec{Z}$	2012

TABLE 10.2
Direction and Sensitivities of the Pylon Accelerometers

Pylon				
Location	**Channel**	**Serial Number**	**Direction**	**Sensitivity (mV/g)**
Y1	1	26183	$-\vec{X}$	2013
	2	26189	$-\vec{Z}$	2008
Y2	3	23435	$-\vec{X}$	2000
	4	26194	$\vec{Z}$	2013
	5	26191	$-\vec{X}$	2010
Y3	6	26185	$-\vec{X}$	2008
	7	26186	$\vec{Z}$	2008
	8	26195	$\vec{X}$	2009

10.5 APPLICATION OF THE NAR ALGORITHM TO THE MONITORED DATA

In this study, a NAR network model is employed to predict the dynamic response of a cable-stayed bridge induced by ambient vibration through the measurements recorded by the SHM system previously described. The data-driven processing has been carried out through the Neural Network Time Series, which is a specific tool implemented in MATLAB. The accelerations elaborated by the algorithm were recorded on July 17, 2017, at 11:00 a.m. The registration has a total length of almost 10 minutes, and it is composed of 59,991 samples since the rate was 100 Hz. A first simulation has been performed by selecting different parameters: the training function *trainlm* that updates weight and bias values according to Levenberg–Marquardt optimization (among the fastest ones), and the network parameters (number of *hidden layers* and *time delay*) (Marquardt 1963; Yu and Wilamowski 2011). Furthermore, the NN modeling approach requires the determination of the percentage proportions of the data, which are divided into a training dataset, a validation dataset, and a test dataset. More specifically, the properties of the NN are the following: (a) a Levenberg–Marquardt backpropagation algorithm; (b) the hidden layers were equal to 10; (c) the selected number of delays was equivalent to 10; and (d) the network training set ratio was equal to 70%, the validation set was equal to 15%, and the network test ratio was equivalent to 15%. The NN's performance results are presented in Figure 10.6 in terms of real (black line) and simulated (red line) acceleration time series related to location X2 (deck) and channel 3 (Y-direction). Results show that the simulated time series is slightly higher than the real time series, resulting in a good match between them. Furthermore, a comparison between frequencies has been performed in the training dataset (the first 70% of the recorded samples) and test dataset (the last 15%) (see Figure 10.6b and 10.6c). From the figures, we observe a good agreement of the frequencies and higher amplitudes between the experimental and simulated results by the network (red line). Figure 10.7 depicts different zoom (ranging

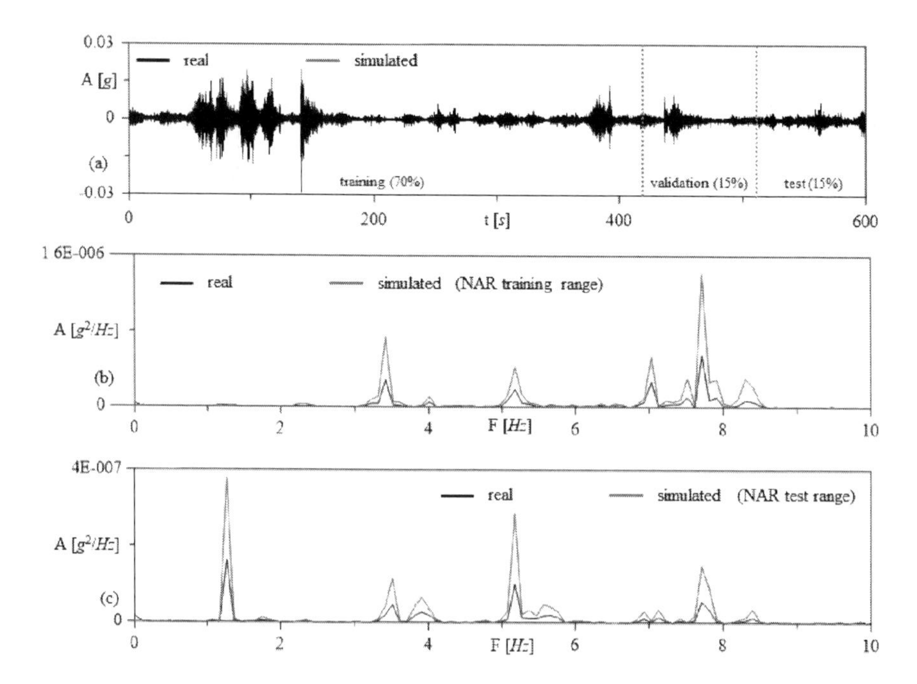

FIGURE 10.6 (a) Time and (b, c) frequency comparisons between recorded and simulated acceleration trained by the NAR algorithm. Training (70%), validation (15%), and test (15%). Hidden layers: 10; and time delay: 10.

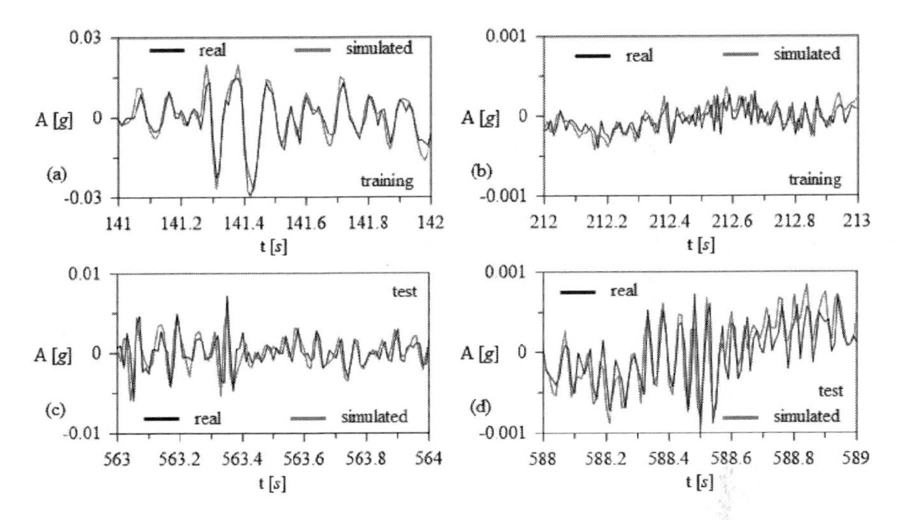

FIGURE 10.7 Zoom of the comparison reported in Figure 10.6: range of relatively (a, c) high and (b, d) low amplitudes.

1 second) of the analyzed results. The ranges have been selected in areas with relatively high and small amplitudes. Results show a better agreement for the higher amplitudes (see Figure 10.7a and 10.7c).

A parametric analysis has been conducted by changing the *hidden layers* and *time delay* network parameters. Specifically, the hidden layers have been set to 10 and 20, while for the time delay, six different values have been chosen (i.e., 10, 20, 50, 100, 1000, and 2000). Results, listed in Tables 10.3 and 10.4, have been compared in terms of the StD, RMSE, and R^2. From the results, it can be claimed that the StD shows a low variability when *hidden layers* and *time delays* are equal to 10 and to 1000 and 2000, respectively. RMSE and R^2 are substantially the same in all cases. However, it is right to highlight that while, for the simulations for the *time delays* 10, 20, 50, and 100, the computation time is up to a few minutes, for *time delays* 1000 and 2000 the computation time is over 10 hours.

TABLE 10.3
Performance of the ANN Algorithm: Hidden Layers 10

Time Delay (Sample)	StD Real (g)	StD Simulated (g)	StD Difference (%)	RMSE	R^2
10	0.00092	0.00081	13.58	0.00060	0.57
20	0.00092	0.00084	9.52	0.00061	0.56
50	0.00092	0.00083	10.84	0.00060	0.57
100	0.00092	0.00086	6.98	0.00059	0.59
1000	0.00092	0.00087	5.75	0.00060	0.58
2000	0.00092	0.00087	5.75	0.00061	0.56

TABLE 10.4
Performance of the ANN Algorithm: Hidden Layers 20

Time Delay (Sample)	StD Real (g)	StD Simulated (g)	StD Difference (%)	RMSE	R^2
10	0.00092	0.00083	10.84	0.00061	0.56
20	0.00092	0.00083	10.84	0.00061	0.56
50	0.00092	0.00084	9.52	0.00060	0.57
100	0.00092	0.00086	6.98	0.00059	0.59
1000	0.00092	0.00084	9.52	0.00059	0.60
2000	0.00092	0.00086	6.98	0.00060	0.58

It is worth highlighting that the training of the algorithm has been performed using only one recorded time series, through which the best result of R^2 (equal to 0.6) has been obtained. Probably, a better value of R may be achieved by using more registrations or other data-driven procedures. Moreover, when such value begins to slightly deviate from a certain reference percentage (e.g., 5%), it is reasonable to hypothesize the activation of an alert threshold.

10.6 CONCLUSIONS

Recently, ANNs have been implemented in time series forecasting applications due to their nonlinear modeling capability. In this context, this chapter aims at evaluating the performance of an ANN in the prediction of the dynamic response of a cable-stayed bridge induced by ambient vibration. The proposed network was formulated to be capable of efficiently predicting relevant dynamic responses associated with the bridge. Accelerations recorded by a SHM system installed in the Tabarly Bridge in Nantes, France, has been used as a case study. Such measurements have been processed by a script generated by the Neural Time Series tool included in the toolbox of MATLAB. The Levenberg–Marquardt solution algorithm was preferred, as it is found to attain faster convergence without compromising the quality of the final network. After the training of NAR, MATLAB software can predict the future value of one time step ahead. The results, even if not completely satisfactory (e.g., the maximum value achieved of R^2 is equal to 0.60), are promising and have margins of improvement. A better value of R^2 could probably be achieved by using more registrations or other data-driven procedures. Indeed, they could be helpful as an adjunct tool to detect the trigger of possible anomalies related to structural health. This research activity aims to provide suggestions to select appropriate values of the parameters used in the procedure and falling in a well-defined range (e.g., the hidden layer and time delay). It is also worth specifying that these indications are evaluated for an experimental structural response induced by ambient noise, showing usually low amplitudes.

Future attempts, including different variables to address missing parameters, such as temperature, that explain the predicted error, would likely attain a higher degree of accuracy. Finally, deep learning models for time series forecasting could be applied to improve the results.

ACKNOWLEDGMENTS

Part of the research leading to these results has received funding from the research project DESDEMONA—Detection of Steel Defects by Enhanced MONitoring and Automated procedure for self-inspection and maintenance (grant agreement no. RFCS-2018_800687), supported by EU Call RFCS-2017. Other funding resources were in part sponsored by the NATO Science for Peace and Security Programme under grant identification no. G5924.

REFERENCES

Bakhary, N., Hao, H., and Deeks, A. J. (2007). "Damage detection using artificial neural network with consideration of uncertainties." *Engineering Structures*, 29(11), 2806–2815.

Bal, C., and Demir, S. (2021). "JMASM 55: MATLAB Algorithms and Source Codes of'cbnet'Function for Univariate Time Series Modeling with Neural Networks (MATLAB)." *Journal of Modern Applied Statistical Methods*, 19(1), 19.

Barai, S., and Pandey, P. (1995). "Vibration signature analysis using artificial neural networks." *Journal of Computing in Civil Engineering*, 9(4), 259–265.

Benmouiza, K., and Cheknane, A. (2016). "Small-scale solar radiation forecasting using ARMA and nonlinear autoregressive neural network models." *Theoretical and Applied Climatology*, 124(3), 945–958.

Cascardi, A., Micelli, F., and Aiello, M. A. (2017). "An Artificial Neural Networks model for the prediction of the compressive strength of FRP-confined concrete circular columns." *Engineering Structures*, 140, 199–208.

Chang, C., Chang, T., and Xu, Y. (2000). "Adaptive neural networks for model updating of structures." *Smart Materials and Structures*, 9(1), 59.

Chow, T., and Leung, C.-T. (1996). "Nonlinear autoregressive integrated neural network model for short-term load forecasting." *IEE Proceedings-Generation, Transmission and Distribution*, 143(5), 500–506.

de Oliveira Dias Prudente dos Santos, J. P., Crémona, C., da Silveira, A. P. C., and de Oliveira Martins, L. C. (2016). "Real-time damage detection based on pattern recognition." *Structural Concrete*, 17(3), 338–354.

Denton, J. W. (1995). "How good are neural networks for causal forecasting?" *The Journal of Business Forecasting*, 14(2), 17.

Di Girolamo, G. D., Smarra, F., Gattulli, V., Potenza, F., Graziosi, F., and D'Innocenzo, A. (2020). "Data-driven optimal predictive control of seismic induced vibrations in frame structures." *Structural Control and Health Monitoring*, 27(4), e2514.

Domaneschi, M., Cimellaro, G., De Iuliis, M., and Marano, G. (2021). "Laboratory investigation of digital image correlation techniques for structural assessment." *Bridge Maintenance, Safety, Management, Life-Cycle Sustainability and Innovations*, CRC Press, 3260–3266.

Dutta, A., and Talukdar, S. (2004). "Damage detection in bridges using accurate modal parameters." *Finite Elements in Analysis and Design*, 40(3), 287–304.

Fang, X., Luo, H., and Tang, J. (2005). "Structural damage detection using neural network with learning rate improvement." *Computers & Structures*, 83(25–26), 2150–2161.

Feng, N., Wang, F., and Qiu, Y. (2006). "Novel approach for promoting the generalization ability of neural networks." *International Journal of Signal Processing*, 2(2), 131–135.

Gattulli, V., Cunha, A., Caetano, E., Potenza, F., Arena, A., and Di Sabatino, U. (2021). "Dynamical models of a suspension bridge driven by vibration data." *Smart Structures and Systems, An International Journal*, 27(2), 139–156.

Gattulli, V., Franchi, F., Graziosi, F., Marotta, A., Rinaldi, C., Potenza, F., and Sabatino, U. D. (2022a). "Design and evaluation of 5G-based architecture supporting data-driven digital twins updating and matching in seismic monitoring." *Bulletin of Earthquake Engineering*, 20(9), 4345–4365.

Gattulli, V., Lepidi, M., Potenza, F., and Di Sabatino, U. (2016). "Dynamics of masonry walls connected by a vibrating cable in a historic structure." *Meccanica*, 51(11), 2813–2826.

Gattulli, V., Lepidi, M., Potenza, F., and Di Sabatino, U. (2019). "Modal interactions in the nonlinear dynamics of a beam–cable–beam." *Nonlinear Dynamics*, 96(4), 2547–2566.

Gattulli, V., Potenza, F., and Piccirillo, G. (2022b). "Multiple tests for dynamic identification of a reinforced concrete multi-span arch bridge." *Buildings*, 12(6), 833.

Hagan, M. T., Demuth, H. B., and Beale, M. (1997). *Neural Network Design*, PWS Publishing Co.

Hasançebi, O., and Dumlupınar, T. (2013). "Linear and nonlinear model updating of reinforced concrete T-beam bridges using artificial neural networks." *Computers & Structures*, 119, 1–11.

Haykin, S., and Lippmann, R. (1994). "Neural networks, a comprehensive foundation." *International Journal of Neural Systems*, 5(4), 363–364.

Ji, W., and Chee, K. C. (2011). "Prediction of hourly solar radiation using a novel hybrid model of ARMA and TDNN." *Solar Energy*, 85(5), 808–817.

Kromanis, R., and Kripakaran, P. (2017). "Data-driven approaches for measurement interpretation: Analysing integrated thermal and vehicular response in bridge structural health monitoring." *Advanced Engineering Informatics*, 34, 46–59.

Lee, J. J., Cho, S., Shinozuka, M., Yun, C.-B., Lee, C.-G., and Lee, W.-T. (2006). "Evaluation of bridge load carrying capacity based on dynamic displacement measurement using real-time image processing techniques." *Steel Struct*, 6, 377–385.

Lee, L. S., Karbhari, V. M., and Sikorsky, C. (2007). "Structural health monitoring of CFRP strengthened bridge decks using ambient vibrations." *Structural Health Monitoring*, 6(3), 199–214.

Lee, J. J., and Shinozuka, M. (2006). "A vision-based system for remote sensing of bridge displacement." *NDT & E International*, 39(5), 425–431.

Liu, C., DeWolf, J. T., and Kim, J.-H. (2009). "Development of a baseline for structural health monitoring for a curved post-tensioned concrete box–girder bridge." *Engineering Structures*, 31(12), 3107–3115.

Li, H., Wang, T., and Wu, G. (2021). "Dynamic response prediction of vehicle-bridge interaction system using feedforward neural network and deep long short-term memory network." *Proc., Structures*, Elsevier, 2415–2431.

Magalhães, F., Cunha, A., and Caetano, E. (2010). "Continuous dynamic monitoring of an arch bridge: Strategy to eliminate the environmental and operational effects and detect damages." *Proc., Proc. 24th International Conference on Noise and Vibration Engineering (ISMA2010), Leuven, Belgium.*

Maier, H. R., and Dandy, G. C. (2000). "Neural networks for the prediction and forecasting of water resources variables: A review of modelling issues and applications." *Environmental Modelling & Software*, 15(1), 101–124.

Maity, D., and Saha, A. (2004). "Damage assessment in structure from changes in static parameter using neural networks." *Sadhana*, 29(3), 315–327.

Markham, I. S., and Rakes, T. R. (1998). "The effect of sample size and variability of data on the comparative performance of artificial neural networks and regression." *Computers & Operations Research*, 25(4), 251–263.

Marquardt, D. W. (1963). "An algorithm for least-squares estimation of nonlinear parameters." *Journal of the society for Industrial and Applied Mathematics*, 11(2), 431–441.

Nguyen, D., and Widrow, B. (1990). "Improving the learning speed of 2-layer neural networks by choosing initial values of the adaptive weights." *Proc., 1990 IJCNN international joint conference on neural networks*, IEEE, 21–26.

Orcesi, A. D., and Frangopol, D. M. (2010). "Inclusion of crawl tests and long-term health monitoring in bridge serviceability analysis." *Journal of Bridge Engineering*, 15(3), 312–326.

Park, H. S., An, J. H., Park, Y. J., and Oh, B. K. (2020). "Convolutional neural network-based safety evaluation method for structures with dynamic responses." *Expert Systems with Applications*, 158, 113634.

Pavelka, A., and Procházka, A. (2004)"Algorithms for initialization of neural network weights." *Proc., In Proceedings of the 12th annual conference, MATLAB*, 453–459.

Phares, B. M., Wipf, T. J., Klaiber, F., and Abu-Hawash, A. (2003). "Bridge load rating using physical testing." *Proc., Proceedings of the 2003 Mid-Continent Transportation Research Symposium, Iowa State University.*

Rafiq, M., Bugmann, G., and Easterbrook, D. (2001). "Neural network design for engineering applications." *Computers & Structures*, 79(17), 1541–1552.

Ribeiro, D., Calçada, R., Ferreira, J., and Martins, T. (2014). "Non-contact measurement of the dynamic displacement of railway bridges using an advanced video-based system." *Engineering Structures*, 75, 164–180.

Tang, C.-W., Chen, H.-J., and Yen, T. (2003). "Modeling confinement efficiency of reinforced concrete columns with rectilinear transverse steel using artificial neural networks." *Journal of Structural Engineering*, 129(6), 775–783.

Wang, C., Ansari, F., Wu, B., Li, S., Morgese, M., and Zhou, J. (2022). "LSTM approach for condition assessment of suspension bridges based on time-series deflection and temperature data." *Advances in Structural Engineering*, 25(16), 3450–3463, 13694332221133604.

Yun, C.-B., and Bahng, E. Y. (2000). "Substructural identification using neural networks." *Computers & Structures*, 77(1), 41–52.

Yun-Lai, Z., and Wahab, M. A. (2017). "Damage detection using vibration data and dynamic transmissibility ensemble with auto-associative neural network." *Mechanics*, 23(5), 688–695.

Yu, H., and Wilamowski, B. M. (2011). "Levenberg-Marquardt training." *Industrial Electronics Handbook*, 5(12), 1.

Yu, J., Zhu, P., Xu, B., and Meng, X. (2017). "Experimental assessment of high sampling-rate robotic total station for monitoring bridge dynamic responses." *Measurement*, 104, 60–69.

Zhang, G. P. (2003). "Time series forecasting using a hybrid ARIMA and neural network model." *Neurocomputing*, 50, 159–175.

Zhao, R., Yan, R., Chen, Z., Mao, K., Wang, P., and Gao, R. X. (2019). "Deep learning and its applications to machine health monitoring." *Mechanical Systems and Signal Processing*, 115, 213–237.

11 Modeling Uncertainties by Data-Driven Bayesian Updating for Structural and Damage Detection

Chiara Pepi, Massimiliano Gioffrè, and Mircea D. Grigoriu

11.1 OVERVIEW

Structural health monitoring (SHM) methods are essential for the identification of system properties and the accurate prediction of the performance and resilience of structural systems subjected to time-dependent actions. It is common to estimate the response of existing structures by numerical models that depend on unknown mechanical and/or geometrical parameters and to characterize probabilistically these parameters from input/output measurements by solving inverse problems in a stochastic setting. The quality of the resulting models depends on the magnitude of the measurement errors and the methods for solving the posed inverse problems. We propose a robust framework for damage identification that is based on the Bayesian representation of the uncertain parameters of the finite element (FE) model for dynamic analysis. The potential of the proposed method is assessed by virtual response samples that are polluted by measurement noise. The method is crucial to quantify the time evolution of structural performance and reliability and to develop suitable damage detection procedures for early warning. A cable-stayed footbridge is used as a case study to demonstrate the implementation and the capability of our method. The proposed method advances the field of data-driven SHM in three directions. First, the efficiency of the classical Bayesian updating procedure is improved by using polynomial chaos. Second, the updating algorithm is viewed as a model prediction error, e.g., the differences between measured and calculated natural frequencies and modal shapes. Third, the measurement errors are explicitly handled in the Bayesian updating procedures for solving the damage identification and localization problems.

11.2 INTRODUCTION

Experimental modal analysis [1–5] and model updating [6–8] have been hot research topics because their combined use can provide useful information for structural identification and structural damage detection [9–11]. Experimental

DOI: 10.1201/9781003306924-11

modal analysis is first carried out in order to estimate the modal parameters of a structural system (e.g., natural frequencies, vibration modes, and damping) from ambient vibration tests (AVTs) using operational modal analysis (OMA) procedures [12–14]. In the OMA technique, the structural input excitation is not measured, and it is assumed to be a stationary white noise [15–17]. Nevertheless, these ambient vibrations can be very far to be stationary because of wind and/or microtremors, and furthermore the extracted modal features can be affected by measurement noise, estimation errors, and changing environmental conditions. Therefore, tools are needed to take into account all these sources of uncertainties. Model updating is then used for estimating the unknown mechanical and/or geometrical parameters of the FE so that the discrepancy between the model-predicted and the experimental modal parameters is minimized. This procedure leads to a constrained optimization problem that is often ill-posed since the existence and the uniqueness of the inverse problem solution are not guaranteed. For this reason, in the last few years, probabilistic model-updating procedures gained growing interest in the scientific community [18–20]. This kind of technique can be grouped into two main classes: classical probabilistic approaches and Bayesian methods based on the well-known Bayes' theorem [21].

In the Bayesian framework, the prior probability density functions (PDFs) of the uncertain model parameters are transformed into posterior PDFs, accounting explicitly for all the sources of errors involved in the process, including measurement and modeling errors [22–24]. This transformation is done through the so-called likelihood function, which represents the contribution of the measured data in the definition of the posterior distribution and practically reflects how well the measurements can be explained from the model when a set of unknown input parameters is defined. The likelihood function is computed by using suitable probabilistic models of the prediction error, i.e., the difference between the FE model and the measured modal parameters. While the definition of suitable prior PDFs of the uncertain model parameters has been extensively documented in literature [25–28], less attention has been devoted to the selection of prediction error models. In most cases, uncorrelated zero-mean Gaussian errors are adopted, meaning that prediction error variances are assumed to be fixed and known [23, 29, 30]. Most studies assume variances to be equal for all modes for natural frequencies and mode shape measurements. This assumption is supported by the principle of maximum entropy [25], but it can significantly affect the results since the measurements and the corresponding prediction errors may be correlated in space and in time [31, 32]. Second, the prediction error variances for natural frequencies and mode shapes can be estimated based on the real measurements when a large number of observations are available [33]. Finally, the prediction error variances can be considered as additional uncertain parameters, and they can be updated together with the uncertain FE model parameters in the Bayesian inference scheme [34–36].

One of the challenges in the application of Bayesian model-updating methods is the case of multidimensional complex systems with several updating parameters and/or large datasets used as reference. In these cases, a complex multidimensional integration problem has to be solved that can be rather time-consuming. The Markov chain Monte Carlo (MCMC) methods are the most widely used technique for such integration and are based on the generation of random sequences of input parameter samples (so-called Markov chains) that are in equilibrium with the target posterior PDFs [37, 38]. These methods require several solutions of the structural problem using the deterministic FE model, one for each occurrence of the input parameters, which can make the Bayesian procedure unfeasible. A commonly encountered method to overcome this drawback is based on a low-order Taylor approximations of the quantities of interest (QoIs), but these formulas are only accurate in very simple situations (e.g., univariate input) [39]. Therefore, surrogate models based on PC expansion [40–42] can be used to solve this issue, dramatically reducing the time needed for the Bayesian updating framework by replacing the numerical FE model solution with the surrogate solution.

The efficient performance of model updating and the prediction of the uncertainties in the results are crucial tasks, especially in the main field of application of AVTs and FE model–updating procedures, which is the vibration-based damage detection method [11, 29, 43, 44]. Although deterministic procedures for FE model updating have been largely used for damage identification in large-scale real structures, there are only a few applications of Bayesian FE model updating to full-scale complex structures due to the existing challenges with dealing with modeling errors and real modal data. In particular, when vibration modes are included in the reference dataset, direct mode shape matching is required, and this is a very complex task since modal data is incomplete and the experimental eigenvectors may be characterized by a significant imaginary component requiring suitable formulation for the modal vector's prediction error probabilistic model.

In this study, the challenges of implementing a Bayesian updating framework for structural model updating and damage detection of full-scale structures using real incomplete dynamic modal data are studied, and a novel framework is proposed. First, estimates of the modal parameters are obtained via AVTs on an actual structural system, and PC-based surrogate models for each natural frequency and for each eigenvector component of the modal vectors are evaluated and calibrated using an initial FE model. An improved version of the MCMC method is used to estimate the posterior marginal PDFs of the selected model parameters: besides the commonly used system natural frequencies, the modal assurance criterion (MAC) is introduced as a constraint for the random walk in order to ensure direct mode shape matching at each step of the Markov chains; furthermore, the deterministic FE model solution at each step of evaluation is replaced with the surrogate solution for reducing the computational costs required to estimate the posterior marginal PDFs; and, finally, the prediction error variances are treated as further updating parameters. The effectiveness of the proposed method is eventually demonstrated using a FE numerical model describing a curved cable-stayed footbridge located in Terni (Umbria Region, Central Italy).

The reduction of the computational costs and the use of a reference dataset including both natural frequencies and modal vectors make the proposed procedure particularly suitable to be used in real-time SHM applications for the detection of the existence and the localization of damage. For this reason, virtual samples of the updated cable-stayed footbridge's numerical model structural response affected by random loads are evaluated. These virtual samples are used as *virtual experimental responses* in order to assess the effect of the signal sampling parameters on the estimation of modal parameters. The information given by the measurement uncertainties is used in the likelihood function formulation, and the Bayesian method for detection of structural damage is then established by calculating the probability distribution of the unknown numerical model parameters using MCMC Metropolis Hastings (MH)-based model updating of the undamaged and damaged structures.

11.3 THE BAYESIAN SETTING FOR INVERSE PROBLEM SOLUTIONS

Let us consider a mathematical representation $\mathcal{M}$ of a physical system, which relates a QoI to a set of measurable variables x and random parameters gathered in Θ. The probabilistic model can be written in the form:

$$\mathcal{M}(x,\Theta)= \mathcal{M}(x,\theta)+\sigma\varepsilon \tag{11.1}$$

In this equation, $\Theta=(\theta,\sigma)$, where $\theta=(\theta_1,\ldots,\theta_N)\in R^N$ denotes the set of unknown model parameters; $\sigma\varepsilon$ is the total model error, in which ε is a random variable with zero mean and unit variance and σ represents the standard deviation of the model error. This probabilistic model is formulated under three main hypotheses: (i) σ does not vary with x (homoskedasticity assumption), (ii) ε follows normal distribution (normality assumption), and (iii) the model error can be added to the model (additive assumption) [36, 45, 46].

The unknown model parameters in the real valued random vector Θ can be estimated using the Bayes theorem [21] as:

$$p\left(\Theta|\bar{D}\right)=\frac{p\left(\bar{D}|\Theta\right)p(\Theta)}{p\left(\bar{D}\right)} \tag{11.2}$$

where $p\left(\Theta|\bar{D}\right)$ is the posterior PDF of the uncertain parameters Θ to be updated, conditional on a data vector of M measurements $\bar{D}=\left\{\bar{D}_1,\ldots,\bar{D}_M\right\}$; $p\left(\bar{D}|\Theta\right)$ is the likelihood function that captures the information from the measurements; and $p(\Theta)$ is the prior distribution, which quantifies the initial knowledge of the vector of parameters Θ. The normalizing factor $p\left(\bar{D}\right)=\int p\left(\bar{D}|\Theta\right)p(\Theta)d\Theta$ is called evidence and is used to ensure that the integration over the parameter space of the posterior PDF in Eq. (11.2) is equal to one.

11.4 THE BAYESIAN SETTING FOR FINITE ELEMENT MODEL INPUT PARAMETER ESTIMATION USING EXPERIMENTAL MODAL DATA

The unknown FE model input parameters are traditionally estimated using deterministic approaches minimizing the distance between the experimental and the FE model–predicted modal data (e.g., natural frequencies and vibration modes), but the problem is ill-posed since a unique solution does not exist. Therefore, being $\mathcal{M}(x,\Theta)$ in Eq. (11.1), the mathematical representation of a FE model, the Bayes theorem in Eq. (11.2) can be set up defining: (i) the FE deterministic numerical model $\mathcal{M}(x,\Theta)$; (ii) the unknown numerical model parameters in the random vector $\Theta=(\theta,\sigma)$ (i.e., Young's modulus E, mass density ρ and boundary conditions); (iii) the set of modal data extracted from measured acceleration time histories in the vector $\bar{D}=\{f_1,\ldots,f_M,\Phi_1,\ldots,\ \Phi_M\}$, where f_k and Φ_k are, respectively, the k-th natural frequency and the k-th mode shape vector, and M is the total number of the observed modes with $k = 1,\ldots, M$.

11.5 PROBABILISTIC CHARACTERIZATION OF THE MODEL AND MEASUREMENT PREDICTION ERROR

A numerical mechanical model is not able to perfectly reproduce the real behavior of the true structural system [18]. Therefore, a modeling error e_M defined as the difference between the real behavior of the true system and the model predictions—that is, $e_M = \boldsymbol{D}-\mathcal{M}(\boldsymbol{x},\Theta)$—is always present. Since the measurements are in practice always disturbed also, a measurement error e_D determines a difference between the true system output and the actual observed data $\bar{D}$, that is, $e_D = \bar{D} - D$. The total prediction error can be obtained as the sum of the modeling and measurement error as

$$r = e_M + e_D = \bar{D} - \mathcal{M}(\boldsymbol{x},\Theta) \tag{11.3}$$

where the dependency on the unknown true system behavior $\boldsymbol{D}$ is eliminated. Using Eqs. (11.1) and (11.3), the prediction error of the k-th natural frequency $r_k^f(\theta)$ can be expressed as

$$r_k^f(x,\Theta) = f_k - \widehat{f_k}(x,\Theta) = \sigma_k^f \varepsilon_k \tag{11.4}$$

where f_k and $\widehat{f_k}(x,\Theta)$ denote the k-th observed and model-predicted natural frequency, respectively. The prediction error of the k-th vibration modes can be defined by means of MAC [47]. The MAC coefficient is defined to measure the degree of correlation between the measured (Φ_k) and the numerically computed ($\widehat{\Phi_k}(x,\Theta)$) modal vectors. Taking into account that the MAC coefficient assumes values between 1 and 0 for, respectively, perfect match and no correlation, its complement *1-MAC* can be considered as the residual error for mode shape; and, using Eqs. (11.1) and (11.3), the probabilistic model error on mode shapes can be expressed as

$$r_k^{ms}(x,\Theta) = 1 - \text{MAC}(x,\Theta) = \sigma_k^{ms}\varepsilon_k \tag{11.5}$$

Gathering all the prediction errors in a vector r having $2 \times M$ components such that

$$r(x,\Theta) = \begin{bmatrix} r^f(x,\Theta) \\ r^{ms}(x,\Theta) \end{bmatrix} \tag{11.6}$$

The likelihood function for a set of $(2 \times M)$-variate observations (i.e., M natural frequencies and M mode shapes) can be expressed as

$$P(\bar{D}|\Theta) = (2\pi|\Sigma|)^{-\frac{1}{2}} \exp\left(-\frac{1}{2} r^T(x,\Theta)\Sigma^{-1} r(x,\Theta)\right) \tag{11.7}$$

where Σ is the covariance matrix of the total model error $\sigma_k\varepsilon_k$. Note that, considering symmetry, Σ includes $2 \times M$ unknown variances σ_k^2 and $M(M-1)/2$ unknown correlation coefficients. A practical issue in this approach is to guarantee the respect of the homoskedastic model. Considering the non-negative nature of the QoIs in Bayesian inference based on dynamic data (natural frequencies and vibration modes to be taken in the form of distance between perfect correlations), a logarithmic transformation is used [48].

11.6 PROBABILISTIC CHARACTERIZATION OF THE MODEL AND MEASUREMENT PREDICTION ERROR

A source of uncertainty related to the unknown input parameters and prior distributions reflects the initial state of knowledge about them. Bayesian prior probabilities can be subjective and objective priors. Subjective priors are chosen depending on expert judgment (e.g., personal belief, visual inspections, and measurements). Their choice can be relevant, since different results of the Bayesian updating framework may be obtained when the dataset used as reference is small or not properly informative. In contrast, objective priors are formulated according to some formal rules like the widely used principle of maximum entropy [25]. When no such information is available, one should use a prior distribution that has minimal influence on the posterior distribution, so that inferences are unaffected by information external to the observations. Following Refs. [36, 45, 46], non-informative priors are selected for the global error standard deviation σ:

$$p(\Sigma) \propto |R|^{-\frac{(2M)+1}{2}} \prod_{i=1}^{2M} \frac{1}{\sigma_i} \tag{11.8}$$

where $|\cdot|$ denotes the determinant, and R is the correlation matrix.

11.7 A MODIFIED VERSION OF THE MCMC METROPOLIS HASTINGS ALGORITHM FOR THE POSTERIOR DISTRIBUTION EVALUATION

Once the Bayesian setting is set up (i.e., the prior PDFs and probabilistic model of error are defined), Eq. (11.2) allows for the updating of the PDFs of the model parameters Θ_i using experimental observations of the structural system $\bar{D}$. If the number of parameters and the data space dimension are large, the multi-dimensional integration cannot be solved analytically, and sampling methods such as the MCMC and its derivatives are used. The MH algorithm—as an MCMC method—is used in this work [37, 38]. This algorithm is based on generating samples from any target distribution of the uncertain parameters Θ_i. The proposed parameter sample Θ^* is generated by a proposal density $p\left(\Theta^* \middle| \Theta^t\right)$ depending on the current state of the chain. To start the Markov chain, draw Θ_0 from the prior distribution. To simulate the next sample Θ^{t+1} from the current sample Θ^t, $t = 1, \ldots, N$, first simulate a candidate state Θ^* from the proposal PDF $q\left(\Theta^* \middle| \Theta^t\right)$. Defining

$$\rho\left(\Theta^*, \Theta^t\right) = \frac{p\left(\Theta^* \middle| \bar{D}\right) p\left(\Theta^t \middle| \Theta^*\right)}{p\left(\Theta^t \middle| \bar{D}\right) p\left(\Theta^* \middle| \Theta^t\right)} \tag{11.9}$$

the candidate sample Θ^* has a probability $min\left\{1, \rho\left(\Theta^*, \Theta^t\right)\right\}$ to be accepted and a probability $1 - min\left\{1, \rho\left(\Theta^*, \Theta^t\right)\right\}$ to be rejected. If accepted, the sample Θ^* will be taken as the next state of the chain $\Theta^{t+1} = \Theta^*$; otherwise, the current state is taken as the next step of the chain $\Theta^{t+1} = \Theta^t$. The process is repeated until N Markov chain samples have been simulated and the specification of the acceptance probability $\rho\left(\Theta^*, \Theta^t\right)$ allows generating a Markov chain with the desired target density.

In FE model–updating procedures, the updating process can cause misleading results when experimental modal data is used as reference because of possible frequency matching associated to different mode shapes. To overcome this problem, the idea proposed in this work is to use the MAC coefficient [47] in order to measure the correlation rate between the experimental and numerical mode shapes. Recalling that the MAC coefficient assumes values ranging from 0 to 1, when the two modes have 0 or perfect correlation, the classical MCMC MH algorithm is modified using the MAC coefficients as a soft constraint so that the total error in Eq. (11.6) at each step of the chain is computed as the difference between the model predicted and the observed natural frequencies only when they correspond to the same mode shape.

11.8 A SURROGATE MODEL–BASED MCMC MH ALGORITHM FOR THE POSTERIOR DISTRIBUTION EVALUATION

Estimation of the posterior distribution using the MCMC MH method is computationally prohibitive, since it requires the computation of the FE model deterministic solution at each step of the chain and usually requires about 10^5 sample generations to have solution convergency. In order to obtain a significant reduction in the computational burden, an effective method based on the functional approximation of the forward model response in Eq. (11.1) is used. To this end, the PC representation method [49–51] is used to obtain an analytical representation of the model itself as a function of the main random input parameters, leading directly to a surrogate model in the form of a response surface. This means that the posterior sampling via MCMC can be carried out directly from the response surface without the need to solve the analytical deterministic model for all the samples.

Assuming that the model response is a finite variance random variable, the PC approximation of the structural response can be expressed in a coordinate system described by independent and identically distributed, zero-mean, unit variance Gaussian RVs $\xi \sim N(0,1)$:

$$\tilde{\mathcal{M}}(x,\theta) = \sum_{\alpha \geq 0}^{N_P - 1} \widehat{a_\alpha} H_\alpha(\xi) \tag{11.10}$$

where $H_\alpha(\xi)$ represents the multivariate Hermite polynomials with a finite multi-index set; $\widehat{a_\alpha}$ are the polynomial coefficients; and N_P is the number of the unknown coefficients in the summation. When the forward model response $\mathcal{M}(x,\theta)$ is a R^d-valued random vector consisting of $d > 1$ independent components with finite variances, the procedure can be applied to each component of $\mathcal{M}(x,\theta)$ without loss of generality.

The coefficients $\widehat{a_\alpha}$ in Eq. (11.10) can be obtained by means of non-intrusive regression-based methods using the deterministic solutions of the input realizations [41]. Further accurate details about the theoretical background of PC-based surrogate models can be found in Ref. [51]. Once the approximation of $\mathcal{M}(x,\theta)$ has an explicit functional form, the forward model response in Eq. (11.3) can be replaced by its approximation $\tilde{\mathcal{M}}(x,\theta)$, obtaining

$$r = \bar{D} - \tilde{\mathcal{M}}(x,\Theta) \tag{11.11}$$

Therefore, the main advantage of the surrogate model–based MCMC MH method is that, if an accurate approximation $\tilde{\mathcal{M}}(x,\Theta)$ is obtained, the posterior density $p(\Theta|\bar{D})$ can be evaluated for a large number of samples without computationally demanding simulations of the forward problem. Accounting for the constraint used in the random walk to ensure that the residual vector is estimated

Algorithm 11.1: Surrogate model–based improved MCMC MH algorithm

Data: Initiate the algorithm with a value $\boldsymbol{\Theta}^0$
for Each $t = 1 \rightarrow N$ **do.**
Draw $\boldsymbol{\Theta}^*$ from the proposal probability distribution density $p(\Theta^* | \Theta^t)$.
Estimate the surrogate model solution for $\tilde{\mathcal{M}}(\boldsymbol{x}, \boldsymbol{\theta}^*)$.
Compute the diagonal MAC coefficient.
Reorder the natural frequencies.
Compute the residual vector r_t in Eq. (11.11).
Compute the acceptance rate $\rho(\Theta^*, \Theta^t)$ in Eq. (11.9).
Draw u from a uniform distribution $u \sim \mathcal{U}(0,1)$.
  if $u < \rho(\Theta^*, \Theta^t)$ **then**
Accept the state $\Theta^{t+1} = \Theta^*$.
  else
Reject the state $\Theta^{t+1} = \Theta^t +$.
  end
end

as the difference between the model predicted and the observed natural frequencies only when they correspond to the same mode shape, an improved version of the classical MCMC MH algorithm based on the surrogate model is obtained and reported in Algorithm 11.1.

11.9 A SURROGATE MODEL–BASED BAYESIAN METHOD FOR DAMAGE DETECTION IN REAL-TIME SHM APPLICATIONS

SHM is the process of equipping a structure with an array of sensors and then extracting features (e.g., modal parameters) for the main purposes of damage detection and structural performance assessment. The main ideas at the basis of SHM is that changes in structural system parameters due to damage (e.g., stiffness, mass, damping, and boundary conditions) can provide changes in the vibration characteristics (e.g., natural frequencies, vibration modes, and damping). Hence, the existence of structural damage can be determined from statistical analysis on the large amount of modal data estimated at every fixed time, while a FE numerical model needs to be developed and calibrated in order to obtain information about the damage localization.

The natural frequencies evolution in time-based damage detection methods has been a hot research topic for assessing the existence of damage and, in some cases, also the severity of damage. Nevertheless, conventional damage localization detection methods are based on deterministic optimization problems able to determine the unknown stiffness distributions by fitting the measured and the FE model–predicted system response. These methods cannot address the effect of uncertainties due to measurement noise and modeling hypotheses and approximation, and furthermore they can be ill-posed and computationally unfeasible, especially in real-time SHM practical applications.

The proposed procedure, based on an improved version of the classical MCMC MH method, is able to overcome the main drawbacks in SHM practical applications.

1. Evaluation of experimental natural frequencies f_k and vibration modes Φ_k using a single ambient vibration measurement
2. Development of an initial numerical FE model $\mathcal{M}(x,\theta)$ driven by a set of uncertain mechanical parameters
3. Definition of an initial PDF for the unknown numerical model parameters Θ
4. Development of high-fidelity PC-based surrogate models for each of the selected QoIs (i.e., $\widehat{f_k}$, $\widehat{\Phi_k}$) and quantification of the model uncertainties (i.e., variances of the error between the deterministic FE numerical model and the surrogate model solutions)
5. Evaluation of the posterior distribution of the uncertain parameters θ using Algorithm 11.1 to obtain an updated undamaged numerical model to be used as reference for damage identification
6. Performing continuous OMA using the acceleration time histories recorded at every fixed time, obtaining a large amount of data to be used to quantify the measurement uncertainties
7. Sum of the measurement and the modeling uncertainties to obtain the total model error explicitly introduced in the likelihood function
8. Continuous updating of the PDFs of the uncertain parameters every time a new measurement (i.e., $\widehat{f_k},\widehat{\Phi_k}$) becomes available to track continuously the main parameter estimates and their associated uncertainties in order to obtain information about the damage existence and localization

11.10 APPLICATIVE CASE STUDY AND DISCUSSION

This section reports the validation of the previously introduced surrogate model–based MCMC MH algorithm for the estimation of the FE model parameters. Specifically, a cable-stayed footbridge (Figure 1 in Ref. [50]) is considered in this study. The footbridge consists of two main parts: a curved part with a total length of 120 m, supported by an asymmetric array of cables connected to a 60 m tripod tower through a pair of circular rings, and a straight 60 m span with two bowstring arches.

11.10.1 Ambient Vibration Tests and Operational Modal Analysis

Ambient-vibration dynamic tests were carried out on the cable-stayed spans with the aim of identifying the modal properties of the cable-stayed footbridge. The ambient vibrations induced by traffic, human, and wind actions in operating conditions were recorded and used to estimate the experimental modal properties

TABLE 11.1
Natural Frequencies Identified from the Measurements

Mode	Φ_1	Φ_2	Φ_3	Φ_4	Φ_5	Φ_6
f (Hz)	0.964	1.693	1.797	2.474	2.591	3.307

by means of the enhanced frequency domain decomposition (EFDD) method [15, 17]. The obtained natural frequencies are summarized in Table 11.1. The corresponding mode shapes are reported in Figure 11.1: three vertical bending modes (denoted as Φ_1, Φ_5, and Φ_6), two lateral modes (Φ_2 and Φ_4), and one mixed-vibration mode (Φ_3) are identified.

11.10.2 Initial FE Model Development

An initial three-dimensional FE model of the footbridge was built using SAP2000 software [49]. Different mechanical characteristics are used for each structural component, while each stay is modeled with a nonlinear element describing both tension stiffening and large deflections so that an iterative solution is required. The dynamic characteristics of the structural system are estimated by performing a pre-stress modal analysis in the dead load and cable pre-tension deformed equilibrium configuration. Further details about the FE model of the footbridge can be found in Refs. [50, 51].

The results of the numerical modal analysis are reported in Table 11.2 with the relative frequency discrepancies $\Delta f = \dfrac{\hat{f}_i - f_i}{\hat{f}_i}$ and the mutual MAC value.

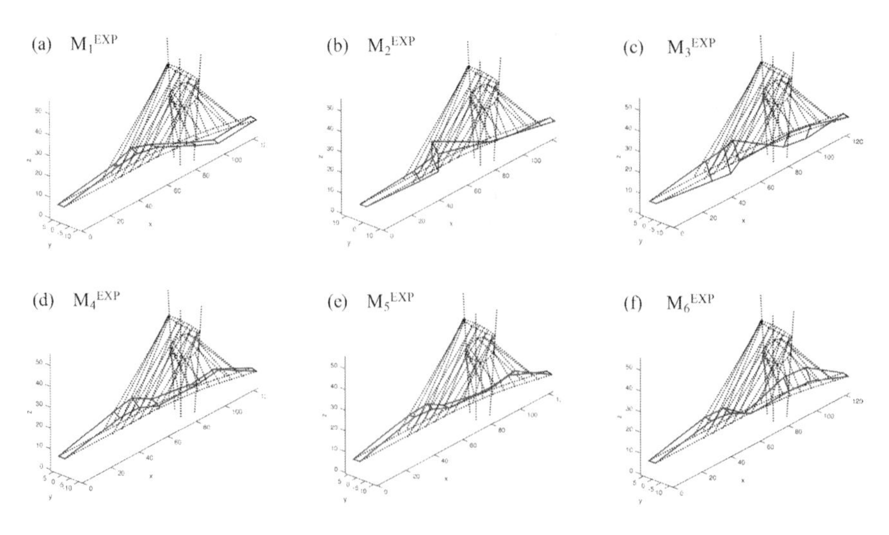

FIGURE 11.1 Vibration modes estimated from experimental data using enhanced frequency domain decomposition (EFDD).

TABLE 11.2
Natural Frequencies Identified from the Numerical Model

Mode	$\widehat{\Phi_1}$	$\widehat{\Phi_2}$	$\widehat{\Phi_3}$	$\widehat{\Phi_4}$	$\widehat{\Phi_5}$	$\widehat{\Phi_6}$
$\hat{f}$(Hz)	1.030	1.514	1.774	2.184	2.365	2.982
Δf(%)	6.40	−10.57	4.56	−13.27	−9.55	−10.89
Diagonal MAC	0.922	0.886	0.734	0.831	0.811	0.920

The relative discrepancies in the natural frequencies range between 4.56% and 13.27%. All the modes display MAC values higher than 0.80, with the exception of the torsional mode shape ($\widehat{\Phi_3}$). Overall, the modal features generally show a fairly good correlation between experimental results and numerical results from the initial FE model. The development of an initial FE model as precisely as possible is crucial to achieve the best model-updating results because it provides a better starting point for a given model-updating problem.

11.10.3 Definition of the Uncertain Parameters' PDF

The uncertain parameter vector Θ contains the set of the uncertain parameters θ and each element of the covariance matrix of the total model error Σ. The set of the uncertain parameters vector $\theta \in R^2$ contains the deck and the cable stiffness described by the steel (i.e., E_{steel}) and the cables (i.e., E_{cables}) elastic modulus $\theta_1 = E_{steel}$ and the cables $\theta_2 = E_{cables}$. The uncertain parameters were selected by means of sensitivity analyses carried out by varying the following: steel and cable elastic modulus, cable tension stiffening, and stiffnesses of rotational/translational springs used to model the soil structure interaction problem. Eigenfrequencies were very sensitive to E_{steel}; eigenvectors were sensitive to both E_{steel} and E_{cables}; and negligible effects on modal properties were found varying in cable tension stiffening and stiffnesses of rotational/translational springs. Further details are reported in Refs. [50, 51]. Since no direct information is available, a normal prior distribution is assumed for both components θ_1 and θ_2. Mean values and the coefficient of variation (cov) of θ_1 and θ_2 PDFs are reported in Table 11.3. Non-informative priors are selected for each component of the random vector $\Sigma \in R^{\frac{M(M-1)}{2}}$ [45, 46].

11.10.4 Development of Polynomial Chaos (PC)-Based Surrogate Models

The natural frequencies $\widehat{f_k}$, $k = 1,\ldots, 6$ and each eigenvector component of the FE model computed mode shapes $\widehat{\Phi_k}$, $k = 1,\ldots, 6$ are set as QoIs and the PC expansion. Eq. (11.10) is used in order to build 90 surrogate models, one for each of the selected QoIs.

TABLE 11.3
Probability Density Functions (PDFs) of the Uncertain Parameters

Parameters	Mean Value (GPa)	COV
θ_1	210	0.15
θ_2	160	0.20

The accuracy of each of the PC-based surrogate models depends on the quality and the quantity of the sampling points. In this work, samples are drawn with the Gaussian quadrature approach using a full tensor grid scheme. The PC expansion is built using two support RVs ξ of polynomials order $p = 3,\ldots, 5$ (typical of the stochastic finite element method [FEM] practical applications). A convergence analysis is carried out in order to select the polynomials order using a two-step procedure. First, the mean and the variance of the error vector $\tilde{e} = \dfrac{\tilde{\mathcal{M}}(x,\theta) - \mathcal{M}(x,\theta)}{\tilde{\mathcal{M}}(x,\theta)}$ are evaluated on the pairs of samples $\{\theta_1, \theta_2\}$ used as the training population for each polynomials order p; and, second, the error vector $\tilde{e}$ is estimated driving the simulation of the parameters $\{\theta_1, \theta_2\}$ in the tail values of the vector θ joint PDF where maximum error is expected. Polynomial order $p = 5$ is found to be the best selection for all the tested pairs of input parameters. Further details about the surrogate model definition and validation can be found in Refs. [50, 51].

Figure 11.2a shows the variances of the errors σ_f and σ_{ms} for each of the first six natural frequencies and associated mode shapes, respectively. Values

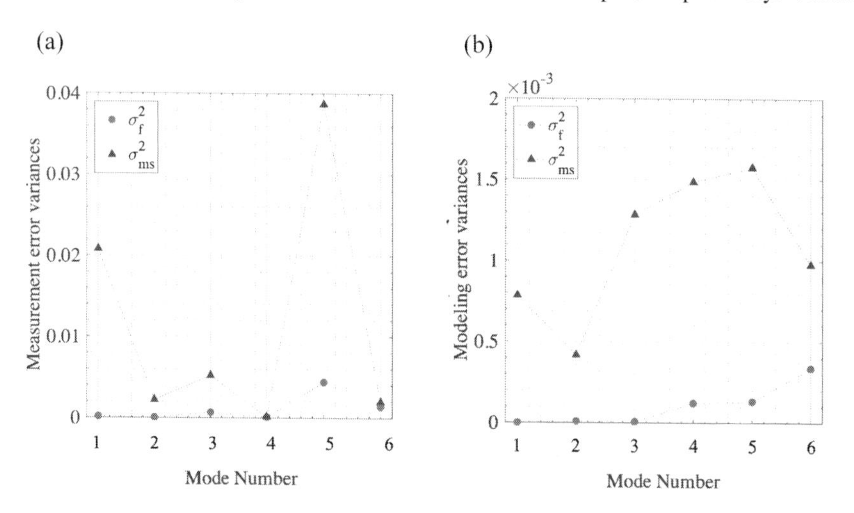

FIGURE 11.2 (a) Error vector variance for the first six natural frequencies and mode shapes with polynomial order 5; (b) measurement error variance for the first six natural frequencies and the first six modal vectors.

lower than 2×10^{-3} are obtained. It is noteworthy that even if the surrogate models are developed for each eigenvector component of all the considered mode shapes, the error between the surrogate and the FE model solution is estimated by using the diagonal MAC coefficient, measuring the degree of correlation between the FE model computed mode shapes and those obtained from the eigenvector response surfaces for each of the pairs of samples $\{\theta_1,\theta_2\}$ used as samples.

11.10.5 Bayesian Updating Framework Using an Improved MCMC MH Algorithm

Bayesian model updating is carried out by following the improved surrogate model–based MCMC MH described in Section 11.9. In the present work, the Bayesian updating is carried out by using three different reference datasets consisting, respectively, of the first measured natural frequency (i.e., $\bar{D}_1=\{f_1\}$), the first six natural frequencies (i.e., $\bar{D}_2=\{f_k\}$, k = 1,…, 6), and the first six natural frequencies and associated vibration modes (i.e., $\bar{D}_3=\{f_k,\Phi_k\}$, k = 1,…, 6) in order to highlight the importance of using the proper informative reference dataset. Results are reported in Figure 11.3.

The posterior marginal PDFs of θ_1 and θ_2 when $\bar{D}_1$ and $\bar{D}_2$ are used as reference datasets are plotted in Figure 11.3a and 11.3d and in Figure 11.3b and 11.3d, respectively. The θ_1 posterior distribution has mean values equal to 190 and 266 GPa, about 0.90 and 1.20 times the mean value of the prior PDF, when datasets $\bar{D}_1$ and $\bar{D}_2$ are used as targets, respectively.

The posterior marginal distribution of θ_2 is very similar to the prior PDF in both cases. This result is expected since the sensitivity analyses showed that the natural frequencies are mainly influenced by the deck stiffness θ_1. It is worth noticing that the use of $\bar{D}_1$ as target provides a reduction in both updating parameters' mean values, while the use of $\bar{D}_2$ as target provides an increase in both updating parameters' mean values. The posterior marginal PDFs of θ_1 and θ_2 when $\bar{D}_3$ is used as reference are plotted in Figure 11.3c and 11.3f. The θ_1 posterior marginal PDF has a mean value equal to 270 GPa, about 1.3 times the prior mean value; the θ_2 posterior distribution has a mean value equal to 183 GPa, about 1.125 times the prior mean value. A significant reduction in the uncertainties of the updated parameters (i.e., a reduction of the PDFs' standard deviation from $\bar{D}_1$ to $\bar{D}_3$) can be observed, meaning that the updating procedure is effective.

11.10.6 Measurement Uncertainties

Since continuous real-time SHM data is not available, the measurement uncertainties quantification is carried out by using simulated data. The acceleration time histories of the updated FE model to different white noise base excitation are used for the system dynamic identification using EFDD by varying signal sampling parameters (i.e., the time length of the signals, the number of frequency lines in the Power Spectral Density (PSD) spectrum). The position and the number of virtual sensors are selected in order to obtain the same configuration of the

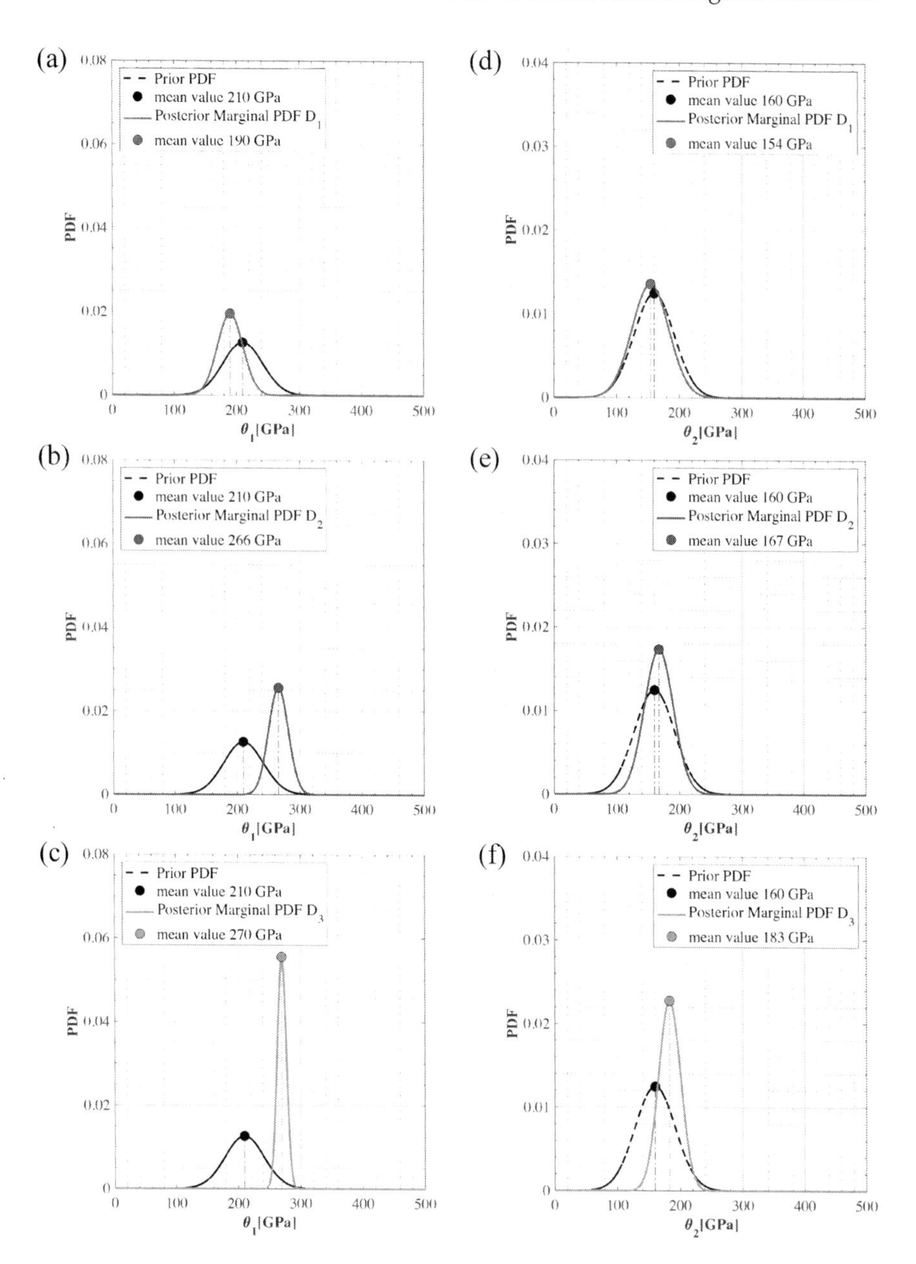

FIGURE 11.3 (Dashed lines) Prior and (solid lines) posterior marginal density function: (a,d) Reference dataset $\bar{D}_1$, (b,e) reference dataset $\bar{D}_2$, and (c,f) reference dataset $\bar{D}_3$. Left panels refer to deck stiffness; right panels refer to cable stiffness.

real experimental data. A total number of dynamic identification runs equal to 48 is performed and used in order to evaluate the variability of the modal parameters, providing a quantification of the variances of the natural frequencies σ_f^2 and the diagonal MAC σ_{ms}^2. Results are shown in Figure 11.2b.

11.10.7 Damage Detection

Before choosing the damage scenario to simulate for testing the potential application of the proposed method, the variations of modal parameters with different damage position and severity were investigated through a parametric analysis in view of the potential use of the proposed procedure for the vibration-based damage detection when highly redundant structures such as cable-stayed footbridges and bridges are of interest. Cable-stayed footbridges are particularly vulnerable to different kinds of events related to tension loss or to a cross section reduction due to cyclic loads or natural corrosion. For this reason, the cable damage was modeled as a reduction of the cable stiffness under different damage scenarios. Effects on eigenfrequencies and eigenvectors due to different damage levels (i.e., a percentage reduction in the cable stiffness), different damage positions, and different damage extensions (i.e., damage diffuse in one or more consecutive cables) were investigated in order to assess the possibility to detect different classes of damage severity. It was found that when different damage levels affect single cables taken alone (damage scenario due to local failure or impulsive high tensile stress in cables), the maximum frequency variations are lower than 3% and the diagonal MAC values variations are very close to 1. In contrast, when damage affects more cables consecutively (damage diffused into a region due to local phenomena affecting the deck), significant frequency and diagonal MAC values variations were observed. It should be pointed out that, also for significant damage severity, the frequency shifts are smaller than the variations due to the operating conditions and signal-processing parameters. In this sense, the diagonal MAC coefficients appear to be a powerful damage indicator, highlighting the importance of developing suitable Bayesian procedures minimizing computational costs to be used in real-time SHM applications. Further details are reported in Ref. [50].

A structural damage scenario characterized by a 60% stiffness reduction of four consecutive cables was numerically simulated. The damaged FE model's continuous response to a Gaussian white noise base excitation is evaluated and used for the dynamic identification process. The evaluated natural frequencies and vibration modes are set as the reference dataset $\overline{D_{DAM}} = \left\{ \widehat{f_k^{DAM}}, \widehat{\Phi_k^{DAM}} \right\}$ with $k = 1,\ldots, 6$ to carry out the Bayesian updating procedures. The real valued random vector $\theta \in R^2$ and each term of the covariance matrix $\Sigma \in R^{21}$ of the total model error are updated. The prior PDFs of θ_1 and θ_2 are assumed to be equal to the posterior PDFs obtained from the Bayesian updating of the initial FE model when both natural frequencies and vibration modes are used as target. The initial covariance matrix of the total model error Σ is assumed to be diagonal with

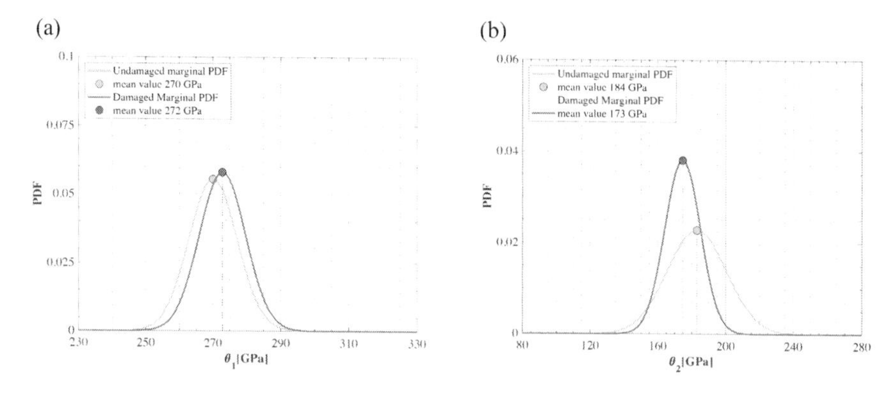

FIGURE 11.4 Marginal posterior probability density functions (PDFs) of the damaged state: (a) deck stiffness θ_1 and (b) cable stiffness θ_2.

components equal to the sum of measurement and the modeling error variances σ_f^2 and σ_{ms}^2 in Figure 11.2.

The results of the proposed updating Bayesian procedure are shown in Figure 11.4. It is observed that the θ_1 posterior distribution has a mean value equal to 273 GPa, about 1.01 times the mean value of the prior PDF θ_1, while the θ_2 posterior distribution has a mean value equal to 173 GPa, about 0.94 times the mean value of the prior PDF θ_2. These obtained results are in good agreement with the simulated damage. It is worth noting that the proposed framework is able to give information also on the localization of damage. In fact, a stiffness loss in the cable consistent with the simulated damage is detected, while the deck stiffness variation is low enough that it can be considered negligible.

11.11 CONCLUSIONS

In this work, a Bayesian method for structural model updating and damage detection was developed. The main idea of the proposed method is to perform a Bayesian model-updating procedure to identify the posterior PDF of the stiffness distributions of undamaged and/or damaged structures. A statically indeterminate cable-stayed footbridge was used to validate the proposed Bayesian method. Ambient vibration tests were carried out to obtain modal data (e.g., natural frequencies and associated vibration modes) to be used as target for the updating of the cable-stayed footbridge model parameters (i.e., deck stiffness and cable stiffness). The posterior marginal PDF of these model parameters was estimated by means of a modified MCMC MH method replacing the solution of the deterministic FE model with the solution given by the surrogate models at each step of evaluation. Three different reference datasets were used in the updating procedure to assess the importance of using adequate datasets for model updating. Results show that the proposed Bayesian method can predict the posterior uncertainties satisfactorily. If all measured modal data is included, the accuracy of the solution is remarkable. The main drawback of the standard Bayesian updating framework in SHM and vibration-based damage detection is related to the excessive computational costs required to find posterior distributions. In contrast, the proposed method based on high-fidelity surrogate models overcomes this problem so that the Bayesian updating framework can be used in SHM practical applications for the identification of damage existence and localization.

In particular, the structural damage detection problem was investigated using simulated data, starting from the results of preliminary parametric numerical analyses carried out to evaluate the effects of different kinds of damage on the structural modal properties. The continuous response simulation of the damaged FE numerical model was evaluated and used to define a virtual reference dataset consisting of both natural frequencies and vibration modes. Setting the obtained modal data as reference, the updating framework was carried out, and the ability of the proposed Bayesian-based SHM procedure to detect the existence and to provide useful information about the localization of severe damage in a statically indeterminate complex structure was discussed. It was found that the proposed Bayesian approach is not only efficient but also accurate.

REFERENCES

1. Heylen, W., Lammens, S., & Sas, P. (1998). *Modal Analysis Theory and Testing*. Katholieke Universiteit Leuven, Faculty of Engineering, Department of Mechanical Engineering, Division of Production Engineering, Machine Design and Automation.
2. Ewins, D. J. (1984). *Modal Testing: Theory and Practice*. Research Studies Press.
3. Rossi, G., Marsili, R., Gusella, V., & Gioffré, M. (2002). Comparison between accelerometer and laser vibrometer to measure traffic excited vibrations on bridges. *Shock and Vibration*, *9*(1–2), 11–18.
4. Rossi, G., Gusella, V., & Gioffré, M. (2008). Performance evaluation of monumental bridges: Testing and monitoring 'Ponte Delle Torri' in Spoleto. *Structure and Infrastructure Engineering*, *4*(2), 95–106.
5. Gioffré, M., Cavalagli, N., Pepi, C., & Trequattrini, M. (2017). Laser doppler and radar interferometer for contactless measurements on unaccessible tie-rods on monumental buildings: Santa Maria Della Consolazione Temple in Todi. *Journal of Physics: Conference Series*, *778*(1), 012008.
6. Ubertini, F. (2014). Effects of cables damage on vertical and torsional eigen properties of suspension bridges. *Journal of Sound and Vibration*, *333*(11), 2404–2421. doi:10.1016/j.jsv.2014.01.027
7. Benedettini, F., & Gentile, C. (2011). Operational modal testing and FE model tuning of a cable-stayed bridge. *Engineering Structures*, *33*(6), 2063–2073. doi:10.1016/j.engstruct.2011.02.046
8. Daniell, W. E., & Macdonald, J. H. G. (2007). Improved finite element modelling of a cable-stayed bridge through systematic manual tuning. *Engineering Structures*, *29*(3), 358–371. doi:10.1016/j.engstruct.2006.05.003
9. Gentile, C., Guidobaldi, M., & Saisi, A. (2016). One-year dynamic monitoring of a historic tower: Damage detection under changing environment. *Meccanica*, *51*, 2873–2889.
10. Masciotta, M. G., Ramos, L. F. T., Lourenco, P., & Vasta, M. (2017). Damage identification and seismic vulnerability assessment of a historic masonry chimney. *Annals of Geophysics*, *60*(4), 252–265.

11. Cabboi, A., Gentile, C., & Saisi, A. (2017). From continuous vibration monitoring to FEM-based damage assessment: Application on a stone-masonry tower. *Construction and Building Materials*, *156*, 252–265. doi:10.1016/j.conbuildmat.2017.08.160
12. Magalhaes, F., Cunha, A., Caetano, E., & Brincker, R. (2010). Damping estimation using free decays and ambient vibration tests. *Mechanical Systems and Signal Processing*, *24*(5), 1274–1290.
13. Pioldi, F., & Rizzi, E. (2017). A refined Frequency Domain Decomposition tool for structural modal monitoring in earthquake engineering. *Earthquake Engineering and Engineering Vibration*, *16*(3), 627–648.
14. Cardoso, R., Cury, A., & Barbosa, F. (2017). A robust methodology for modal parameters estimation applied to SHM. *Mechanical Systems and Signal Processing*, *95*, 24–41.
15. Brincker, R., Zhang, L., & Andersen, P. (2001). Modal identification of output only systems using Frequency Domain Decomposition. *Smart Materials and Structures*, *10*, 441. doi:10.1088/0964-1726/10/3/303
16. Brincker, R., & Zhang, L. (2009). Frequency domain decomposition revisited. *IOMAC 2009 - 3rd International Operational Modal Analysis Conference*, 615–626.
17. Brincker, R., Ventura, C. (2015). *Introduction to Operational Modal Analysis*. Wiley-Blackwell.
18. Simoen, E., De Roeck, G., & Lombaert, G. (2015). Dealing with uncertainty in model updating for damage assessment: A review. *Mechanical Systems and Signal Processing*, *56–57*, 123–149.
19. Marwala, T. (2010). *Finite-Element-Model Updating Using Computational Intelligence Techniques*. Springer-Verlag, London, UK.
20. Tarantola, A. (2005). *Inverse Problem Theory and Methods for Model Parameter Estimation*. doi:10.1137/1.9780898717921
21. Bayes, T. (1763). An essay towards solving a problem in the doctrine of chances. *Philosophical Transactions of the Royal Society*, *53*, 370–418.
22. Beck, J. L., & Katafygiotis, L. S. (1998). Updating models and their uncertainties. I: Bayesian statistical framework. *Journal of Engineering Mechanics*, *124*(4), 455–461.
23. Vanik, M. W., Beck, J. L., & Au, S. K. (2000). Bayesian probabilistic approach to structural health monitoring. *Journal of Engineering Mechanics*, *126*(7), 738–745.
24. Cheung, S. H., & Beck, J. L. (2009). Bayesian model updating using hybrid Monte Carlo simulation with application to structural dynamic models with many uncertain parameters. *Journal of Engineering Mechanics*, *135*(4), 243–255.
25. Jaynes, E. T. (1957). Information theory and statistical mechanics. *Physical Review*, *106*, 620–630.
26. Jeffreys, H. (1946). An invariant form for the prior probability in estimation problems. *Proceedings of the Royal Society of London A: Mathematical, Physical and Engineering Sciences*, *186*(1007), 453–461. doi:10.1098/rspa.1946.0056
27. Lindley, D. V. (1972). *Bayesian Statistics, A Review*. Society for Industrial and Applied Mathematics.
28. Berger, J. O., Bernardo, J. M., & Sun, D. (2009). The formal definition of reference priors. *The Annals of Statistics*, *37*(2), 905–938.
29. Goller, B., & Schueller, G. I. (2011). Investigation of model uncertainties in Bayesian structural model updating. *Journal of sound and vibration*, *330*(25–15), 6122–6136.
30. Bartoli, G., Betti, M., Facchini, L., Marra, A. M., & Monchetti, S. (2017). Bayesian model updating of historic masonry towers through dynamic experimental data. *Procedia Engineering*, *199*, 1258–1263. doi:10.1016/j.proeng.2017.09.267

31. Simoen, E., Papadimitriou, C., & Lombaert, G. (2013). On prediction error correlation in Bayesian model updating. *Journal of Sound and Vibration, 332*(18), 4136–4152. doi:10.1016/j.jsv.2013.03.019
32. Asadollahi, P., Li, J., & Huang, Y. (04 2017). *Prediction-error variance in Bayesian model updating: a comparative study*. 101683P.
33. Behmanesh, I., & Moaveni, B. (2014). Probabilistic identification of simulated damage on the Dowling Hall footbridge through Bayesian finite element model updating. *Structural Control and Health Monitoring, 22*(3), 463–483.
34. Behmanesh, I., Moaveni, B., Lombaert, G., & Papadimitriou, C. (2015). Hierarchical Bayesian model updating for structural identification. *Mechanical Systems and Signal Processing, 64–65*(3849), 360–376.
35. Zhang, E. L., Feissel, P., & Antoni, J. (2011). A comprehensive Bayesian approach for model updating and quantification of modeling errors. *Probabilistic Engineering Mechanics, 26*(4), 550–560.
36. De Falco, A., Mori, M., & Sevieri, G. (2018). Bayesian updating of concrete gravity dams model parameters using static measurements. *Proceedings 6th European Conference on Computational Mechanics - ECCM 6, Glasgow, UK.*
37. Gamerman, D., & Lopes, H. F. (2015). *Markov Chain Monte Carlo: Stochastic Simulation for Bayesian Inference*. Chapman & Hall, CRC2006.
38. Hastings, W. K. (1970). *Monte Carlo Sampling Methods Using Markov Chains and Their Applications* (p. 57, pp. 97–109). Oxford University Press, Biometrika Trust.
39. Nagel, J. B. (2017). *Bayesian techniques for inverse uncertainty quantification*. PhD dissertation, ETH Zurich.
40. Konakli, K., & Sudret, B. (2016). Polynomial meta-models with canonical low-rank approximations: Numerical insights and comparison to sparse polynomial chaos expansions. *Journal of Computational Physics, 321*, 1144–1169. doi:10.1016/j.jcp.2016.06.005
41. Choi, S.-K., Canfield, R., Grandhi, R., & Pettit, C. (2004). Polynomial chaos expansion with Latin hypercube sampling for estimating response variability. *AIAA Journal, 42*, 1191–1198.
42. Sudret, B., Marelli, S., & Wiart, J. (2017). Surrogate models for uncertainty quantification: An overview. *2017 11th European Conference on Antennas and Propagation (EUCAP)*, 793–797.
43. Ferrari, R., Froio, D., Rizzi, E., Gentile, C., & Chatzi, E. N. (2019). Model updating of a historic concrete bridge by sensitivity- and global optimization-based Latin Hypercube Sampling, *Engineering Structures, 179*, 139–160.
44. Au, S. K., Zhang, F. L., & Ni, Y. C. (2013). Bayesian operational modal analysis: Theory, computation, practice. *Computers and Structures, 126*, 3–14.
45. Gardoni, P. (2002). *Probabilistic models and fragility estimates for structural components and systems*. PhD dissertation, Univ. of California, Berkeley, Berkeley, California
46. Gardoni, P., Der Kiureghian, A., & Mosalam, K. M. (2002). Probabilistic capacity models and fragility estimates for reinforced concrete columns based on experimental observations. *Journal of Engineering Mechanics, 128*(10), 1024–1038. doi:10.1061/(ASCE)0733-9399(2002)128:10(1024)
47. Allemang, A. J. (2003). The Modal Assurance Criterion (MAC): Twenty years of use and abuse. *Journal of Sound and Vibrations, 37*(8), 14–21.
48. Box, G. E. P., & Cox, D. R. (1964). An analysis of transformations. *Journal of the Royal Statistical Society Series B: Statistical Methodology*, 26(2): 211–243.
49. SAP2000. Static and dynamic finite element of structures. (2009). *Computers and structures, Inc Berkeley CA USA.*

50. Pepi, C., Gioffré, M., & Grigoriu, M. D. (2019). Parameters identification of cable stayed footbridges using Bayesian inference. *Meccanica, 54*(9), 1403–1419.
51. Pepi, C., & Gioffré, M. (2019). Vibaration based Bayesian inference for finite element model parameters estimation and damage detection. *Proceedings of the AIMETA 2019 XXIV - Conference of the Italian Association of Theoretical and Applied Mechanics Rome, Italy, 15–19 September 2019.*

12 Image Processing for Structural Health Monitoring

The Resilience of Computer Vision–Based Monitoring Systems and Their Measurement

Nisrine Makhoul, Dimitra V. Achillopoulou, Nikoleta K. Stamataki, and Rolands Kromanis

12.1 OVERVIEW

In today's Information Age, the use of emerging technologies and techniques, such as machine learning (ML), has made structural health monitoring (SHM) systems reliable, and their measurements accurate and recoverable. Computer vision (CV)-based systems offer remote measurement of structures' responses (e.g., vertical deflections, accelerations) or damage patterns on their surface (e.g., cracks). In comparison to measurements collected with contact sensors (e.g., displacement sensors and accelerometers), CV-based measurements, such as from cameras, can be fairly easily recovered in case of sensor malfunctions, or easily predicted using novel measurement interpretation techniques (e.g., ML algorithms). The quality of the measurements depends on environmental and camera-specific factors. The replacement of CV-based systems is effortless once they become defective due to their resilient dimension. The adaptable and sustainable nature of the smart, sophisticated CV-based systems, which also include measurement interpretation tools, emphasizes their suitability for monitoring various external stressors, hazards, and risks due to climate change. This chapter (i) defines and explains the resilience dimension of the CV-based systems and their measurements, and (ii) informs the quantification of the resilience curves for structures being monitored (i.e., by updating the fragility functions and enriching the recovery function library).

DOI: 10.1201/9781003306924-12

12.2 INTRODUCTION

Our aging civil infrastructure must be kept at a certain level of structural integrity and safety with an optimal usage of resources to enable societal activities. This is especially true under the current and upcoming economically challenging times, associated with the post-pandemic era and societal unrest, in order to allow economic growth using technological novelties and fulfilling sustainability criteria (Ivankovic et al., 2019).

Bridges are crucial transportation links of civil infrastructure joining two or more areas by crossing one or more obstacles. These assets facilitate product and civilian travel; therefore, their proper functioning has a great impact on the economy and society. The maintenance and detection of damage or faults in a timely manner are crucial since the safety of the bridges is important. Not only do poorly maintained and aged bridges collapse, such as the Morandi Bridge (Calvi et al., 2019) in Italy, but also newly constructed bridges fail. For example, in Norway, the timber bridges Perkolo (Solberg, 2016) and the Tretten (NRK, 2022) failed after operating for two and 10 years, respectively.

To keep bridges safe, regular and detailed inspections are carried out every two and six years, respectively. This, however, is not frequent enough to detect early defects. SHM has been introduced to aid bridge management. It offers sensors and measurement interpretation techniques to monitor conditions of bridges periodically and continuously (Brownjohn, 2007), as well as in real time. A range of sensors and sensing technologies are available. The most frequently used SHM systems consist of contact sensors, such as electric strain gauges as well as fiber-optic and temperature sensors. Contact sensors are exposed to the harsh environment and may be damaged or vandalized, or they may malfunction. Their replacement usually is expensive and labor-intensive. Also, interpretation of the measurements can be challenging. Technologically emerging approaches such as CV-based measurement are promising. The chaotic concept of resilience is also projected to SHM systems that are needed to be smart, sustainable, and adaptive but not necessarily complicated. Modern technology introduces a great range of innovative methods and techniques and has enhanced (or combined already existing) systems, such as CV-based SHM systems.

A CV-based SHM system consists of digital camera(s), a computer, an image-processing toolbox, and some mechanical components, such as a tripod or a pan-tilt platform (Kromanis & Forbes, 2019). The main advantages of CV-based systems are their easy application and low cost, making them an attractive option for asset managers. CV-based systems usually are not installed on the bridges, but at a suitable distance away from them. This way, they can be easily replaced without disrupting data collection.

CV-based systems quantify resilience at the micro and macro scales and are thought to be resilient themselves, since they are low cost, can be replaced, are adaptive, and can be paired with emerging technologies such as ML and artificial intelligence (AI) to interpret and predict bridge responses. Therefore, combining such systems with the novelties of the 5G informs resilience at any level and can make reliable predictions in order to update the current codes for natural and human-induced hazards.

However, the resilience of CV-based monitoring systems has not been debated before. For this purpose, Sections 12.3 through 12.7 introduce CV-based SHM systems and their measurements, techniques, and limitations. Sections 12.8 through 12.11 debate the importance of resilience and its components, the quantification of resilience, and how resilience is informed by modern SHM systems. Then Sections 12.12 through 12.14 discuss emerging technologies in monitoring, specifically AI and ML with a focus on SHM. And, finally, Section 12.15 concludes the chapter.

12.3 CV-BASED SHM SYSTEMS

Modern digital cameras offering high image resolution (the number of pixels in an image frame) and high frame rate (the number of image frames captured in a second) are available at low costs. The main element of a camera is an image sensor that enables capturing images. Today, image sensors for affordable cameras are manufactured in a wide range of sizes and quality. Some modern smartphones have four rear cameras and at least one front camera. In the field of civil SHM, each pixel of an image frame can present information relevant to the condition of a structure. Still, images are used predominantly to perform monitoring at a local level, in which cracks, spalling, and buckling, for example, can be detected, characterized, quantified, and further analyzed and interpreted using software or algorithms. Multiple consecutively collected images are turned into a video, creating a time history of the changes in image frames. Videos allow monitoring at a global level, in which bridge dynamic, static, and quasi-static response under applied loads such as traffic and temperature is captured. This chapter focuses on CV-based SHM systems that capture the response of structures at a global level.

12.4 MEASUREMENTS

In a CV-based measurement approach, a target(s) is selected in the reference frame; usually, it is the first image frame of the video. The location of the target(s) is then detected in the consecutive image frames. Pixel motions of targets are then converted to displacements in engineering units (e.g., millimeters), or their derivatives such as strains and inclination angles. Figure 12.1 shows the application of the CV-based measurement for estimating bridge vibration parameters. Multiple cameras, smartphones, and a modified GoPro camera with a telescopic lens are used for measurement collection (Figure 12.1a). The cameras are placed at suitable and safe distances from the bridge and focused on an area of interest. The first image frame capturing the field of view of the GoPro action camera is given in Figure 12.1b. This figure also shows a selected region of interest (ROI) within which lies a target or object of interest. The motion of the target is extracted from each image frame, forming a time history of pixel motion (see Figure 12.1c). Parameters corresponding to the bridge dynamic response, such as

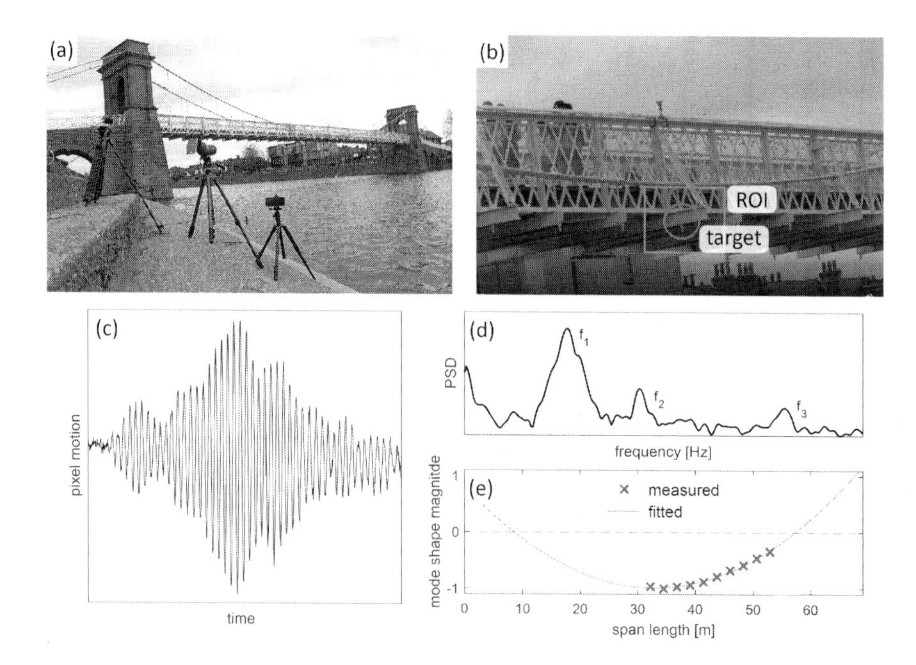

FIGURE 12.1 The Wilford Suspension Bridge monitoring: (a) Camera setup, (b) image frame with a region of interest (ROI) and a target, (c) target pixel motion, (d) power spectrum density (PSD) from the pixel motion (where f_i is the frequency at ith mode), and (e) estimated mode shape at the first frequency.

its vibration frequencies and corresponding mode shape, can then be estimated (see Figure 12.1d and 12.1e).

12.5 CV-BASED TECHNIQUES

Although the main advantage of CV-based measurement is the ability to collect distributed measurements, in some applications CV-based systems serve for measuring displacements of a single target. In general, four techniques are the most commonly deployed for target tracking in civil SHM applications. These techniques are briefly discussed below and in-depth in Xu and Brownjohn (2018) and Dong and Catbas (2021).

1. *Template matching.* A target is usually defined as a rectangular area within the ROI(s). The area that is most closely matching with the template is sought in the ROI(s) of the consecutive image frames. The technique is quite frequently deployed in the SHM field (Feng, Feng, Ozer, & Fukuda, 2015; Xu, Brownjohn, & Kong, 2018). This technique offers high measurement resolution, but it comes at a cost—much time is needed for the computation of target displacements.

2. *Feature point (or keypoint) matching.* In this technique, salient features such as corners, edges, and blobs are first detected (in the reference image frame) and then sought in the consecutive image frames in the defined ROI(s) (Khuc & Catbas, 2017; Lydon et al., 2018). The main challenge is to make the algorithm robust to missing feature points in the image frames. This could be due to the image being blurred (due to the vibration of the structure) or a change of brightness (environmental conditions).
3. *Optical flow.* Similarly to template matching, the target is defined as an area in the ROI(s) in which the optical flow of pixels is estimated. Optical flow is defined as the velocity of movement resulting from the brightness shift of the pattern or target in image frames (Beauchemin & Barron, 1995). Usually in structural monitoring, the Kanade–Lucas–Tomasi method, which assumes the brightness consistency and small displacement with neighboring points or pixels, is deployed (Lydon et al., 2018; Zhu, Lu, & Zhang, 2021). The application of the technique can also be extended to measuring full-field displacements (Bhowmick, Nagarajaiah, & Lai, 2020).
4. *Shape-based tracking.* In the shape-based-tracking approach, a known shape is sought, and its location is tracked in the ROI(s). The shape is defined by the user and usually is attached to the surface of the structure. The shapes can be, for example, Aruco markers (Kromanis & Forbes, 2019; Kalybek, Bocian, & Nikitas, 2021), individual blobs (circles) (Kromanis, Lydon, Martinez del Rincon, & Al-Habaibeh, 2019; Kromanis & Kripakaran, 2021), or their ensemble (e.g., four spots) (Lee & Shinozuka, 2006). The disadvantage of this technique is that, usually, a target needs to be installed on the structure. However, the fact that the target has known dimensions makes the conversion of displacement from pixels to engineering units much easier compared to the other techniques.

12.6 LIMITATIONS

The main element of CV-based SHM systems is the camera, which is attached, most frequently, to a tripod or, sometimes, to a fixed support system. The camera is, therefore, also the most vulnerable element of the system. The main external factors affecting the quality of the images and, consequently, the measurement accuracy and resolution are (i) ground motions induced by surrounding infrastructure (e.g., trains and vehicles); (ii) wind shaking the camera system; and (iii) rain, fog, haze, birds, and so on obstructing targets or making the field of view unclear or blurry. Also, camera-related factors such as internal camera drift can affect the quality of measurement. Providing that the quality of images is good, the other cause of poor measurement accuracy can be related to image-processing techniques. For example, a change in brightness (day to night) may affect the detection of the target in the new images.

Further parameters that affect the response or the functionality of CV-based systems, apart from hardware issues (e.g., decaying wiring cables), are the interruption of the network due to internet connection or even software issues in processing or storing data.

12.7 RESILIENCE

Natural and manmade hazards are becoming increasingly frequent, inflicting severe consequences on the infrastructure, societies, and the environment. Thus, preserving the functionality and operability of the critical infrastructure is crucial during and after hazard events. This allows for minimizing the losses (i.e., direct and indirect) and mitigating the aftermath of disasters. Recently, the new philosophy of resilience thinking emerged in the assessment and design of infrastructure, along with urban planning, while dealing with external acute shocks and chronic stresses. Resilience was defined by the Federal Highways Administration (FHWA) Order 5520 as "the ability to anticipate, prepare for, and adapt to changing conditions and withstand, respond to, and recover rapidly from disruptions."

Within this context, the concept of resilience-based engineering goes beyond risk-based engineering. Thus, it aims to assess the community and/or system adaptation, and recovery from shocks in addition to classical mitigations. National and local disaster mitigation strategies have shifted their focus from disaster response to prevention and preparedness (Makhoul, 2015; Makhoul & Argyroudis, 2018; Makhoul & Argyroudis, 2019).

12.8 IMPORTANCE

The importance of resilience comes from the capabilities of structures, infrastructures, and so on to adequately accommodate certain events or stressors (e.g., earthquake, tsunami, or climate change) throughout their life. Those capabilities are (i) to prepare and plan before an event or stressor, (ii) to absorb shocks and stresses generated by an event, (iii) to recover from this incident or hazard, and (iv) to adapt to the endured risks and retain improvement lessons for the future. Traditional resilience aims to be studied at variate levels (i.e., material, component, structural, network, infrastructure, system, the system of systems, and city levels). Resilience considers interdependencies between multiple levels and hazards. Community resilience encompasses all the aforementioned, and considers the socio-economic interactions, targeting thus a more realistic representation of the phenomenon and an amelioration of the decision-making process.

However, quantitative metrics and tools for the resilience investigation of socio-technical systems are not yet well established, and standards and procedures are still emerging (Bruno & Clegg, 2015; Lloyd's Register Foundation, 2015). The complex issue of quantifying the resilience of civil infrastructure can be addressed on many levels, starting from a component scale by measuring engineering data (e.g., cracks, deflections, and material properties) up to the structure or network scale by measuring functional or operational data. Certain types of SHM such as CV-based SHM systems help create records containing large amounts and diverse kinds of data throughout the lifetime of the bridge. However, in the current increased digitalization era, until now research has not addressed the resilience of SHM systems as well as CV-based SHM. Effectively, the built environment is becoming more reliant on SHM (including CV-based SHM) data and information for the prediction of hazards, performances, and losses. Therefore, it becomes of utmost importance to ensure the resilience of those monitoring systems.

12.9 COMPONENTS

Resilience quantification is a data-based procedure accounting for proper criteria throughout the distinct time periods of an asset: (a) the initial state, (b) absorption, (c) the idle period, (d) recovery, and (e) adaptation. The quantification includes four main components: (i) structural integrity, (ii) functionality, (iii) operations, and (iv) resources, each one containing a group of parameters (Achillopoulou, Stamataki, Psathas, Iliadis, & Karabinis, 2022). Thus, in the case that the asset is a reinforced concrete bridge, the structural integrity is indicated by a group of parameters such as the material properties (e.g., strength), the width of a crack, strains, and deflections. The component of functionality can be described by parameters such as the changes in traffic load, the number of lanes that are not in function, and the traffic regulations. Regarding an asset—for instance, a critical transport infrastructure asset such as a bridge—the operations include but are not limited to parameters such as the type of monitoring system, analytics, data management, and storage. Resources can be estimated both in monetary terms and in staff needed to run, manage, maintain, and repair the asset. Lately, due to climate change, extreme events happen more and more often, having tremendous and catastrophic consequences on the built environment. The occurrence of a natural hazard is indirectly accounted for in each component of resilience. For instance, in the design process, seismic, wind, or snow loads are taken into consideration depending on the probability of occurrence and the return period or the location characterized by an increased reliability level.

Each parameter is taken into account depending on a scale of importance ranging from 1 to 4, with 4 corresponding to a high importance level. For instance, at the initial and absorption states, the importance level of the material's strength is considered higher, while in the idle period, recovery, and adaptation, the importance level is lower since decaying is expected and it is not necessarily the most crucial parameter upon which to base decisions (or for the decision-making process). Another important aspect is the reliability level of each quantity. For example, the reliability level of a chosen material's strength is considered to range from 95% to 99% at the initial design. After this period, depending on the damage type and monitoring data available, the material's reliability level decreases, especially close to the end of the life of an asset. The reliability level throughout the life cycle of the asset is included by applying a reducing factor. The average number for all parameters consists of the resilience value. The overall resilience is the sum of the components, each one multiplied with a participation factor that depends on the age of the structure and the criticality of the components of this period (Achillopoulou, Stamataki, Psathas, Iliadis, & Karabinis, 2022).

The estimation of the participation factors of each component results from a sensitivity analysis based on a specific or targeted resilience level. The more accurate the data and the technology used are, the more correspondent the resilience value is to the performance of the structure. For instance, during preparedness periods, the targeted resilience level can be informed by various systems, directly or indirectly, each one having a particular accuracy depending on the equipment, acquiring method, and conditions. For example, in the case of visual inspections, which is the oldest technique for the evaluation of a bridge's condition, there are some limitations and difficulties. In cases where components of a bridge are inaccessible, this technique is considered unsuitable. During this kind of inspection, the functionality of the bridge may be disrupted. Therefore, digital and CV-based systems, such as CV-based SHM systems, are proven to be efficient for the collection of data with high accuracy without affecting the functionality of the bridge, enhancing the interpretation and validation of the condition of the bridge in an easy, expedient, and objective manner (Spencer, Hoskere, & Narazaki, 2019).

As such, technologically advanced methods assist in a better understanding of the capacity and the response of the structure by informing resilience. Especially in the climate change era during extreme events or unpredicted conditions, emerging technologies used in assessment methods aim at the evaluation of response and updating of existing performance and recovery models (Spencer, Hoskere, & Narazaki, 2019; Argyroudis et al., 2022). In order to do that, the methods themselves need to enable a certain level of resilience.

12.10 QUANTIFICATION

Over the past decade, assessment of the resilience of critical infrastructure facing natural and extreme hazards and events attracted great research efforts. Numerous available resilience assessment methods were investigated, such as (i) quantitative and qualitative assessment; (ii) deterministic and probabilistic techniques; and (iii) components, systems, networks, and systems of systems.

Those methods suggest different infrastructure and system phases for responding to hazards. Those phases are characterized by four abilities, which are to (i) absorb the consequences of the hazardous event, (ii) adapt to the changing situation, (iii) quickly recover and restore the functionality of the infrastructure or system under consideration, and (iv) successfully respond to disorder caused by the hazard event by efficiently using the available resources (e.g., manpower, equipment, and critical resources for operations of response and recovery).

1. The *attribute-based method* (Fisher & Norman, 2010; Pettit, Fiksel, & Croxton, 2010) focuses on the system's resilience properties (i.e., what makes it resilient), which are robustness, resourcefulness, adaptivity, and recoverability. This is done by assessing the degree of presence of these properties within the studied system. These methods help assess resilience qualitatively or semi-quantitatively. However, they lack practical assessment of the extent of the system's ability to operate properly while

facing disruptions, or the efficiency of prospective resilience improvement and investment, hence the need for the performance-based method.
2. The *performance-based approach* (Bruneau et al., 2003; Rose, 2007; Alderson, Brown, & Carlyle, 2015) quantifies the resilience of an infrastructure or system. It offers the capability of interpreting and formulating quantitatively the infrastructure resilience metrics and thus facilitates comparative analyses. Furthermore, those methods are suitable for cost–benefit and planning analyses, as the metrics allow calculating the profits and costs related to the recommended resilience improvements and investments.

The performance-based methods are more complex and time-consuming, as they depend on computational models and require more resources. Those two methods (i.e., attribute and performance-based) are complementary, as they have their limitations and advantages. However, if combined, they provide the stakeholders with an enhanced understanding of infrastructure resilience. Other resilience methods are also available, such as agent-based among others which (a) considers complex networks of systems, (b) represent users, (c) control, and (d) approximative reasoning including ML fuzzy logic (Afrasiabi, Tavana, & Di Caprio, 2022).

Furthermore, numerous assessment procedures for resilience are offered. The basic resilience quantification models for a single structure are the resilience triangle (Bruneau et al., 2003), availability-based resilience (Ayyub, 2014), and the simplified model (Ayyub, 2015). This resilience quantification basis increases in sophistication conditional on the resilience needs or scales, such as (i) a single facility or system; (ii) systems, by using a simplified approach; (iii) a system of systems utilizing a practical framework; and (iv) infrastructure networks.

12.10.1 A Simple Resilience Metric

Many have offered resilience metric formulations with varying levels of sophistication (Cimellaro, Reinhorn, & Bruneau, 2006; Bruneau & Reinhorn, 2007; Reed, Kapur, & Christie, 2009; Cimellaro, Reinhorn, & Bruneau, 2010; Ouyang & Duenas-Osorio, 2014; Nan & Sansavini, 2016). The initial and most basic formulation is presented in Eq. (12.1) as a simple formula for a single event:

$$R = \int_{t_A}^{t_B} Q(t)/T_{AB}\, dt \tag{12.1}$$

where $Q(t)$ is the functionality, which is a continuous function that ranges from 0 to 100%. The 0% is the total performance loss, and the 100% is the no-performance reduction. The time is represented by t, which varies between the initial time t_A of an extreme event E under consideration and the recovery time t_B from this event E. The total recovery time is given by $T_{AB} = t_B - t_A$.

The functionality depends on the initial and indirect losses produced by a certain catastrophic event (E). When E occurs, the functionality of the asset drops and is expressed by a loss function. Losses are divided into structural and nonstructural. The nonstructural losses are time-dependent (i.e., they will keep decreasing until recovery measures start taking place). The losses involve the casualties as well as economic losses (direct and indirect), which in turn encompass social and economic components, and thus are more difficult to quantify.

The structural losses are the easiest ones to compute and are thus widely available. They are articulated as a percentage of the structural repair and replacement costs and computed by involving the structural fragility function. Fragility functions are defined as the probability of exceeding a limit state performance given the occurrence of an extreme event. They are expressed by $P(LS/IM = Y)$, where P is the probability of exceeding a limit state given an intensity measure, LS is the structure or structural component limit state (denoted sometimes by DS, the damage state), IM is the intensity measure (e.g., ground motion), and Y is the level of realization of the intensity measure.

Finally, the recovery functions represent the functionality function after catastrophic event E and the losses endured. It can take a variety of shapes (Cimellaro, Reinhorn, & Bruneau, 2006; Cimellaro, Reinhorn, & Bruneau, 2010; Bocchini, Decò, & Frangopol, 2012), which depend on society's preparedness and recovery capabilities.

12.11 HOW MODERN SHM SYSTEMS INFORM RESILIENCE

Generally, SHM (short- or long-term) can inform infrastructure resilience at many analysis levels. It can vary from the local to global levels (e.g., material, component, structure, network, system, and regional levels). It can also inform before, during, and after the hazard event. Sousa, Santos, and Makhoul (2022) reported the foremost SHM contributions to infrastructure resilience qualities: (i) SHM information allows for improved maintenance strategies and optimization, which help increase infrastructure **robustness**. (ii) The identification of alternative tracks, for example in case of damaged network components, enhances the infrastructure network **redundancy**. (iii) In emergency situations, SHM can be used to detect failures, identify priorities, and activate resources (i.e., enhancing **resourcefulness**). (iv) SHM supports fast recovery by alerting first responders in the after-event (i.e., enhancing **rapidity**).

Lately, studies have pointed out that monitoring systems are proven to enhance resilience (Achillopoulou, Mitoulis, Argyroudis, & Wang, 2020; Argyroudis et al., 2022). The various systems permit the acquisition of data regardless of the condition, the position, and the type of the asset, without necessarily disrupting its function. The massive quantity of real data and measurements helps in making immediate, direct, and indirect conclusions about the function of the asset, and it is strong documentation for further assessment. Especially when the information gathered is cross-checked by data acquired from various systems, the reliability level and accuracy are enhanced. Lastly, this kind of data-driven assessment facilitates decision making for end users and stakeholders.

A very characteristic example of those kinds of modern systems is CV-based SHM systems, which enable measurements that can contribute to making conclusions and quantifying all distinct components of resilience. The architecture of modern CV-based SHM systems introduces a different dimension of resilience in a more local sense, which is the monitoring system level. This means that a resilient monitoring system better supports the quantification of the resilience of an asset (Borah, Al-Habaibeh, Kromanis, & Kaveh, 2020).

Such systems need both hardware (e.g., sensors, cameras, and mobile phones) and software parts (e.g., databases and process algorithms) as well as an internet network (e.g., cloud storage) (Borah, Al-Habaibeh, Kromanis, & Kaveh, 2020; Dong & Catbas, 2021). The resilience at the SHM level can also be described with an approach that is analogous to the one for an asset. However, the distinct periods (Figure 12.2 nomenclature) and their duration do not necessarily coincide with the resilience periods on a more global level (e.g., bridge resilience). Obviously, it is system dependent to quantify local resilience (sensors, wiring, internet, software, algorithms, output, time, environmental conditions, maintenance of the system, access, and users). For example, in cases of extreme environmental conditions such as strong wind or high temperature, the sensor can be dislocated or even fall; thus, the lens of a camera of a CV-based SHM can possibly break down, and the measurement can be lost. Therefore, the system's functionality is affected, and an immediate replacement is needed. Similarly, the case of a power shutoff, which results in the disconnection of the network or even a bug in the algorithm used, affects the resilience quantification or further model updating at both the local and global levels, increasing uncertainty.

At this local level, many factors affect differently the form of the resilience curve. For instance, for the CV-based SHM systems, the main factors can be (i) the interruption of the network or the CV-based SHM system's aging (e.g., restart of the computer, internet disconnection, a need for cable replacement, or a bug in the AI algorithm), as well as (ii) the occurrence of cumulative events (e.g. fault measurements, or the breakdown of the camera or camera lenses) and (iii) the environmental conditions (e.g., wind, fog, rain, and light conditions). Therefore, depending on the occurrence of an extreme event and the exposure conditions, especially nowadays with climate change events (e.g., floods, wildfires, and hurricanes), there is an impact on the resilience components of operation and resources. For example, the absorption of an occurrence when monitoring with a CV-based SHM system can have two forms (Figure 12.2): either (1) an abrupt change, for example, the interruption of the system and hardware damage; or (2) an exponential form, when cumulative events that affect the CV-based SHM system's response occur, such as the need for system upgrading, the decaying of wiring or cables, and damage to the system's components due to environmental exposure and conditions such as wind, rain, and fog. Additionally, the recovery period depending on the mitigation strategy follows (a) a steep form (immediate recovery), (b) a linear form, or (c) a trigonometric form, resulting in adapting the system's hardware and software in future time.

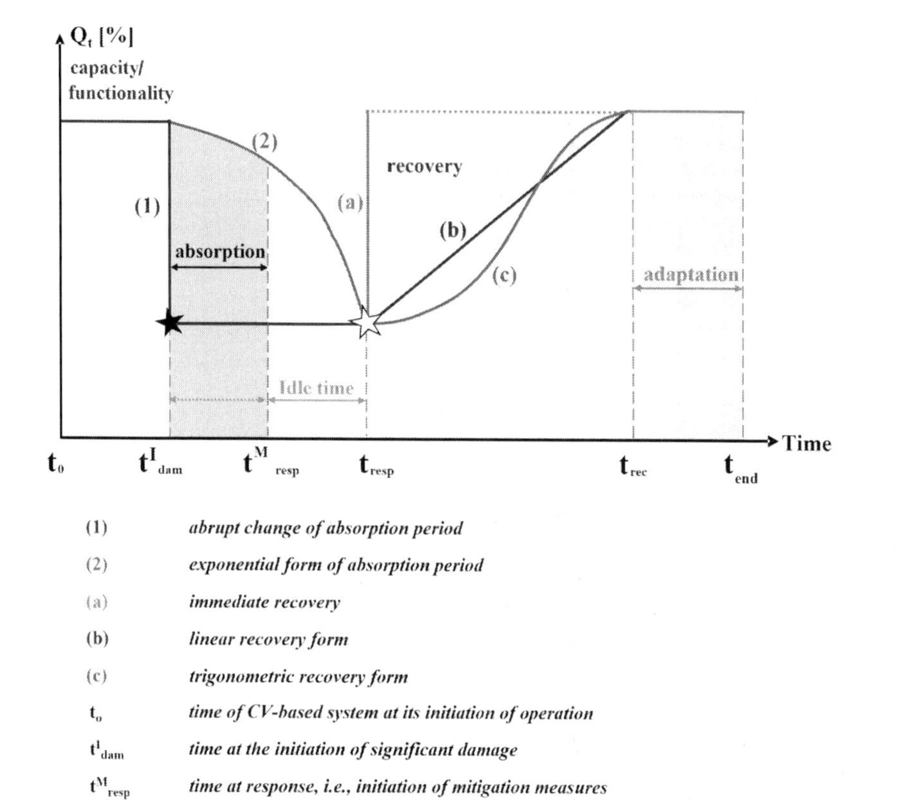

FIGURE 12.2 Different forms of computer vision (CV)-based structural health monitoring (SHM) systems' resilience curve.

12.12 EMERGING TECHNOLOGIES IN MONITORING

Precaution, prevention, and management of infrastructures before, while, and after the occurrence of an extreme event have become critical issues in the past decade supported by novel technological achievements. Hurricanes, wildfires, thermal extremes, heavy rain, snow, and earthquakes are only some examples of hazards that were amplified or became more frequent due to climate change worldwide.

Nowadays, emerging technologies have offered unique solutions for the monitoring of structures either with the use of contact sensors (e.g., SHM systems such as accelerometers, geophones, and thermometers) or remotely (e.g., CV-based SHM systems such as cameras, mobile phones, and drones), combining AI and ML. Thus, tools such as AI and ML are necessary for collecting, analyzing, predicting, and managing data in order to extract useful information about the asset's condition and functionality as well as to inform resilience, assist in taking decisions, and manage the asset (prevent, repair, and restore). Based on real data, machine learning models (MLMs) can make future predictions and inform or update modeling, resulting in reliable information regarding the probability of the occurrence of extreme events. Therefore, AI algorithms and MLMs contribute to a safer design of structures.

12.13 ARTIFICIAL INTELLIGENCE WITH A FOCUS ON SHM

AI has proven to be an effective tool for solving complex problems while considering embedded uncertainties, and it has been used to express cognitive functions. In order to achieve this, algorithms describing many logical operations are developed for increasing the speed of data processing, decreasing errors, and improving computational efficiency. Those AI algorithms have a great impact on the field of structural engineering, and various AI methods have been applied in the different fields of civil engineering through the present. Fuzzy logic, neural networks, autoregressive models, and ML are the methods that have gathered attention in the past decade. The focus on these methods is due to their main advantage of being reliable and efficient tools in the field of structural engineering (Salehi & Burgueño, 2018; Tyrtaiou, Papaleonidas, Elenas, & Iliadis, 2020; Thai, 2022).

Moreover, AI methods have been widely applied in SHM systems to face the challenge of interpreting complex or big data acquired by sensors. Among the potentials of combining AI and SHM systems are reductions in diagnosis time and error rate in monitoring. AI assists in the condition assessment of an asset by identifying the critical parameters, making reliability analyses, identifying load patterns or extreme loads on structures (e.g., traffic load, extreme temperature, wind, or snow), analyzing load effect, early-warning assisting in decision making, and recognizing the fault measurements of sensors (Sun, Shang, Xia, Bhowmick & Nagarajaiah, 2020).

Focusing on the modern CV-based SHM systems, AI algorithms can enable the detection of ROIs by using object detection algorithms, image-labeling processes (e.g., different structural elements or materials, crack labeling, and damage detection) with the utilization of data-clustering algorithms, the interpretation of sensors' data (e.g., strain values and acceleration), optical flow estimation (e.g., traffic flow estimation), and future predictions based on big data. Although some critical parameters can affect the reliability of the AI algorithms, including image blur or brightness level, MLMs can assist in enhancing reliability by identifying errors (e.g., fault measurements) and updating algorithms (Spencer, Hoskere, & Narazaki, 2019; Dong & Catbas, 2021; Nian, 2021).

12.14 MACHINE LEARNING WITH A FOCUS ON SHM

ML is one of the most famous AI methods that has been developed and improved in the past few years. Those models utilize AI learning algorithms and monitoring data from SHM and CV-based SHM systems to build an MLM. The MLM is a trained computer system that generates data, makes accurate predictions, and is able to identify errors or fault measurements to update itself (Salehi & Burgueño, 2018; Thai, 2022).

Based on the utilized algorithms, the MLM decision is made according to the three available learning methods, which are (i) supervised learning (e.g., classification or regression), (ii) unsupervised learning (e.g., clustering), and (iii) reinforcement learning (Derras & Makhoul, 2022; Thai, 2022), and it follows the data pre-processing. After the MLM decision, the pre-processed data is divided into training and testing sets in order to begin an iterative process of developing, training, and validating the ML algorithm. Once the algorithm is trained, evaluation criteria are used for the MLM. The evaluation criteria for a well-trained MLM performance can either be a sensitivity analysis or an accuracy analysis as well as evaluation indices (e.g., root mean square error [RMSE], mean absolute error [MAE], and R square [R^2]). However, a crucial parameter of the MLM design is the imbalanced data classes that can lead to fault results, and thus a greater range of data should be secured for reliable predictions (Kostinakis, Morfidis, Demertzis, & Iliadis, 2022; Psathas et al., 2022).

Today, the application of MLMs is popular in SHM and CV-based SHM systems, as they can make reliable predictions for the structure's response by acquiring those monitoring data (e.g., strain values, acceleration, displacement, and deflections). Also, MLMs can interpret sensor data, identify damages and label them, make risk analyses, estimate fragility functions and resilience curves, as well as identify fault measurements by utilizing and combining the proper AI algorithms and setting engineering criteria (Lazaridis et al., 2022; Psathas et al., 2022). Therefore, the MLM enhances the reliability of the forecast through the optimization of the algorithm and assists in decision making as well as in model updating for designing and assessment of the structures. Finally, regarding the CV-based SHM system, those models are also used for the detection of either objects (e.g., vehicles or cracks) or ROI, the automatic classification using images, and the interpretation of sensor data (Salehi & Burgueño, 2018; Thai, 2022).

12.15 CONCLUSIONS

In the past decades, bridge engineering and its management have become increasingly reliant on monitoring systems (sensors and sensing techniques) and measurement interpretation techniques. This chapter takes interest specifically in image processing for structural health monitoring (SHM) and discusses the resilience of computer vision–based (CV-based) monitoring systems and their measurements. Effectively, SHM and CV-based SHM are very interesting technologies and have proven their benefits; however, are they resilient enough to withstand likely shocks and stresses?

To investigate this idea, this chapter discusses first the CV-based SHM systems and their measurements, techniques, and limitations. Then it examines the importance of resilience, its components, and its quantification, and explains how resilience is informed by modern SHM systems. The analysis shows that CV-based SHM systems are interesting in terms of resilience, as they can be easily replaced and recovered along with their measurements in the event of an unfortunate damaging hazard. All of this can be done at low costs and in a short recovery time. Finally, this chapter presents emerging technologies in monitoring, specifically artificial intelligence and machine learning with a focus on CV-based SHM, as a perspective for future enhanced monitoring solutions and improved resilience.

REFERENCES

Achillopoulou, D. V., Mitoulis, S. A., Argyroudis, S. A., & Wang, Y. (2020). Monitoring of transport infrastructure exposed to multiple hazards: A roadmap for building resilience. *Science of the Total Environment 746*, pp. 1–25.

Achillopoulou, D., Stamataki, N., Psathas, A., Iliadis, L., & Karabinis, A. (2022). Resilience Quantification based on Monitoring & Prediction data Using Artificial Intelligence (AI). *IABSE Congress Nanjing 2022 -Bridges & Structures: Connection, Integration and harmonisation (accepted).*

Afrasiabi, A., Tavana, M., & Di Caprio, D. (2022). An extended hybrid fuzzy multi-criteria decision model for sustainable and resilient supplier selection. *Environmental Science & Pollution Research, 29*, 37291–37314. https://doi.org/10.1007/s11356-021-17851-2

Alderson, D., Brown, G., & Carlyle, W. (2015). Operational models of infrastructure resilience. *Risk Analysis*, *35*(4), pp. 562–586.

Argyroudis, S., Mitoulis, S., Chatzi, E., Baker, J., Brilakis, I., Gkoumas, K., & Linkov, I. (2022). Digital technologies can enhance climate resilience of critical infrastructure. *Climate Risk Management*, *35*, p. 1003812.

Ayyub, B. (2014). Systems resilience for multi-hazard environment: Definition, metrics and valuation for decision making. *Risk Analysis.*, *34*(2), pp. 304–355.

Ayyub, B. (2015). Practical resilience metrics for planning design, and decision making. *ASCE-ASME Journal of Risk & Uncertainty in Engineering Systems Part A Civil*, *1*(3). doi:10.1061/AJRUA6.0000826

Beauchemin, S., & Barron, J. (1995). The computation of optical flow. *ACM Computing Surveys (CSUR)*, *27*(3), pp. 436–466.

Bhowmick, S., Nagarajaiah, S., & Lai, Z. (2020). Measurement of full-field displacement time history of a vibrating continuous edge from video. *Mechanical Systems & Processing*, *144*, p. 1068412.

Bocchini, P., Decò, A., & Frangopol, D. (2012). Probalistic functionality recovery model for resilience analysis. In Biondini, F., & Frangopol, D. (eds.), *Bridge maintenance, safety, management, resilience and sustainability. Proceedings of the Sixth International IABMAS Conference, Stresa, Lake Maggiore, Italy* (pp. 1920–1921).

Borah, S., Al-Habaibeh, A., Kromanis, R., & Kaveh, B. (2020). The Resilience of Vision-Based Technology for Bridge Monitoring: Measuring and Analysing. *In:* S. BORAH, *ed., Proceedings of the Joint International Resilience Conference 2020.*

Brownjohn, J. (2007). Structural health monitoring of civil infrastructure. *Philosophical Transactions of the Royal Society A: Mathematical, Physical and Engineering Sciences*, *365*(1851), pp. 589–622.

Bruneau, M., Chang, S., Eguchi, R., Lee, G., O'Rourke, T., & Reinhorn, A. (2003). A framework to quantitatively assess and enhance the seismic resilience of communities. *Earthquake Spectra*, *19*(4), pp. 733–752.

Bruneau, M., & Reinhorn, A. (2007). Exploring the concept of seismic resilience for acute care facilities. *Earthquake Spectra*, *23*(1), pp. 41–62. doi:10.1193/1.2431396

Bruno, M., & Clegg, R. (2015). *A foresight review of resilience engineering*. Lloyd's Register Foundation Report Series.

Calvi, G., Moratti, M., O'Reilly, G., Scattarreggia, N., Monteiro, R., Malomo, D., ... Pinho, R. (2019). Once upon a time in Italy: The tale of the Morandi bridge. *Structural Engineering International*, *29*(2), pp. 198–2112.

Cimellaro, G., Reinhorn, A., & Bruneau, M. (2006, April 18–22). Quantification of Seismic Resilience. *Proceedings of the 8th U.S. National Conference on Earthquake Engineering*, p. 1094.

Cimellaro, G., Reinhorn, A., & Bruneau, M. (2010). Framework for analytical quantification of disaster resilience. *Engineering Structures*, *32*(11), pp. 3639–3649. doi:10.1016/j.engstruct.2010.08.008

Derras, B., & Makhoul, N. (2022, July 5–7). An Overview of the Infrastructure Seismic Resilience Assessment Using Artificial Intelligence and Machine-Learning Algorithms. *ICONHIC2022 - 3rd International Conference on Natural Hazards & Infrastructures.*

Dong, C., & Catbas, F. (2021). A review of computer vision-based structural health monitoring at local and global levels. *Structural Health Monitoring*, *2*(20), pp. 692–743.

Feng, D., Feng, M., Ozer, E., & Fukuda, Y. (2015). A vision-based sensor for noncontact structural displacement measurement. *Structural Control & Health Monitoring*, *25*(5), pp. 16557–16575.

Fisher, R., & Norman, M. (2010). Developing measurement indices to enhance protection and resilience of critical infrastructures and key resources. *Journal of Business Continuity & Emergency Planning*, *4*(3), pp. 191–206.

Ivankovic, M., Strauss, A., & Soussa, H. (2019). European review of performance indicators towards sustainable road bridge management. *Proceedings of the Institution of Civil Engineers-Engineering Sustainability*, *173*(3), pp. 109–124.

Kalybek, M., Bocian, M., & Nikitas, N. (2021). Performance of optical structural vibration monitoring systems in experimental modal analysis. *Sensors*, *24*(4), p. 1239.

Khuc, T., & Catbas, F. (2017). Completely contactless structural health monitoring of real-life structures using cameras and computer vision. *Structural Control & Health Monitoring*, *24*(1), p. e1852.

Kostinakis, K., Morfidis, K., Demertzis, K., & Iliadis, L. (2022). Classification of buildings' potential for seismic damage by means of artificial intelligence techniques. *arXiv preprint arXiv:2205.01076*.

Kromanis, R., & Forbes, C. (2019). A low-cost camera system for accurate collection of structural response. *Inventions*, *4*(3), p. 412.

Kromanis, R., & Kripakaran, P. (2021). A multiple camera position approach for accurate displacement measurement using computer vision. *Journal of Civil Structural Health Monitoring*, *11*, pp. 661–678.

Kromanis, R., Lydon, D., Martinez del Rincon, J., & Al-Habaibeh, A. (2019). Measuring structural deformations in the laboratory environment using smartphones. *Frontiers in Built Environment*, *5*, p. 44.

Lazaridis, P., Kavvadias, I., Demertzis, K., Iliadis, L., & Vasiliadis, L. (2022). Structural damage prediction of a reinforced concrete frame under single and multiple seismic events using machine learning algorithms. *Applied Sciences*, *12*(8), p. 3845.

Lee, J., & Shinozuka, M. (2006). A vision-based system for remote sensing of bridge displacement. *NDT & E International*, *39*(5), pp. 425–431.

Lloyd's Register Foundation. (2015). Foresight review of resilience engineering. *Report Series: No. 2015.2.*

Lydon, D., Lydon, M., del Rincon, J., Taylor, S., Robinson, D., O'Brien, E., & Catbas, F. (2018). Development & field testing of a time-synchronized system for multi-point displacement calculation using low-cost wireless vision-based sensors. *IEEE Sensors Journal*, *18*(23), pp. 9744–9754.

Makhoul, N. (2015, September 23–25). From Sustainable to Resilient and Smart Cities. *IABSE Conference, Structural Engineering: Providing Solutions to Global Challenges.*

Makhoul, N., & Argyroudis, S. (2018, June 18–21). Loss Estimation Software: Developments, Limitations and Future Needs. *16ECEE - Thessaloniki 16th European Conference on Earthquake Engineering.*

Makhoul, N., & Argyroudis, S. (2019, June 23–26). Tools for Resilience Assessment: Developments, Limitations and Future Needs. *ICONHIC20199 - 2nd International Conference on Natural Hazards & Infrastructure.*

Nan, C., & Sansavini, G. (2016). A quantitative method for assessing resilience of interdependent infrastructures. *Reliability Engineering & System Safety*, *157*, pp. 35–53.

Nian, H. (2021). Civil engineering stability inspection based on computer vision and sensors. *Microprocessors & Microsystems*, *82*, p. 103838.

NRK. (2022). *Use collapse at Tretten.* Retrieved September 1, 2022, from NRK: https://www.nrk.no/nyheter/brukollaps-pa-tretten-1.16067817

Ouyang, M., & Duenas-Osorio, L. (2014). Multi-dimensional hurricane resilience assessment of electric power systems. *Structural Safety*, *48*, pp. 15–24.

Pettit, T., Fiksel, J., & Croxton, K. L. (2010). Ensuring supply chain resilience: Development of a conceptual framework. *Journal of business logistics*, *31*(1), pp. 1–21. doi:10.1002/j.2158-1592.2010.tb00125.x

Psathas, A., Iliadis, L., Achillopoulou, D., Papaleonidas, A., Stamataki, N., Bountas, D., & Dokas, I. (2022). Autoregressive Deep Learning Models for Bridge Strain Prediction. *International Conference on Engineering Applications of Neural Networks*, pp. 150–164.

Reed, D. A., Kapur, K. C., & Christie, R. D. (2009). Methodology for assessing the resilience of networked infrastructure. *IEEE Systems Journal*, *3*(2), pp. 174–180.

Rose, A. (2007). Economic resilience to natural and man-made disasters: Multidisciplinary origins and contextual dimensions. *Environmental Hazards*, *7*(4), pp. 383–398.

Salehi, H., & Burgueño, R. (2018). Emerging artificial intelligence methods in structural engineering. *Engineering Structures*, *171*, pp. 170–189.

Solberg, M. G. (2016, February 19). *Found the cause of the bridge collapse.* Retrieved September 1, 2022, from TU-Teknisk Ukeblad Media AS: https://www.tu.no/artikler/prosjekteringsfeil-ga-brukollapsen/277233

Sousa, H., Santos, L., & Makhoul, N. (2022, July 5–7). Next Generation of Monitoring Systems Towards Infrastructure Resilience. *ICONHIC2022 - 3rd International Conference on Natural Hazards & Infrastructure.*

Spencer Jr, B., Hoskere, V., & Narazaki, Y. (2019). Advances in computer vision-based civil infrastructure inspection and monitoring. *Engineering*, *5*(2), pp. 199–222.

Sun, L., Shang, Z., Xia, Y., Bhowmick, S., & Nagarajaiah, S. (2020). Review of bridge structural health monitoring aided by big data and artificial intelligence: From condition assessment to damage detection. *Journal of Structural Engineering*, *146*(5), p. 04020073.

Thai, H. T. (2022). Machine learning for structural engineering: A state-of-the-art review. *Structures*, *38*, pp. 448–491.

Tyrtaiou, M., Papaleonidas, A., Elenas, A., & Iliadis, L. (2020). Accomplished Reliability Level for Seismic Structural Damage Prediction Using Artificial Neural Networks. *In International Conference on Engineering Applications of Neural Networks*, pp. 85–98.

Xu, Y., Brownjohn, J., & Kong, D. (2018). A non-contact vision-based system for multi-point displacement monitoring in a cable-stayed footbridge. *Structural Control & Health Monitoring*, *25*(5), p. e2155.

Xu, Y., & Brownjohn, J. (2018). Review of machine-vision based methodologies for displacement measurement in civil structures. *Journal of Civil Structural Health Monitoring*, *8*, pp. 91–110.

Zhu, J., Lu, Z., & Zhang, C. (2021). A marker-free method for structural dynamic displacement measurement based on optical flow. *Structure & Infrastructure Engineering*, *18*(1), pp. 84–96.

13 Automatic Structural Health Monitoring of Road Surfaces Using Artificial Intelligence and Deep Learning

Andrea Ranieri, Elia Moscoso Thompson, and Silvia Biasotti

13.1 OVERVIEW

Road infrastructure is one of the pillars of modern societies: it allows efficient transport of people and goods where other transport modalities could not compete in terms of cost efficiency. However, road asphalt tends to deteriorate over time, with use, and due to atmospheric and environmental phenomena. It is precisely the capillarity of road infrastructure, its greatest value, which also makes it very difficult and expensive to monitor and maintain, thus making road monitoring an important part of structural health monitoring (SHM). To date, in most countries, the detection of damage to the road surface is done manually, with specialized operators in the field operating expensive equipment, greatly limiting the effectiveness of monitoring by maintenance bodies. The recent breakthroughs in computer vision, neural networks, and artificial intelligence make low-cost monitoring of infrastructure possible and scalable, reducing the reliance on monitoring performed by more expensive and invasive methods that require human intervention to just a small number of critical cases. In this chapter, we overview and discuss a number of methods for pothole and crack detection on the road surface, focusing on the recent achievements in semantic segmentation of road images to identify both potholes and cracks using deep learning (DL) techniques. In addition, we present some of the devices for data acquisition currently available on the market, focusing in particular on depth cameras, and we introduce the most important open datasets and benchmarks currently available in the literature. Finally, we discuss future research directions regarding the use of RGB-D (Red Green Blue-Depth) technology and depth cameras as an additional source of information to be provided to the neural network during training or to set up self-supervised learning pipelines.

DOI: 10.1201/9781003306924-13

13.2 INTRODUCTION

In the field of transportation engineering, the importance of road infrastructure is obvious, as wheeled vehicles remain relevant in both the industrial and private transportation of goods and people. The road network is an efficient way to connect different locations that is also cheap due to the low cost of its main constructive component, the asphalt. However, the latter tends to deteriorate to the point of becoming a threat to road users or, at least, a warning sign of a future failure of part of the infrastructure. Indeed, atmospheric events are the main cause of damage to the road surface, followed by landslides of the ground below the asphalt and by the passage of bulky vehicles. The continuous use of the infrastructure itself makes driving on the road either dangerous or, in the case of severe damage, impossible. This creates inconvenience to users proportional to the importance of the road, especially for those with heavy traffic or connecting small towns that usually can rely on one or two communication routes for connections to suppliers and other commodities.

For these reasons, it is essential to keep the state of the road surface under constant scrutiny, preventing where possible its deterioration or intervening to mitigate any damage that could cause injuries to drivers or damage to vehicles. This is usually done via qualitative manual detection, resulting in a tedious inspection with subjective evaluations of the damage. As a side effect, this forces the monitoring to be less frequent and efficient, decreasing the possibility of acting preventively on roads that are soon to be in need of intervention, and triggering repairs only in critical scenarios.

Moreover, the raw size of the road infrastructure is an issue in itself, which raises the cost of road maintenance in both economic and practical terms. Indeed, this job is done by specialized workers, trained to know and operate heavy machinery (Du et al., 2021). The amount of money spent in this maintenance process testifies to the issue: in the USA alone, considering the costs of fixing damage to both drivers' cars and the road itself, "billions USD/year" are spent (Bumpy Road Ahead, 2018). More importantly, the cost of human lives caused by this issue is high: approximately one-third of road fatalities in the USA are victims of poor road surface conditions (Fan et al., 2022). Finally, it is also worth mentioning the importance of this topic for building safe and efficient self-driving cars, which have to be aware of road damage to avoid them by dodging them at the maximum "safe speed" possible, stopping the vehicle only in cases of extreme danger.

The need for efficient road maintenance sparked great interest in the research world. Starting with the detection of potholes and extending to other damages and entities, a number of methods and datasets to test and/or train such methods are being developed (Figure 13.1). From an algorithmic point of view, the approach may differ according to the task and/or availability of public datasets (for classification, object detection, and semantic segmentation); however, the final goal is to find an automatic and efficient way to perform such a task, reducing both human and economic efforts.

In the scientific literature, efforts in this topic began in the early 1990s (Mahler et al., 1991) with computer vision methods on either 2D or 3D data. More recently,

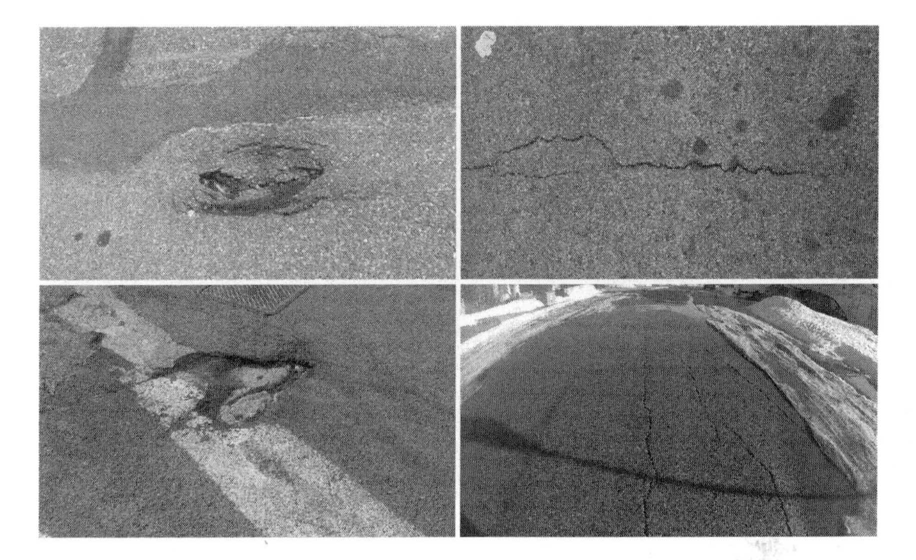

FIGURE 13.1 Examples of potholes and cracks within the Pothole Mix dataset.

with the renewed interest in DL-based methods, the trend shifted toward the development of models able to learn how to perform these tasks, creating the need for large amounts of data to train ever-larger models, which usually involve ever-increasing handmade annotations.

There is a wide variety of road damage that is worth detecting. However, most of the efforts in the current literature are devoted to pothole detection. By *pothole*, we mean a portion of asphalt that is missing or crumbled to the point of having a significant displacement on the surface (i.e., the inside of a pothole is lower than the rest of the road surface) and/or the terrain under the road surface is clearly visible. These are certainly the most immediate threats for road users, albeit not the most common. Besides potholes, cracks are the most frequent sign of road degradation. By *cracks*, we mean one or multiple fractures in the road surface that are not, however, wide enough to be considered potholes. The length of cracks tends to always exceed their width by orders of magnitude. It is necessary to distinguish between cracks without structural significance and cracks that, for instance, imply a pattern of landslide or structural failure. For this reason, some types of cracks are extremely important to detect and report because they allow municipalities to plan preventative interventions that are much less expensive than repairing landslide damage.

In this chapter, we explore the current trends regarding road damage detection, with emphasis on the approaches based on artificial intelligence and DL. Before that, we present the types of data on which these models are trained, showcasing the obstacles in their acquisition both in the tools used to acquire them and in how the ground truth, which is the key for supervised learning methods, is generated. In the final remarks, we recap the content of the chapter and outline possible future developments, together with a possible way to overcome one of the current bottlenecks of supervised learning, that is, the time-consuming and tedious hand-labeling of huge amounts of data.

13.3 A SYNOPSIS OF THE CHALLENGES AND TRENDS IN THE AUTOMATIC DETECTION OF ROAD DAMAGE

In this section, we look at the main elements involved in addressing the problem of automatic detection of road damage: the process of data acquisition, their arrangement in annotated or unannotated datasets, and, finally, an overview of the methods for the automatic classification, detection, and semantic segmentation of road damage. Regarding the last part, in particular, we also briefly overview some classic computer vision methods that preceded the current trend based on DL techniques.

13.3.1 Data Acquisition

It is worth spending a few words on the tools used to acquire the data necessary for the field of road surface monitoring, since their characteristics, especially when combined with data fusion approaches, can lead to the creation of pipelines for self-supervised learning capable of increasing overall learning efficiency. Depending on their nature, mainly image-like or 3D-like, two groups can be identified: cameras and range-imaging devices. The first group is the easiest to work with based on both the tools required, from professional cameras to even smartphone cameras (Maeda et al., 2018), and the speed of acquisition. This translates directly into the possibility of acquiring a large amount of data with relative ease. However, the cost of this efficiency is that the geometry of the road is not always clear due to perspective reasons.

Moreover, there are high variations in the colors of the images, mainly due to weather conditions and lighting. Since the information contained in an RGB image is heavily influenced by colors, models risk being severely affected by this problem and inheriting the biases resulting from datasets acquired too homogeneously (e.g., it is important not to acquire outdoor datasets only in cloudy weather, because otherwise the resulting models will not be able to handle hard shadows). Hence, there is interest in the second group of tools, which mainly contains laser scanners and depth camera sensors. However, the first of these tools has to be paired with ad-hoc vehicles that are not widely used. This, together with the actual cost of the scanner itself, makes the use of laser scanners very expensive and sometimes impractical.

More variety can be seen in depth camera-like sensors (Figure 13.2): they can be used either to reconstruct the geometry of the road (like with Microsoft Kinect) or to add a fourth channel of information to images (like RGB-D images acquired with depth cameras). While these are still more complex to set up than 2D-image-only sensors, it is possible to install these tools on common vehicles and/or on simpler structures (e.g., a stereo camera on a dolly-like cart). However,

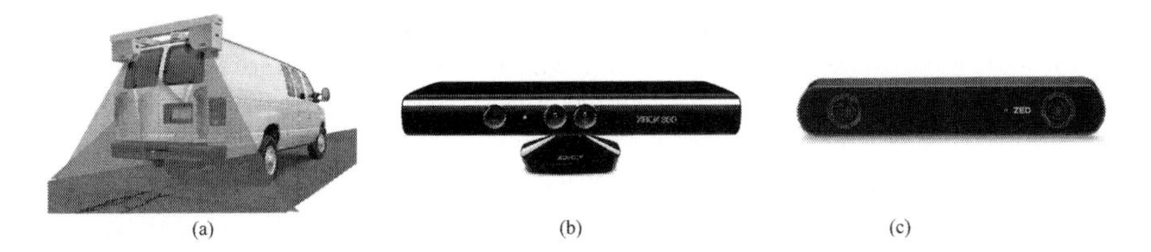

FIGURE 13.2 Commonly used sensors for 3D road data acquisition: (a) a laser scanner (Laurent et al., 2012), (b) Microsoft Kinect (Banerjee et al., 2017), and (c) a stereo camera (StereoLabs, 2022). (From https://arxiv.org/pdf/2204.13590.pdf.)

the resolution of the reconstruction varies significantly, affecting the quality and, consequently, its effectiveness.

Moreover, when assembling an annotated dataset, the acquisition is not the most relevant part of the work: finding and labeling the areas of interest in the acquired data are tedious tasks usually done by hand. This is especially true when talking about segmentation, in which the authors of the dataset carefully draw segmentation masks—a label much richer in terms of information than simple categorical variables or bounding boxes for object detection—in an error-prone activity that can undermine a model's training process.

13.3.2 Datasets and Benchmarks

Given that the current literature on the semantic segmentation of road damage, as we will see, revolves almost only around DL methods, significant amounts of data are required. Indeed, in this task but in general with learning-based methods, choosing a good architecture is only part of the work, as a good dataset can significantly increase the effectiveness of the model that learns on its training set. To train such models, huge amounts of data are required. Currently, there are many freely available datasets to train models, suited to perform a widely different range of tasks. To cite just a few examples in the field of road monitoring, Viren (2019) created an unannotated dataset for road image classification that counts around 700 images, half representing healthy roads and half representing potholes. Similar in size and with the same scope is the work presented by Atulya Kumar (2019). Regarding annotated datasets for object detection, Sovit Ranjan Rath (2020) created a collection of road pothole images provided with a list of bounding boxes of each pothole in each image, for a total of around 4400 images. Similar but smaller (around 670 images) is the dataset in Chitholian et al. (2020). In Maeda et al. (2018), we can find a dataset of around 15,000 images of road surfaces, with bounding boxes labeled with a larger number of road damage and other road entities, like crosswalk blur or white-line blur.

The image segmentation task is the most demanding in terms of work regarding the creation of labels, which are usually handmade and take significantly more time than drawing bounding boxes (which is already a tedious task to do on large amounts of images). Yantra IIT Bombay (2021) created a dataset of Indian roads with a number of segmentation annotations, namely, road, pothole, footpath, shallow path, and background. The dataset is made of around 3000 samples; each sample counts the image plus a mask for each annotation.

While working on a challenge for methods developed for autonomous driving, Roberto Guzmán et al. (2015) created a dataset of ~96,000 high-resolution stereo images, with inertial measurement units and Global Positioning System (GPS) data, of roads in developing countries, which tend to have roads in poor condition.

One of the most complete datasets, in terms of the kind of data associated with a single "shot," is in Fan, Ozgunalp, et al. (2020), in which a color image, a subpixel disparity image, a transformed disparity image, and pothole annotation are associated to each element of the dataset.

Another relevant dataset for pothole and crack segmentation is in Ranieri et al. (2022), which is a collection of five publicly available datasets for semantic segmentation. The dataset counts a total of 4340 image pairs (image+mask). The type of images in the dataset can be separated into two kinds: *top-down*, which are taken with the camera pointing vertically above the road, and *wide-view*, in which the camera points ahead. Examples can be seen in Figure 13.3. The semantic masks that come as ground truth were converted to RGB (to help users become familiar with the dataset using any image visualizer) and made uniform in terms of colors across the five datasets. However, there is no change in resolution and/or image size, and, where possible, the original splits have been kept. In addition, the dataset includes 797 unannotated RGB-D clips of road damage, from which additional images (especially useful for CutMix-like data augmentations) can be extracted. The videos were taken with a depth camera that was held by hand on top of the road damage. The camera used is a Luxonis OAK-D camera connected via USB-C to an Android mobile phone using a Unity app; therefore, it is the operator feedback loop that keeps the camera at the optimal height and orientation so that the disparity map is as clean as possible. A sample of three frames is shown in Figure 13.4.

The benchmark in Moscoso Thompson et al. (2022) was developed as a contest, allowing the participants to submit their methods to semantically segment the

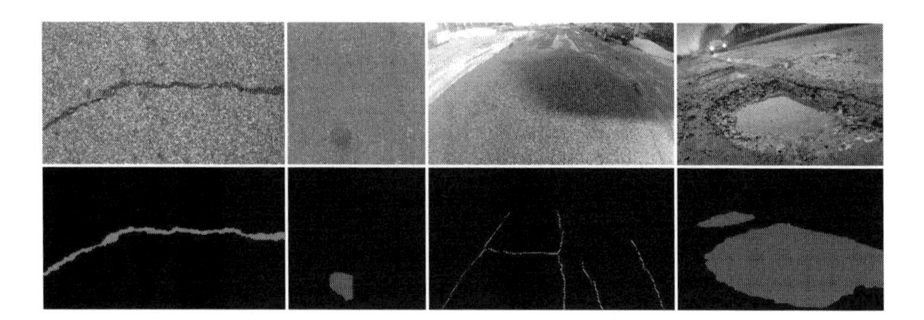

FIGURE 13.3 Example of images and segmentation masks (ground truth) within the Pothole Mix dataset.

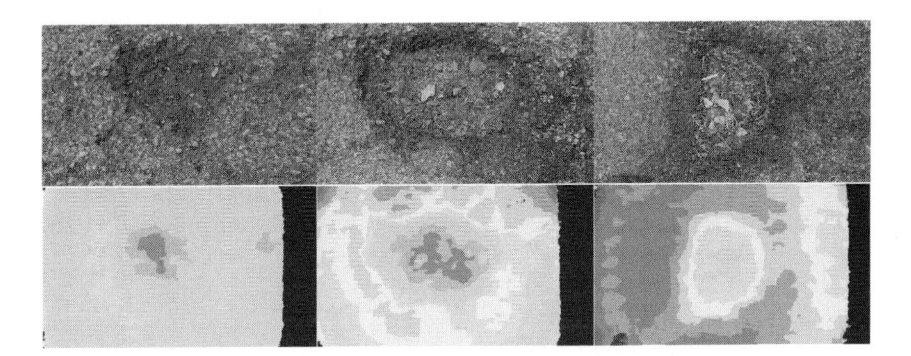

FIGURE 13.4 Examples of frames captured from RGB-D video clips inside the Pothole Mix dataset.

damage in the road images. A quick overview of the results of this benchmark can be found in Section 13.7. Moreover, the code behind this work is open source and available on GitHub (https://gitlab.com/4ndr3aR/pothole-mix-segmentation/), and the related paper was awarded the Replicability Stamp.

Finally, it is worth mentioning that there are other ways to assemble datasets by exploiting already acquired images of roads. Indeed, like Koch et al. (2013), the authors built a dataset partially composed of images from the Google search engine. A similar approach was taken by Majidifard et al. (2020), which proposed a road object detection dataset consisting of more than 14,000 samples created using the Google Street View application programming interface (API).

13.3.3 Classic Computer Vision Methods: An Overview

Around the early 2000s, the problem of automatic road damage monitoring using optical devices gained increasing interest, thanks to the significant development reached by computer vision methods. Mainly two kinds of data were used: RGB or grayscale images, sometimes with an additional disparity/depth map, and 3D road point clouds. For the first kind of data, the general approach (Ma et al., 2022) is to first pre-process the image to remove noise and unwanted components, then to segment the images and, finally, to extract the regions of interest. Examples of the pre-processing step are median filtering, Gaussian filtering, morphological filtering, and so on. Then, tools like histograms are used together with thresholds to segment a color image into damaged and undamaged regions. For example, in Koch and Brilakis (2011), the authors made a binary segmentation (damaged/not damaged) of road color images using histograms and thresholds, further refined with morphological thinning and elliptic regression. Finally, the texture inside a potential pothole is extracted and compared to that of an undamaged road. Similar to Koch and Brilakis (2011) in spirit is Buza et al. (2013), in which in particular Otsu's thresholding method was used to binarize a color image and extract potholes via spectral clustering.

In general, after thresholding, the last step is based on a set of assumptions on the geometry and color that usually characterize a pothole (Fan et al., 2022). More details on classic computer vision methods that work on images can be found in Koch et al. (2015).

This problem was also faced by using depth images, exploiting the distribution of the depth values. For example, Moazzam et al. (2013) studied azimuthal and elevation angles to characterize potholes, extracting depth images using the Kinect depth camera.

Regarding 3D road point clouds, classic computer vision methods focus on exploiting the change in height. This can be done, for example, by fitting a surface to the captured point cloud of the road and checking the discrepancy between the fitted model and the point cloud. The criteria by which the interpolating surface (a plane, a quadric, etc.) is chosen is not unique, and in the literature it is possible to find multiple approaches to this sub-problem (Fan et al., 2018; Fan, Ozgunalp, et al., 2020; Li et al., 2018). It is worth mentioning that, since real roads are not even, this class of approach results is less efficient in general, and results are far more efficient when used in combination with images (Fan, Ozgunalp, et al., 2020).

13.3.4 Deep Learning Methods: An Overview

The rapid spread of learning-based methods such as machine learning (ML) and DL is revolutionizing the world of computer vision. Their success is due to not only the effectiveness of the methods (well known since the 1990s and before) (LeCun et al., 1989), but also the increasing availability of consumer hardware such as graphics processing units (GPUs) capable of performing parallel computing at a fraction of the cost of what was possible a decade ago, and of software libraries that make them efficient and easy to use. Just as with computer vision as a whole, the recent application of these methods to infrastructure monitoring is having an equally significant impact. This revolution began with the increasingly frequent application of learning-based methods that we now call "classic": for example, support vector machines (SVMs) (Cortes & Vapnik, 1995) were used to classify patches of road images (Gao et al., 2020; Lin & Liu, 2010; Pan et al., 2018). In a short time, however, their performance was quickly surpassed by that of convolutional neural networks (CNNs) (Fan et al., 2019) by virtue of the latter's superior ability to extract and "learn" features from the combination of datasets and tasks of the problem being studied.

Thanks to this breakthrough, in recent years, numerous works have been published demonstrating how the classification and object detection of damage to road surfaces are no longer challenges today (Maeda et al., 2018). However, this is no longer true if we attempt to address the problem in terms of semantic segmentation: the inherent complexity of the problem (labeling each pixel of the image) and the scarcity of public datasets labeled with semantic segmentation information make this an extremely challenging problem today. It is also worth remembering how important it is to have as accurate information as possible about damage to the road surface, because, for example, applications such as self-driving cars require the greatest possible accuracy in order to navigate the road surface safely at the maximum possible speed for that road section.

Another reason for the great success of DL-based methods is the possibility of exploiting the transfer learning phenomenon: by pre-training large CNNs on ImageNet (Krizhevsky et al., 2017), they develop feature extractors in the lowest convolutional layers that are effective even for more specific and potentially very different subproblems. It is possible to replace the last layer of these pre-trained networks, called backbones, with a different classification layer, a layer for object detection, or a decoder for semantic segmentation.

In the context of object detection, Yebes et al. (2021) proposed four models: a faster region CNN with Inception-v2 (Szegedy et al., 2016), Inception-ResNet-v2, or ResNet-101 (He et al., 2016) as the backbone and a single-shot detector (SSD) with MobileNet-v2 (Sandler et al., 2018) as the backbone, with the implementation with the ResNet-101 performing better than the others. An extension to other kinds of damage, like cracks, exists in the work (Kortmann et al., 2020). However, in the current literature, the you only look once (YOLO) networks outperformed the others (Suong & Jangwoo, 2018).

The problems of semantic segmentation, however, are much more complex than simple classification or object detection, and they require an entire "segmentation head" capable of performing convolution–upsampling sequences. The segmentation head is connected to the backbone in a classic encoder–decoder structure and is the one that, starting from the feature maps extracted from the backbone, generates the semantic segmentation mask in the inference phase. A further level of complexity is given by the dataset, which is no longer required to have a simple categorical variable for each image, as in the case of classification, or a categorical variable plus four coordinates for each of the bounding boxes of objects in the image, as in the case of object detection. In semantic segmentation, each image is associated with another image (typically in grayscale) in which the value of each pixel represents the attribute class of the object to which that pixel belongs. Indeed, a semantic segmentation approach for road images usually (Ma et al., 2022) falls into one of the following categories: single-modal or data fusion. The single-modal approach relies only on RGB images, while data fusion allows the training of models typically on two (or more) types of data, such as color and a depth map (Hazirbas et al., 2017), normals (Fan, Wang, Cai, et al., 2020; Wang et al., 2021), or transformed disparity images (Fan, Wang, Bocus, et al., 2020). The introduction of a further information channel helps the model to generalize better against the problem, and this generates an inductive bias in the model that allows it to at least partially learn the actual geometry of the road depicted in an image, improving the overall performance, for example, in heterogeneous light conditions. Also, it has been shown that better performance can be obtained with semi-supervised learning techniques: in Chun and Ryu (2019), the authors exploited this approach to generate pseudo-labels and fine-tune a pre-trained fully connected network. Masihullah et al. (2020) also showed the effectiveness of few-shot learning for the problem of pothole segmentation.

In general, the use of attention-based approaches seems to be effective in dealing with this task (J. Fan et al., 2021; R. Fan et al., 2021; Masihullah et al., 2020). The wide participation (for a total of 12 initial registrations) in the 2022 3D Shape Retrieval Contest (SHREC) in pothole recognition (Moscoso Thompson et al., 2022) shows there is a wide interest in the problem, and this in turn confirms it is not a trivial task (also because only two groups were able to submit their solutions).

Both groups proposed solutions mostly based on CNNs to address the problem (with a notable exception of one SegFormer model), albeit with significantly different approaches. Here, we summarize the best methods from each of the two groups. The first method exploits a loss function based on the linear combination between the cross-entropy in function for true and predicted pixels and a loss function based on learned active contours (Chan & Vese, 2001), in a UNet++ architecture with an EfficientNet as the backbone.

The second noteworthy method is based on a semi-supervised method called cross pseudo-supervision. Moreover, it exploits the mosaic data augmentation (Figure 13.5), in which parts of multiple images are aggregated into a new one, increasing both the amount of data and the amount of damage per image, which usually represents just a tiny portion of the image itself, therefore also performing a balancing of the dataset classes. The contest organizers also trained a baseline model based on the DeepLabv3+ architecture with a ResNet-101 as the backbone. Details for these three methods can be found in Moscoso Thompson et al. (2022). The results were evaluated both quantitatively and qualitatively. The quantitative evaluation is done by exploiting classic evaluation measures such as the *weighted pixel accuracy*, the *Dice multiclass coefficient*, and the *intersection-over-union* (IoU). All these measures evaluate the quality of a predicted mask, weighing different properties of the mask differently (for details, see Moscoso Thompson et al., 2022). Despite the variety of the approaches, the evaluation showed that the methods perform quite similarly. Finally, the methods are qualitatively evaluated by testing their performance on the frames of a set of videos that have been withheld by the organizers and not shared with the participants. In this evaluation, it can be observed that the baseline method was pretty conservative in its

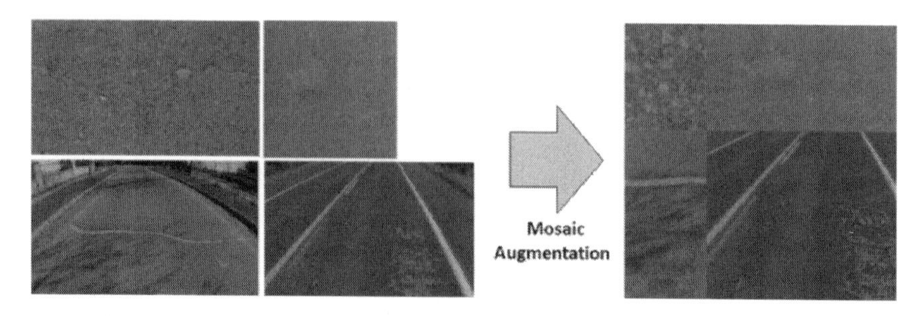

FIGURE 13.5 Example of mosaic data augmentation. (From Moscoso Thompson et al., 2022).

predictions, with a low rate of detection but also a few false positives. However, it incorrectly classified dark areas in images (like the back of a dark street sign; see Figure 13.6). The method that bases its loss function on *active contours* performs better on top-down images; however, its performance drops significantly on wide-view ones. Interestingly, the opposite is true for the method based on the semi-supervised method, as it does much better in wide-view images (or frames). We show examples supporting our observation in Figure 13.6; for more details, we suggest checking Moscoso Thompson et al. (2022).

FIGURE 13.6 Example of segmentation results obtained by models submitted to the 2022 3D Shape Retrieval Contest (SHREC). From top to bottom, left to right: A PUCP–MAnet false-positive prediction on dark asphalt and a CNR–DeepLabv3+ (baseline) false-positive prediction on a road sign. Next to them is a baseline false-negative prediction (several potholes are clearly visible but not detected). Below is a pretty accurate HCMUS–SegFormer prediction of a large pothole (only the bottom part of the pothole is missing). Last row: Very good predictions of PUCP–MAnet on (left) potholes and (right) a large crack.

13.4 DISCUSSIONS AND CONCLUSIONS

In this chapter, we summarized the current state of the art and the main resources available in literature when dealing with the SHM of road surfaces, focusing on artificial intelligence and DL techniques.

Given the complexity of the problem of segmenting potholes and cracks, in this analysis, we identified the labeling of large amounts of data, more than anything else, as one of the key challenges to be faced to effectively tackle the problem of SHM of the road pavement.

We also recognize that the use of more sophisticated instrumentation, such as depth and time-of-flight cameras or LIDARs, can greatly help in data collection and can in part provide the labeling needed for training neural networks. However, the process of extracting labels from this type of structured data is not currently fully automatable (due to image registration and synchronization problems, and difficulties in the data fusion of the different sources) and always requires human intervention in the loop, so the construction or training of pipelines of models for pre-processing could greatly benefit the efficiency of labeled data collection. To this end, our current efforts are aimed at the exploitation of RGB-D data to both ease the creation of semantically segmented datasets and/or exploit the depth map directly to train four-channel neural networks capable of superior performance in semantic segmentation.

Unsupervised learning techniques (contrastive learning, diffusion models) are also very promising and should be exploited more, as they work with unlabeled data and therefore allow researchers to largely bypass the problem of image labeling. Furthermore, the use of particular data augmentation seems to significantly increase performance, as in Moscoso Thompson et al. (2022), where the mosaic data augmentation boosted the segmentation accuracy of some of the models. Indeed, it created not only more data but also more *challenging and balanced* data, with more damage to detect in a single image. From our side, we are also exploring a further improvement in this direction, called *Precise CutMix* data augmentation, in which only the damage is transferred into another image.

Lastly, it must also be taken into account that the discipline of computer vision is undergoing radical transformations that appear to be proceeding at an exponential rate, and therefore radically new approaches (and the related pre-trained models) are announced roughly every six months. These models are, in general, getting larger and trained on ever larger internet-scraped datasets, but a further step of domain-specific experimental research and fine-tuning is needed before they can be successfully applied to very specific datasets and problems such as those belonging to the world of SHM.

REFERENCES

Banerjee, T., Yefimova, M., Keller, J. M., Skubic, M., Woods, D. L., & Rantz, M. (2017). Exploratory Analysis of Older Adults' Sedentary Behavior in the Primary Living Area Using Kinect Depth Data. *Journal of Ambient Intelligence and Smart Environments*, 9(2). https://doi.org/10.3233/AIS-170428

Bumpy Road Ahead, https://tripnet.org/wp-content/uploads/2019/03/Urban_Roads_TRIP_Report_October_2018.pdf, 2018

Buza, E., Omanovic, S., & Huseinovic, A. (2013). Pothole Detection with Image Processing and Spectral Clustering. *Recent Advances in Computer Science and Networking Pothole.*

Chan, T. F., & Vese, L. A. (2001). Active Contours Without Edges. *IEEE Transactions on Image Processing, 10*(2), 266–277. https://doi.org/10.1109/83.902291

Chitholian, A. R., Mustafa, R., & Hossain, M. S. (2022). Real-time pothole detection and localization using convolutional neural network. In *Proceedings of the International Conference on Big Data, IoT, and Machine Learning: BIM 2021* (pp. 579–592). Springer Singapore.

Chun, C., & Ryu, S. K. (2019). Road Surface Damage Detection Using Fully Convolutional Neural Networks and Semi-Supervised Learning. *Sensors (Switzerland), 19*(24). https://doi.org/10.3390/s19245501

Cortes, C., & Vapnik, V. (1995). Support-Vector Networks. *Machine Learning, 20*(3). https://doi.org/10.1023/A:1022627411411

Du, Z., Yuan, J., Xiao, F., & Hettiarachchi, C. (2021). Application of Image Technology on Pavement Distress Detection: A Review. *Measurement: Journal of the International Measurement Confederation, 184.* https://doi.org/10.1016/j.measurement.2021.109900

Fan, J., Bocus, M. J., Hosking, B., Wu, R., Liu, Y., Vityazev, S., & Fan, R. (2021). Multi-Scale Feature Fusion: Learning Better Semantic Segmentation for Road Pothole Detection. *ICAS 2021 - 2021 IEEE International Conference on Autonomous Systems, Proceedings.* https://doi.org/10.1109/ICAS49788.2021.9551165

Fan, R., Bocus, M. J., & Dahnoun, N. (2018). A Novel Disparity Transformation Algorithm for Road Segmentation. *Information Processing Letters, 140.* https://doi.org/10.1016/j.ipl.2018.08.001

Fan, R., Bocus, M. J., Zhu, Y., Jiao, J., Wang, L., Ma, F., Cheng, S., & Liu, M. (2019). Road crack detection using deep convolutional neural network and adaptive thresholding. *IEEE Intelligent Vehicles Symposium, Proceedings, 2019-June.* https://doi.org/10.1109/IVS.2019.8814000

Fan, R., Ozgunalp, U., Hosking, B., Liu, M., & Pitas, I. (2020). Pothole Detection Based on Disparity Transformation and Road Surface Modeling. *IEEE Transactions on Image Processing, 29.* https://doi.org/10.1109/TIP.2019.2933750

Fan, R., Ozgunalp, U., Wang, Y., Liu, M., & Pitas, I. (2022). Rethinking Road Surface 3-D Reconstruction and Pothole Detection: From Perspective Transformation to Disparity Map Segmentation. *IEEE Transactions on Cybernetics, 52*(7). https://doi.org/10.1109/TCYB.2021.3060461

Fan, R., Wang, H., Bocus, M. J., & Liu, M. (2020). We Learn Better Road Pothole Detection: From Attention Aggregation to Adversarial Domain Adaptation. Lecture Notes in Computer Science (Including Subseries Lecture Notes in Artificial Intelligence and Lecture Notes in Bioinformatics), *12538 LNCS.* https://doi.org/10.1007/978-3-030-66823-5_17

Fan, R., Wang, H., Cai, P., & Liu, M. (2020). SNE-RoadSeg: Incorporating Surface Normal Information into Semantic Segmentation for Accurate Free space Detection. *Lecture Notes in Computer Science (Including Subseries Lecture Notes in Artificial Intelligence and Lecture Notes in Bioinformatics), 12375 LNCS.* https://doi.org/10.1007/978-3-030-58577-8_21

Fan, R., Wang, H., Wang, Y., Liu, M., & Pitas, I. (2021). Graph Attention Layer Evolves Semantic Segmentation for Road Pothole Detection: A Benchmark and Algorithms. *IEEE Transactions on Image Processing, 30.* https://doi.org/10.1109/TIP.2021.3112316

Gao, M., Wang, X., Zhu, S., & Guan, P. (2020). Detection and Segmentation of Cement Concrete Pavement Pothole Based on Image Processing Technology. *Mathematical Problems in Engineering, 2020*. https://doi.org/10.1155/2020/1360832

Guzmán, R. et al. (2015). Towards Ubiquitous Autonomous Driving: The CCSAD Dataset. *International Conference on Computer Analysis of Images and Patterns*, 582–593. Springer.

Hazirbas, C., Ma, L., Domokos, C., & Cremers, D. (2017). FuseNet: Incorporating Depth Into Semantic Segmentation Via Fusion-Based CNN Architecture. Lecture Notes in Computer Science (Including Subseries Lecture Notes in Artificial Intelligence and Lecture Notes in Bioinformatics), *10111 LNCS*. https://doi.org/10.1007/978-3-319-54181-5_14

He, K., Zhang, X., Ren, S., & Sun, J. (2016). Deep Residual Learning for Image Recognition. *Proceedings of the IEEE Computer Society Conference on Computer Vision and Pattern Recognition, 2016-December*. https://doi.org/10.1109/CVPR.2016.90

Koch, C., & Brilakis, I. (2011). Pothole Detection in Asphalt Pavement Images. *Advanced Engineering Informatics*, *25*(3). https://doi.org/10.1016/j.aei.2011.01.002

Koch, C., Georgieva, K., Kasireddy, V., Akinci, B., & Fieguth, P. (2015). A Review on Computer Vision Based Defect Detection and Condition Assessment of Concrete and Asphalt Civil Infrastructure. *Advanced Engineering Informatics*, *29*(2). https://doi.org/10.1016/j.aei.2015.01.008

Koch, C., Jog, G. M., & Brilakis, I. (2013). Automated Pothole Distress Assessment Using Asphalt Pavement Video Data. *Journal of Computing in Civil Engineering*, *27*(4). https://doi.org/10.1061/(asce)cp.1943-5487.0000232

Kortmann, F., Talits, K., Fassmeyer, P., Warnecke, A., Meier, N., Heger, J., Drews, P., & Funk, B. (2020). Detecting Various Road Damage Types in Global Countries Utilizing Faster R-CNN. *Proceedings - 2020 IEEE International Conference on Big Data, Big Data 2020*. https://doi.org/10.1109/BigData50022.2020.9378245

Krizhevsky, A., Sutskever, I., & Hinton, G. E. (2017). ImageNet Classification with Deep Convolutional Neural Networks. *Communications of the ACM*, *60*(6). https://doi.org/10.1145/3065386

Kumar, A. (November 2019). *Pothole detection dataset*. URL: shortu rl.at/blBJK.

Laurent, J., Hébert, J. F., Lefebvre, D., & Savard, Y. (2012). Using 3D Laser Profiling Sensors for the Automated Measurement of Road Surface Conditions. RILEM Bookseries, *4*. https://doi.org/10.1007/978-94-007-4566-7_16

LeCun, Y., Boser, B., Denker, J. S., Henderson, D., Howard, R. E., Hubbard, W., & Jackel, L. D. (1989). Backpropagation Applied to Handwritten Zip Code Recognition. *Neural Computation*, *1*(4), 541–551. https://doi.org/10.1162/neco.1989.1.4.541

Lin, J., & Liu, Y. (2010). Potholes Detection Based on SVM in the Pavement Distress Image. *Proceedings - 9th International Symposium on Distributed Computing and Applications to Business, Engineering and Science, DCABES 2010*. https://doi.org/10.1109/DCABES.2010.115

Li, Y., Papachristou, C., & Weyer, D. (2018). Road Pothole Detection System Based on Stereo Vision. *Proceedings of the IEEE National Aerospace Electronics Conference, NAECON, 2018-July*. https://doi.org/10.1109/NAECON.2018.8556809

Maeda, H., Sekimoto, Y., Seto, T., Kashiyama, T., & Omata, H. (2018). Road Damage Detection and Classification Using Deep Neural Networks with Smartphone Images. *Computer-Aided Civil and Infrastructure Engineering*, *33*(12). https://doi.org/10.1111/mice.12387

Ma, N., Fan, J., Wang, W., Wu, J., Jiang, Y., Xie, L., & Fan, R. (2022). *Computer Vision for road imaging and pothole detection: A state-of-the-art review of systems and algorithms*. https://doi.org/10.1093/tse/tdac026

Mahler, D. S., Kharoufa, Z. B., Wong, E. K., & Shaw, L. G. (1991). Pavement Distress Analysis Using Image Processing Techniques. *Computer-Aided Civil and Infrastructure Engineering*, *6*(1). https://doi.org/10.1111/j.1467-8667.1991.tb00393.x

Majidifard, H., Jin, P., Adu-Gyamfi, Y., & Buttlar, W. G. (2020). *PID: A new benchmark dataset to classify and densify pavement distresses.* (arXiv:1910.11123v1 [cs.CV]). *ArXiv Computer Science, October.*

Masihullah, S., Garg, R., Mukherjee, P., & Ray, A. (2020). Attention Based Coupled Framework for Road and Pothole Segmentation. *Proceedings - International Conference on Pattern Recognition.* https://doi.org/10.1109/ICPR48806.2021.9412368

Moazzam, I., Kamal, K., Mathavan, S., Usman, S., & Rahman, M. (2013). Metrology and Visualization of Potholes Using the Microsoft Kinect Sensor. *IEEE Conference on Intelligent Transportation Systems, Proceedings, ITSC.* https://doi.org/10.1109/ITSC.2013.6728408

Moscoso Thompson, E., Ranieri, A., Biasotti, S., Chicchon, M., Sipiran, I., Pham, M.-K., Nguyen-Ho, T.-L., Nguyen, H.-D., & Tran, M.-T. (2022). *SHREC 2022: Pothole and crack detection in the road pavement using images and RGB-D data.*

Pan, Y., Zhang, X., Cervone, G., & Yang, L. (2018). Detection of Asphalt Pavement Potholes and Cracks Based on the Unmanned Aerial Vehicle Multispectral Imagery. *IEEE Journal of Selected Topics in Applied Earth Observations and Remote Sensing, 11*(10). https://doi.org/10.1109/JSTARS.2018.2865528

Ranieri, A., Moscoso Thompson, E., & Biasotti, S. (2022), "Pothole Mix". *Mendeley Data, V1*, doi: 10.17632/kfth5g2xk3.1 (also available at: https://data.mendeley.com/datasets/kfth5g2xk3/2)

Rath, S. R. (September 2020). *Road pothole images for pothole detection.* URL: shorturl.at/sxKUX.

Sandler, M., Howard, A., Zhu, M., Zhmoginov, A., & Chen, L. C. (2018). MobileNetV2: Inverted Residuals and Linear Bottlenecks. *Proceedings of the IEEE Computer Society Conference on Computer Vision and Pattern Recognition.* https://doi.org/10.1109/CVPR.2018.00474

StereoLabs. *ZED stereo camera.* URL: Stereolabs.com/z ed, April 2022.

Suong, L. K., & Jangwoo, K. (2018). Detection of Potholes Using a Deep Convolutional Neural Network. *Journal of Universal Computer Science*, *24*(9), 1244–1257.

Szegedy, C., Vanhoucke, V., Ioffe, S., Shlens, J., & Wojna, Z. (2016). Rethinking the Inception Architecture for Computer Vision. *Proceedings of the IEEE Computer Society Conference on Computer Vision and Pattern Recognition, 2016-December.* https://doi.org/10.1109/CVPR.2016.308

Viren. (December 2019). *Pothole and plain road images.* URL: https://www.kaggle.com/datasets/virenbr11/pothole-and-plain-rode-images

Wang, H., Fan, R., Cai, P., & Liu, M. (2021). SNE-RoadSeg+: Rethinking Depth-Normal Translation and Deep Supervision for Freespace Detection. *IEEE International Conference on Intelligent Robots and Systems.* https://doi.org/10.1109/IROS51168.2021.9636723

Yantra IIT Bombay. (November 2021). *Semantic segmentation datasets of Indian roads.* URL: shorturl.at/coyzB.

Yebes, J. J., Montero, D., & Arriola, I. (2021). Learning to Automatically Catch Potholes in Worldwide Road Scene Images. *IEEE Intelligent Transportation Systems Magazine, 13*(3). https://doi.org/10.1109/MITS.2019.2926370

14 Computer Vision–Based Intelligent Disaster Mitigation from Two Aspects of Structural System Identification and Local Damage Detection

Ying Zhou, Shiqiao Meng, Shengyun Peng, and Abouzar Jafari

14.1 OVERVIEW

Intelligent disaster mitigation is one of the most important parts of intelligent construction, which is of great significance to maintaining the safety and stability of buildings. To evaluate a building's performance under disasters, it is necessary to obtain its structural dynamic characteristic degeneration through structural dynamic response analyses, which can then be used to assess the degeneration of structural stiffness, and ultimately evaluate the damage degree of the structure. Among them, the degeneration of structural dynamic characteristics and the detection of structural damage can be realized through system identification and local damage detection, respectively. With the development of computer vision and deep learning, many state-of-the-art techniques such as convolutional neural networks are used to automatically detect a building's dynamic characteristics and local damage. Besides, the unmanned aerial vehicle (UAV) has good maneuverability and wide detection range, so deploying algorithms on UAVs can achieve more efficient intelligent detection of buildings. This chapter mainly introduces the research progress of intelligent disaster mitigation based on computer vision from two aspects of system identification and local damage detection.

14.2 INTRODUCTION

Intelligent disaster mitigation is a crucial component within the realm of intelligent construction, playing a vital role in ensuring the safety and stability of buildings. To effectively assess a building's resilience during disasters, it becomes

DOI: 10.1201/9781003306924-14

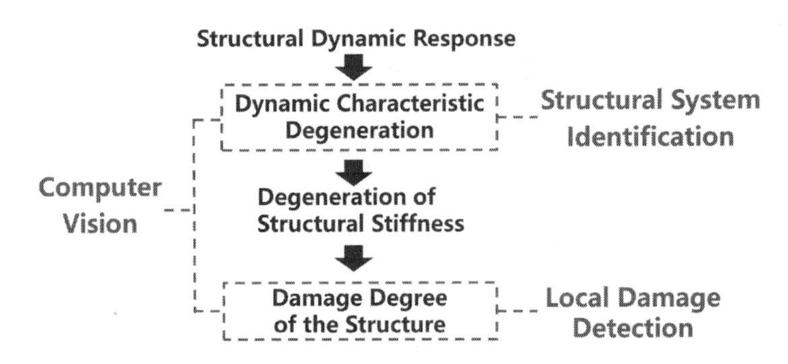

FIGURE 14.1 The schematic diagram of intelligent structural damage detection.

imperative to comprehend its structural dynamic characteristics' degeneration. This necessitates conducting structural dynamic response analyses to acquire pertinent data that can subsequently be utilized to evaluate changes in structural stiffness, thereby determining the extent of damage incurred by the structure. System identification facilitates an understanding of the degeneration in structural dynamic characteristics, while local damage detection methods are employed to detect and assess specific areas of structural deterioration. These approaches collectively contribute to the comprehensive evaluation of a building's performance under disaster scenarios. The realization of structural system identification and local damage detection through computer vision technology can effectively carry out intelligent detection of building damage. The schematic diagram is shown in Figure 14.1.

In traditional detection methods, the identification of structural systems for the degeneration of structural dynamic properties is mainly through the identification of structural displacement responses. The displacement measurement systems can be divided into two types: contact and non-contact systems. The contact systems generally include high-precision transducers, like linear variable differential transformer (LVDT) and cable types. These systems require a stationary reference point to measure the relative displacement between the structure and the fixed point [1]. Meanwhile, installing the measuring system on structures that are super-high-rises or located over a watercourse is difficult or almost impossible [2]. The non-contact systems generally encompass laser, radar, GPS, and camera systems. The measurement of the laser technique is very accurate, but the high-intensity laser beam is detrimental to human health, and the cost of the whole system is too high for regular structural health monitoring (SHM) [3].

Local damage detection for structures mainly includes crack detection, spalling detection, corrosion detection, and water-leaking detection. Among them, crack detection has become one of the most critical detection targets in structural damage detection due to the large degree of damage to building structures caused by structural cracks. Traditional crack detection methods mainly include manual detection and detection by sensors [4]. These methods have low detection efficiency and pose safety risks for the detection personnel.

With the development of computer vision and artificial intelligence technology, the ability to process images and videos has dramatically improved. Since damage to buildings has clear visual characteristics, researchers have proposed a variety of structural system identification [5, 6] and local damage detection [7, 8] techniques based on computer vision methods. Besides, in order to automate damage detection, it is necessary to implement algorithms on multiple hardware devices. Due to their excellent maneuverability and wide detection range, UAVs have become suitable carriers for structural damage detection [9].

This chapter introduces a deep-learning-based SiamSDN algorithm [10] that can effectively achieve structural system identification and a three-stage deep-learning-based method [11] for high-precision crack detection. In addition, this chapter introduces a UAV-based structural system identification and a real-time crack detection method based on a drone [12]. The above method is quantitatively validated on several large shaking-table test buildings and public datasets, such as the first International Project Competition for Structural Health Monitoring [13]. The experimental results show that structural system identification and damage detection based on computer vision can realize automatic damage detection and effectively improve detection efficiency and accuracy.

14.3 STRUCTURAL SYSTEM IDENTIFICATION BASED ON COMPUTER VISION AND UNMANNED AERIAL VEHICLES

In this section, we introduce a vision-based displacement measurement system. The method mainly consists of four parts: (1) original video taken by commercial cellphone cameras and pre-processed video taken by the UAV; (2) camera system setups, including calibrating cameras and building reference coordinates; (3) a deep neural network (DNN)-based video object-tracking algorithm; and (4) displacement measurements and system identifications. The technical flowchart is shown in Figure 14.2.

Original videos can be taken by consumer-grade cameras and UAVs, such as an iPhone 6 rear camera and a Da-Jiang Innovations (DJI) UAV. A UAV has small degrees of vibration, even if it is stationary at one point during its flight. The slight vibrations lead to a huge error while calculating the motions of structures. Therefore, the UAV videos need to be stabilized to eliminate the small and random drifts. Then, videos are divided into subsequent frames for future processing. Since regular commercial cameras use wide-angle lenses, the captured images encompass distortions. It is crucial to calibrate the lens distortion and rectify the video frames. The intrinsic and extrinsic parameters should be calculated, and the world coordinate system (WCS) and local coordinate system (LCS) should be settled to transform the pixel level into a metric system. The pre-processed video frames are the input of the end-to-end DNN object-tracking pipeline, named SiamSDN. The tracking object is identified solely in the first frame. Given a sequence of input images, SiamSDN utilizes a class-specific detector to accurately predict the object's motion state (location, size, and orientation) in each frame. Finally, the pixel displacement time histories are obtained and transformed into metric displacements via the calculated camera parameters.

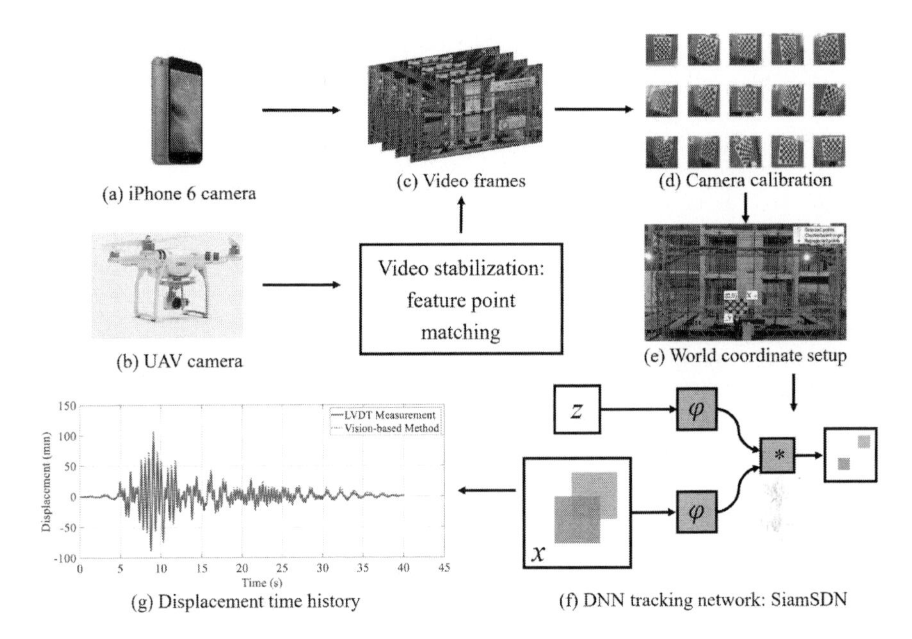

FIGURE 14.2 The flowchart of a vision-based displacement measurement system. Original videos are taken by (a) commercial cameras and (b) unmanned aerial vehicles (UAVs). UAV videos are stabilized to eliminate the vibrations of the UAV itself. Then, (c) video frames need to rectify lens distortions through (d) camera calibration. Camera parameters and (e) world coordinate systems are also established. Raw video images are sent to our firstly proposed (f) deep neural network (DNN) tracking network, SiamSDN. Finally, we transform the (g) displacement time histories from the pixel level to a metric system.

The details of the SiamSDN framework are shown in Figure 14.3, including five typical parts: input image pairs, a feature extraction subnetwork, a matching function, a decoder, and an output score map.

In the pre-processing method for image pairs, the resizing of the image is not included to ensure that all the image pairs are reshaped into a fixed size. Instead, the original image without padding is treated as an instance. Then, exemplar images are upsampled to the maximum size in a batch, which means the exemplar size is randomly altered according to each batch. The feature extraction subnetwork is fully convolutional with 3×3 convolution filters. The network can achieve the same receptive field with a low burden of parameters. Two branches compose the subnetwork. The two feature extraction branches share the same parameters. Thus, the same types of features can be compared in the following network. The correlation operator is a batch-processing function that compares the Euclidean distance or similarity metric between $\phi(z)$ and $\phi(x)$. Here, $\phi(z)$ and $\phi(x)$ denote the outputs of the template and search branches. Combining deep features in a higher dimension is equivalent to dense sampling around the bounding box and evaluating similarity after each feature extraction. However, the former method is more efficient due to the smaller scale of high-dimension features.

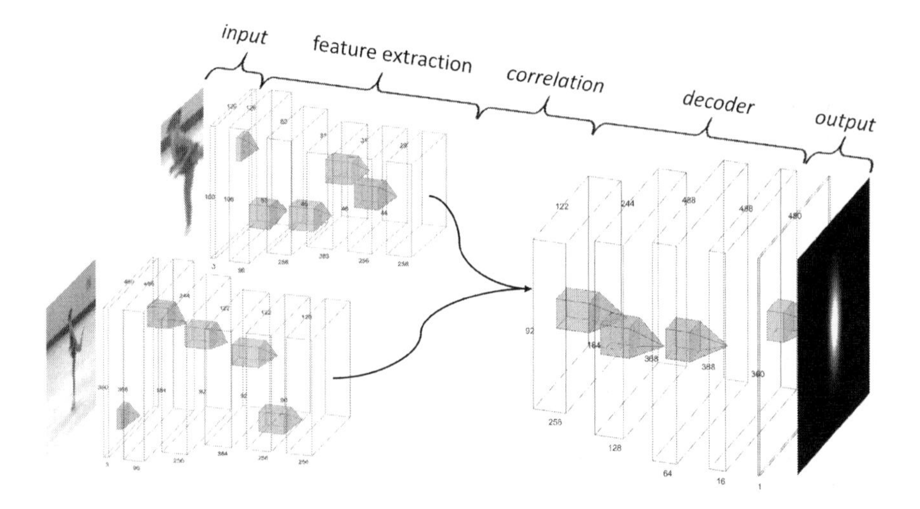

FIGURE 14.3 Main framework of SiamSDN. (Left to right) Input image pair, feature extraction subnetwork, correlation operator, decoder, and output score map. Full-scale exemplar and instance raw images are fed into the template and search branches, respectively. A single decoder is used for restoring both position and scale information.

For convenience, let $u(\phi(z), \phi(x))$ denote the output of the correlation function. The output is a multi-layer two-dimensional (2D) raw map. Since the correlation operator is adopted into the network, the decoder is needed to interpret the comparison results, namely raw maps, from higher into lower dimensions. A decoder with three interpolating layers and three deconvolution layers is adopted, as illustrated in Figure 14.3. The output score map shares the same size as the instance image. In order to combine the classification and bounding box regression tasks of tracking, the label is designed to obey a 2D normal distribution. The mean value is the center of the bounding box. The standard deviation in the label is one-sixth of the width and height. The area outside of the bounding box is set to zero. The response value intensifies with the increase of overlapping area between the exemplar and the instance. Hence, the score around the edge of the bounding box should be lower than that of the center part. Finally, the loss function for each paired instance computes the distance between the label and the score map, which is precisely defined by Eq. (14.1). The parameters in the network are obtained by applying stochastic gradient descent (SGD).

$$L(y, y_{pred}) = \frac{1}{|D|} \sum_{i \in D} (y_i - y_{pred,i})^2 \tag{14.1}$$

where y_{pred} is the response value in the score map, y represents the label, and D indicates the map of scores produced by the decoder.

As a UAV vibrates while hovering over a stationary point, small motions or rotations of the camera will lead to a huge error when measuring the structural displacements. Hence, the raw videos taken by the UAV still have to undergo digital video stabilization (DVS). Motions between two video frames can be divided into global motions and local motions. In the research scene of this study, global motions are vibrations of the UAV camera, whereas local motions are the vibrations of the structural model being shot. Since DVS aims to eliminate the jitter of the camera, its main task includes global motion estimation, motion compensation, and image generation. In the 2D video stabilization algorithm, the estimation of camera global motion parameters mainly includes gray projection, block matching, and feature point matching. During video stabilization, the first frame is used as a reference. Scale-invariant feature transform (SIFT) point detection is carried out for two consecutive image frames. Subsequently, feature descriptions and feature point matching are conducted. Consistent refinement is employed to estimate motion models. During motion estimation, it is assumed that the images satisfy the refined model, the matrix of which can be expressed as:

$$H = \begin{bmatrix} a_1 & a_2 & 0 \\ a_3 & a_4 & 0 \\ t_x & t_y & 1 \end{bmatrix} \tag{14.2}$$

where a_i $(i = 1, 2, 3, 4)$ is related to scaling, rotation, and cropping in image transformation; and t_x and t_y are translation components. Motion compensation for each image is achieved through calculated motion models. The projection transformation matrix of the ith frame is obtained by successive multiplication of the projection transformation matrices of the previous $i - 1$ frames, as shown in Eq. (14.3).

$$H_{cum,i} = \prod_{j=0}^{i-1} H_j \tag{14.3}$$

14.4 CRACK DETECTION BASED ON COMPUTER VISION AND UAVS

14.4.1 A Three-Stage Deep-Learning-Based Method for Crack Detection of High-Resolution Images

Since high-resolution images contain more details, it is necessary to perform geometric edge detection for high-resolution crack images to improve the accuracy of crack detection. Among the crack detection methods based on computer vision, the deep-learning-based method has become one of the most commonly used methods because of its robustness and anti–environmental interference ability. However, limited by computational resources and memory, it is difficult to detect cracks directly on high-resolution images through deep-learning-based methods. This section mainly introduces a high-precision deep-learning-based crack detection method for high-resolution images.

The method comprises three stages: crack classification, segmentation, and morphological post-processing. For a high-resolution crack image, it should be converted into some sub-images first. A crack classification model is then used to identify whether the crack exists in each sub-image, effectively locating the crack's region. Then, by utilizing a segmentation model on the area where the crack exists, the corresponding geometric edges of the crack will be detected efficiently. In the third stage, using a dilate operation and outlier elimination algorithm to eliminate the noise in the prediction results can significantly improve the accuracy of crack geometry edge extraction classification. The flowchart of the method is shown in Figure 14.4. It should be noted that the convolutional neural network model used in the first two stages can be replaced by any classification and segmentation model according to the actual engineering situation. The remainder of this section will individually describe the three stages of the method in detail.

In the first stage of the crack detection process, the image is first divided into several overlapping sub-images. For a high-resolution image, we assume the patch size is $s \times s$, while the overlap length is $s/2$. The overlapping patch aims to improve the accuracy of judging the fracture area. After acquiring the small image, each pixel should be normalized to between 0 and 1 by dividing the value of each pixel in the image by 255 and standardized to between –1 and 1.

The convolutional block attention module (CBAM) ResNet-50 was used to identify the presence or absence of cracks in sub-images. Among them, the role of the CBAM module is to quickly filter out high-value information from a large amount of data, which refers to the information on cracks. Therefore, the ability of the model to identify cracks can be significantly improved after adding this module. The network structure of CBAM ResNet-50 is similar to that of the ResNet network, which is formed by stacking multiple bottlenecks; a schematic diagram is shown in Figure 14.5. A schematic diagram of the network structure of channel attention and spatial attention in the bottleneck is shown in Figure 14.6.

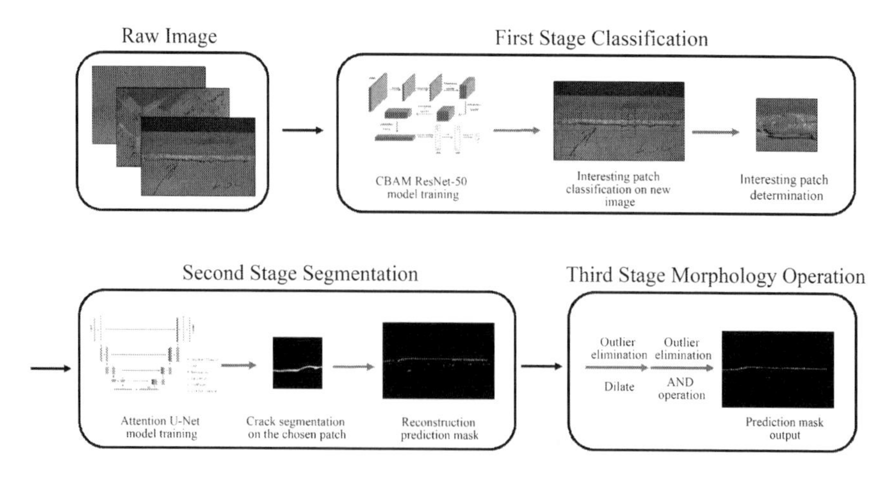

FIGURE 14.4 The flowchart of the crack detection method.

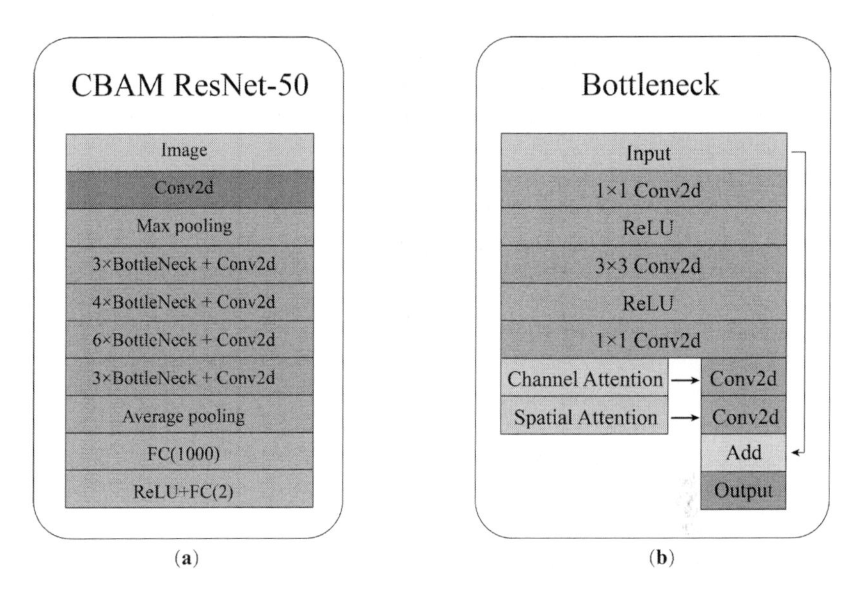

FIGURE 14.5 The architecture of convolutional block attention module (CBAM) ResNet-50: (a) The detailed structure of CBAM ResNet-50, and (b) the detailed structure of the bottleneck. *Conv2d* represents the convolutional layer with a convolution kernel size of 3×3, *Max Pooling* and *Average Pooling* represent the max-pooling layer and average-pooling layer, *FC* represents the fully connected layer, and *ReLU* represents the rectified linear unit.

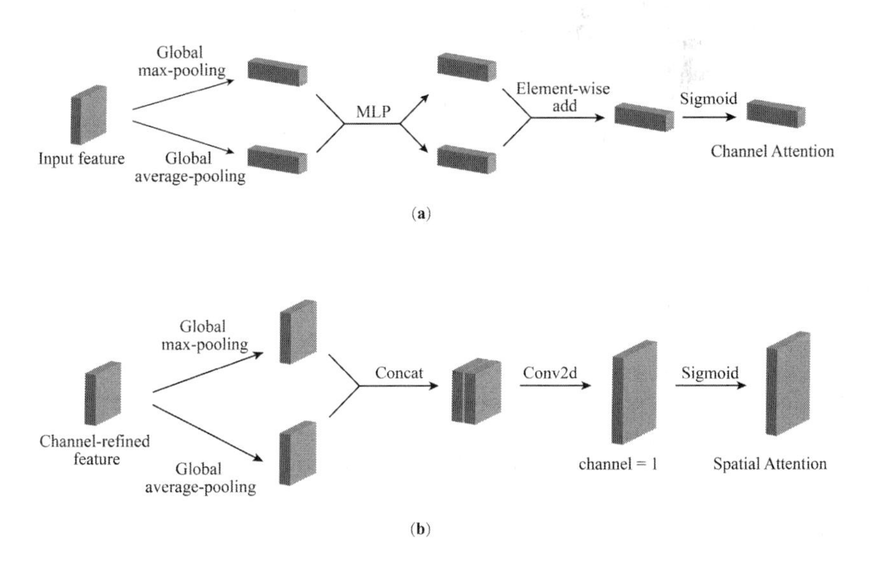

FIGURE 14.6 Diagrams of modules in the convolutional block attention module (CBAM): (a) Diagram of channel attention module, and (b) diagram of spatial attention module. *MLP* represents multi-layer perceptron, and *Sigmoid* represents the sigmoid activation function.

In the second stage of crack detection, since the regions where the crack exists were located, the segmentation model only needs to identify sub-images from crack regions, which means that only the sub-image judged to contain cracks in the first stage should be processed. The Attention U-Net model was implemented in the second stage to segment the geometry edges of cracks. The Attention U-Net model added a self-attention mechanism based on the U-Net network. The schematic diagram of its network structure is shown in Figure 14.7. The Attention CNNLayer adds a channel attention layer and a spatial attention layer based on the CNNLayer, which is shown in Figure 14.8. The spatial attention module used in the Attention U-Net model is the same as in CBAM ResNet-50, and the structure of the channel attention module of the Attention U-Net model is shown in Figure 14.8.

In the third stage of crack detection, several morphological processing methods were used to eliminate errors in crack detection results, including errors caused by cracks' discontinuities and errors caused by wrongly identified outliers. The flowchart of the processing method of the third stage is shown in Figure 14.9.

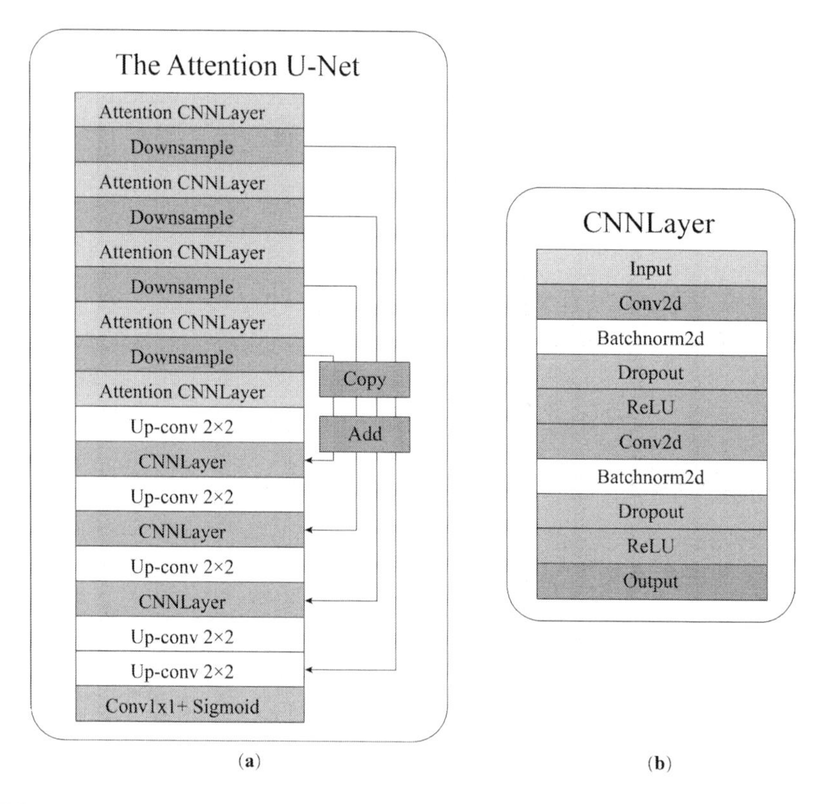

FIGURE 14.7 The overall architecture of the Attention U-Net model: (a) The detailed structure of the Attention U-Net model, and (b) the detailed structure of the CNNLayer. *Batchnorm2d* represents the batch normalization layer, *Dropout* represents the dropout layer, *Up-conv-2 × 2* represents a bilinear interpolation layer, and *Conv1 × 1* represents a convolutional layer with a kernel size of 1 × 1.

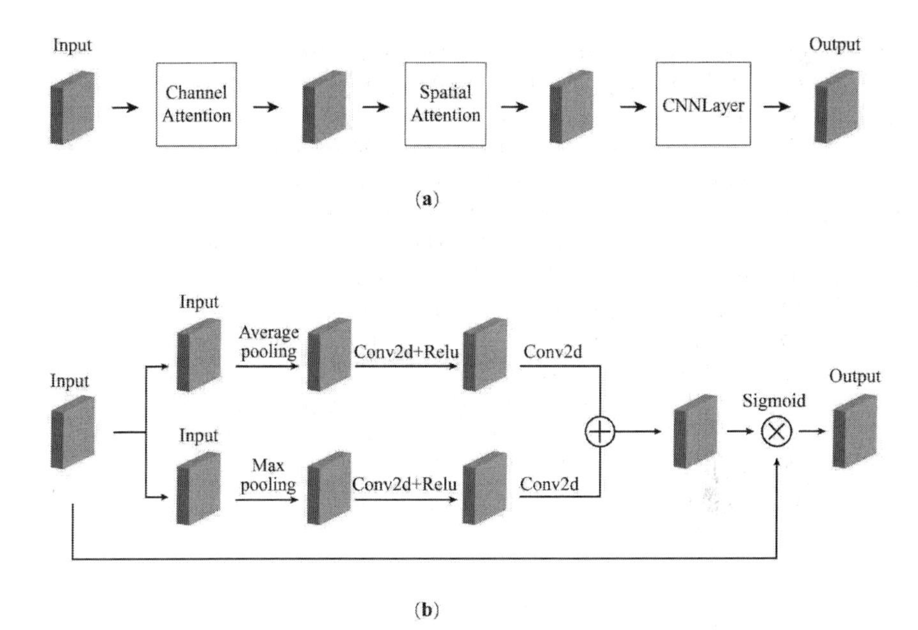

FIGURE 14.8 The architecture of the attention module: (a) The detailed structure of the attention module, and (b) the detailed structure of the channel attention module employed in the attention module.

First, adjust the threshold to a low level and eliminate small outliers through the outlier elimination algorithm. Then, implement the morphologically dilate operation to the entire image to connect the discontinuous cracks, and increase the threshold of the outlier elimination algorithm to eliminate large isolated points. Finally, it is necessary to perform a logical AND operation between the generated image and the original image, guaranteeing that the edge detection results

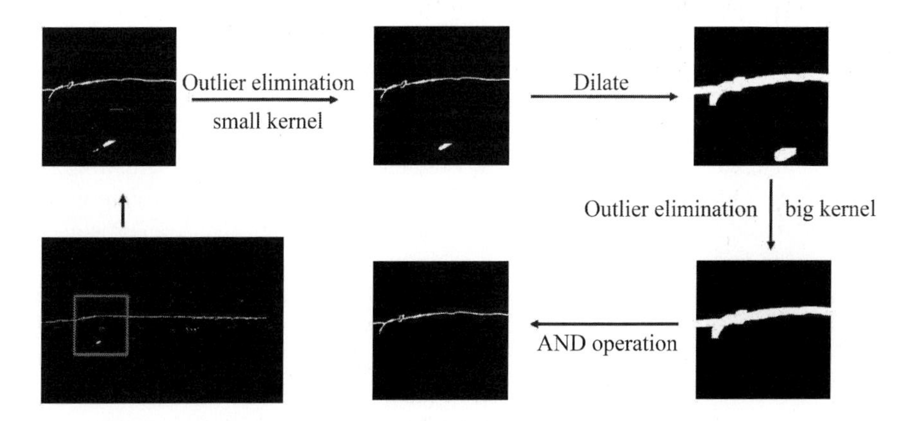

FIGURE 14.9 The flowchart of the third stage.

of cracks are not affected by morphological processing. After processing these operations, the refined results without isolated points will be obtained.

14.4.2 A Real-Time High-Precision Crack Detection Method Based on UAVs

Compared with crack detection through fixed cameras, image acquisition using drones as a carrier can more efficiently and comprehensively obtain all crack information on the building surface. However, there are several problems with using drones for crack detection:

1. Crack width measurement needs to reach the millimeter level in the SHM of concrete structures and in rapid post-disaster damage assessment. However, it is challenging to achieve millimeter-scale accuracy for the crack width obtained by shooting with a drone camera far away from the building.
2. It is difficult to calibrate the shooting distance for crack detection based on drones in practical situations.
3. Drones have turbulence in flight, and there are many environmental interference factors in the captured images.
4. To ensure drone flight safety, it is difficult to control the drone to maintain a short distance from the building for a long time through fixed-point navigation or manual operation of the drone, especially in large-scale crack detection.

The existence of the above problems greatly restricts the application of UAV-based crack detection. Therefore, this section introduces a real-time UAV-based crack detection method that can solve the above ubiquitous problems. The method introduced in this section is mainly applied in drone-based automatic real-time crack detection, and the UAV must carry an onboard computer and a binocular camera. The information from crack detection can be used to assist drones in path planning to achieve higher-precision crack geometry edge extraction and more accurate maximum crack width measurement, and to obtain the spatial position of the crack. The schematic diagram of the overall process is shown in Figure 14.10.

In the first part of the crack detection process, the image of the binocular camera is obtained and downsampled when the drone detects cracks. Then, a lightweight crack classification model image is used to classify the image to determine whether there is a crack. The lightweight crack segmentation algorithm processes the resized image to obtain a rough crack area. By combining the crack position and the depth image generated by the binocular camera, the coordinates of the crack in the drone coordinate system can be calculated, thereby providing critical information for the crack information-assisted drone flight automatic control algorithm. The lightweight crack classification model and the lightweight crack segmentation model need to have a high processing speed to provide enough key information for the flight of the drone, such as the use of ShuffleNet V2 [14]

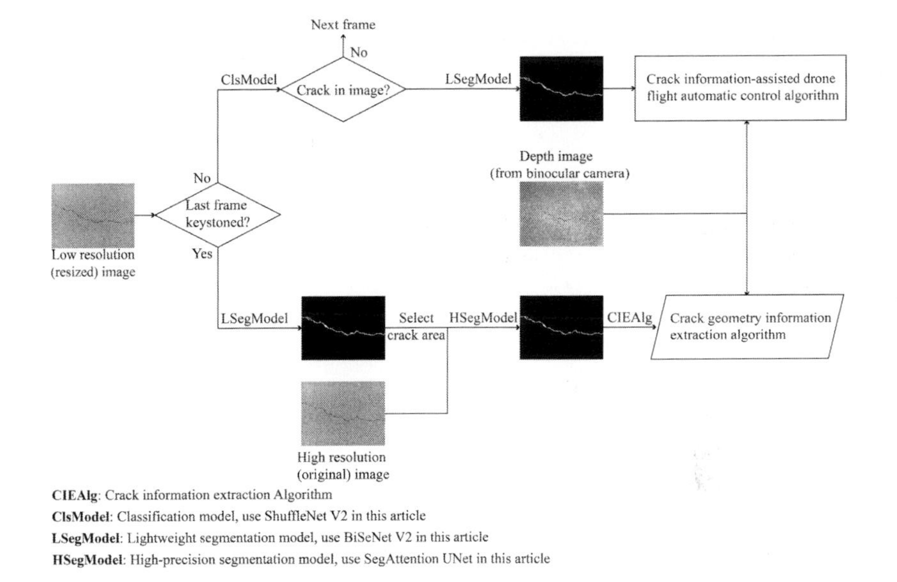

FIGURE 14.10 The flowchart of the UAV-based real-time crack detection method.

and BiSeNet V2 [15]. The crack detection algorithm inevitably has an error rate. However, if the overlapping multi-frame images are repeatedly detected to fully use the information in the captured images, the probability of missing cracks can be significantly reduced.

The second part of the crack detection process is achieved by the crack information-assisted drone flight automatic control algorithm. By using the crack information-assisted drone flight automatic control algorithm, the drone will automatically approach the crack and hover smoothly in the vicinity of the crack. The automatic control algorithm is successfully completed once the drone flies to the target position and hovers smoothly. When the control algorithm is successfully completed, the image captured by the drone at that moment is defined as the keystone. By using the binocular camera, the coordinate of the crack pixels detected from the lightweight crack segmentation model in the world coordinate system can be calculated through the triangulation principle and a gyroscope. We define the moving target vector $\boldsymbol{T} \in R^{3\times1}$ formed by where the camera points to the crack, the adjusted moving target vector $\boldsymbol{D} \in R^{3\times1}$, and the displacement vector $\boldsymbol{I} \in R^{3\times1}$ of the drone between two key frames calculated by the inertial measurement unit (IMU), where the keyframe is defined as the image that successfully calculates the crack position through the image of the binocular camera combined with the crack detection algorithms. The moving target vector $\boldsymbol{T}$ can be calculated through Eqs. (14.4)–(14.6).

$$T' = \frac{1}{N}\left(\sum_{i=1}^{N} x_{w,i}, \sum_{i=1}^{N} y_{w,i}, \sum_{i=1}^{N} z_{w,i}\right) \tag{14.4}$$

$$T = \frac{R - d_t}{R}(T'_x, T'_y, T'_z) \tag{14.5}$$

$$R = \sqrt{T'^2_x + T'^2_y + T'^2_z} \tag{14.6}$$

where R denotes the calculated distance between the drone and the crack, and d_t denotes the target shortest distance between the drone and the crack. The calculation formula of the drone's adjusted moving target vector $\boldsymbol{D}_i$ is shown in Eq. (14.7).

$$D_i = \frac{T_i + \sum_{j=1}^{n-1} \alpha^j \left(T_{i-j} - \sum_{k=1}^{j} I_{i-k} \right)}{1 + \sum_{j=1}^{n-1} \alpha^j} \tag{14.7}$$

where n is the number of the selected keyframes, the value of n is smaller than i, and the specific value of n can be dynamically changed during the flight of the drone since i increases with time. α is a value between 0 and 1. The closer α is to 1, the more $\boldsymbol{D}_i$ tends to rely on the calculation results of the IMU.

The whole process of the crack information-assisted drone flight automatic control algorithm is shown in Algorithm 14.1, where $\boldsymbol{P}_1$ represents the drone's position at the starting point.

The third part of the crack detection process is that when the keystone is obtained after utilizing the crack information-assisted drone flight automatic control algorithm, the same lightweight segmentation model is used to process the low-resolution image and determine the location of the crack in the image. The original high-resolution image is then divided into several image patches with fixed sizes using the patch-based method. The patches containing cracks are selected using the results of the previous lightweight crack segmentation algorithm. After that, a high-precision segmentation model with a large number of parameters and higher precision is used to segment cracks in the selected image patches with higher accuracy, such as the Attention U-Net. Then, all the image patches are combined into a crack prediction mask with the same size as the original image. Finally, a crack geometry information extraction algorithm is used to obtain accurate geometric information about the cracks. It should be emphasized that this part is also completed quickly on the onboard computer of the drone, but since it will not affect the flight trajectory of the drone, it does not need to be completed in milliseconds. The detailed steps of the crack geometry information extraction algorithm are shown in Algorithm 14.2. Among them, C means image set, and each image C_i in set C contains only one crack. S represents the crack skeleton line, which can be realized by the Zhang–Suen algorithm [16]. $N_4(p)$ refers to the four pixels adjacent to a pixel p, B indicates a pixel set, the $dist(S_i, B_j)$ is Euclidean distance, M denotes a large value for initializing $w_{min,i}$, and W represents the maximum crack width. η is the pixel resolution, which can be calculated through Eq. (14.8).

Algorithm 14.1: Pseudocode for the crack information-assisted drone flight automatic control algorithm

1: Set c_{in} and c_{out} to 0.
2: Compute the original location $\boldsymbol{P}_1$, $\boldsymbol{T}_1$.
3: **for** i=2:i_{max} **do**
4: **if** this frame is a keyframe **then**
5: Get the latest $\boldsymbol{T}_i$.
6: Compute $\boldsymbol{I}_{i-1}$.
7: Adjust n and compute $\boldsymbol{D}_i$.
8: **if** $|\boldsymbol{D}_i| < Th_1$ **then**
9: $c_{in} = c_{in} + 1$
10: **else**
11: **if** $c_{in} \neq 0$ **then**
12: $c_{out} = c_{out} + 1$
13: **end if**
14: **end if**
15: **if** $c_{out} > c_{outmax}$ **then**
16: Set c_{in} and c_{out} to 0.
17: **end if**
18: **if** $c_{in} > c_{inmax}$ **then**
19: Terminate the algorithm and obtain images at this moment for final crack detection.
20: break
21: **end if**
22: Adjust the flight direction of the drone to the adjusted moving target vector $\boldsymbol{D}_i$ and move.
23: **end if**
24: **end for**

$$\eta = \frac{1}{N+M}\left(\sum_{i=1}^{N}\left|\frac{x_{1,i}-x_{2,i}}{p_{x1,i}-p_{x2,i}}\right| + \sum_{i=1}^{M}\left|\frac{y_{1,i}-y_{2,i}}{p_{y1,i}-p_{y2,i}}\right|\right) \tag{14.8}$$

where N represents the total number of rows of two crack edge pixels in a row in the crack edge image, and, correspondingly, M represents the total number of columns that contains two crack edge pixels. $x_{1,i}$ and $x_{2,i}$ are the coordinates of two crack edge pixels in a row in the world coordinate system, and $y_{1,i}$ and $y_{2,i}$ are the coordinates of two crack edge pixels in a column in the world coordinate system. $p_{x1,i}$ and $p_{x2,i}$ are the coordinates of the two crack edge pixels in a row in the image coordinate system, and $p_{y1,i}$ and $p_{y2,i}$ are the coordinates of the two crack edge pixels in a column in the image coordinate system. The coordinates of the crack edge pixels in the world coordinate system need to be obtained through the calculation results of the binocular camera.

Algorithm 14.2: Pseudocode for the maximum crack width measurement algorithm

1: Process P using the connected-component labeling algorithm.
2: **for** each $C_i \in C$ **do**
3: **for** each $p \in C_i$ **do**
4: **if** $N_4(p) \neq 1$ **then**
5: Add p into B.
6: **end if**
7: **end for**
8: Set $w_{max} = 0$.
9: **for** each $S_i \in S$ **do**
10: Set $w_{min,i} = M$.
11: **for** each $B_j \in B$ **do**
12: $w_{min,i} = min(dist(S_i, B_j), w_{min,i})$
13: **end for**
14: $w_{max} = max(w_{min,i}, w_{max})$
15: **end for**
16: $W = 2w_{max\eta}$
17: **end for**

14.5 EXPERIMENTS OF THE COMPUTER VISION–BASED METHODS

14.5.1 Experiment of the Structural System Identification Method

A five-story structure is designed and deployed in the test. The video-recording equipment is a DJI UAV with a frame rate of 30 frames per second (fps) and a resolution of 3968 × 2976 pixels. The LVDT displacement transducers used for reference have a sampling rate of 256 Hz and are placed on each floor and at the base of the model. The experimental setup is shown in Figure 14.11. The test conditions are listed in Table 14.1.

Figure 14.12 is the final tracking result of testing scenario 1. The original UAV video-tracking results are also displayed. As shown in the figure, the UAV camera vibrates as it is hovering over a fixed point, which greatly affects the displacement analysis adversely. And the proposed DVS method has eliminated global motions caused by the camera. Errors between the introduced method and LVDT sensors are listed in Table 14.2. For the vision method, both SIFT point matching and the SiamSDN tracking network are considered for comparison. As shown in Table 14.2, all the tracking results of SiamSDN are better than the SIFT method.

FIGURE 14.11 Setup of the UAV test.

TABLE 14.1
Test Conditions

No.	Condition	Amplitude (mm)	Frequency (Hz)	Duration (s)
1	Sine	5	8	60
2	Sine	5	11	60

The normalized root mean squared error (NRMSE) of SiamSDN has improved from 22.70% (3F) to 91.38% (2F) in testing scenario 1 and also increased from 1.01% (3F) to 85.36% (2F) in testing scenario 2. The average improvement is 57.54%. Figure 14.13 manifests the power spectrum density (PSD) computed from the displacements in testing conditions 1 and 2. Comparing the results of the vision and LVDT methods, we can observe that the frequency component obtained from both displacements agrees well.

14.5.2 Experiment of the Three-Stage Deep-Learning-Based Method for Crack Detection in High-Resolution Images

The dataset used in this experiment is from the first International Project Competition for Structural Health Monitoring [13], which is composed of 360 images taken in a steel box girder. A total of 180 images were randomly selected as the training set, and 20 images were prescribed as a validation set, while the remaining 160 images were selected as a test set. The dataset consists of RGB images with pixel resolution of 3264 × 4928 and a corresponding mask. The experimental result shows that *Recall* of the first stage of the

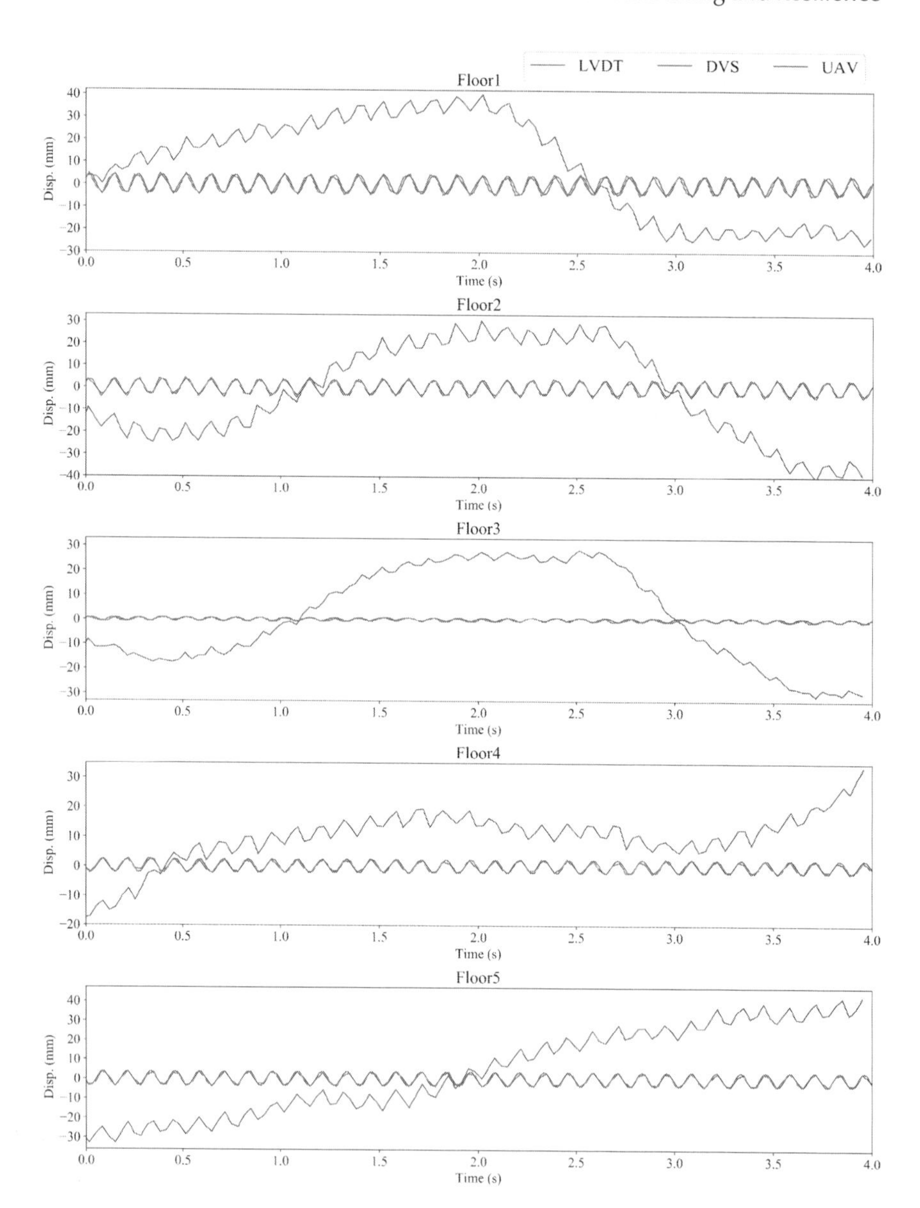

FIGURE 14.12 Testing scenario 1: Comparison of displacements by the SiamSDN system and linear variable differential transformer (LVDT) transducers. The curve that has the highest maximum displacement values in all plots (30 or 40mm). The other two curves that overlap include the plot obtained from the video after digital video stabilization (DVS) processing and the LVDT displacement.

TABLE 14.2
Measurement Errors of Each Floor in the Shaking-Table Test with UAV Cameras

Condition	Algorithm	Error Type	1F	2F	3F	4F	5F
Scenario 1	SiamSDN	Root mean squared error (RMSE) (mm)	0.38	0.09	0.05	0.03	0.16
		Normalized RMSE (NRMSE) (%)	4.52	1.32	4.12	0.70	2.22
	SIFT	RMSE (mm)	1.18	1.05	0.06	0.31	0.29
		NRMSE (%)	13.91	15.32	5.33	6.65	4.01
		Improvement (%)	67.51	91.38	22.70	89.47	44.64
Scenario 2	SiamSDN	RMSE (mm)	0.20	0.19	0.44	0.04	0.27
		NRMSE (%)	6.30	2.44	5.91	1.69	3.33
	SIFT	RMSE (mm)	0.25	1.32	0.45	0.21	0.82
		NRMSE (%)	8.07	16.67	5.97	8.42	9.94
		Improvement (%)	21.93	85.36	1.01	79.93	66.50

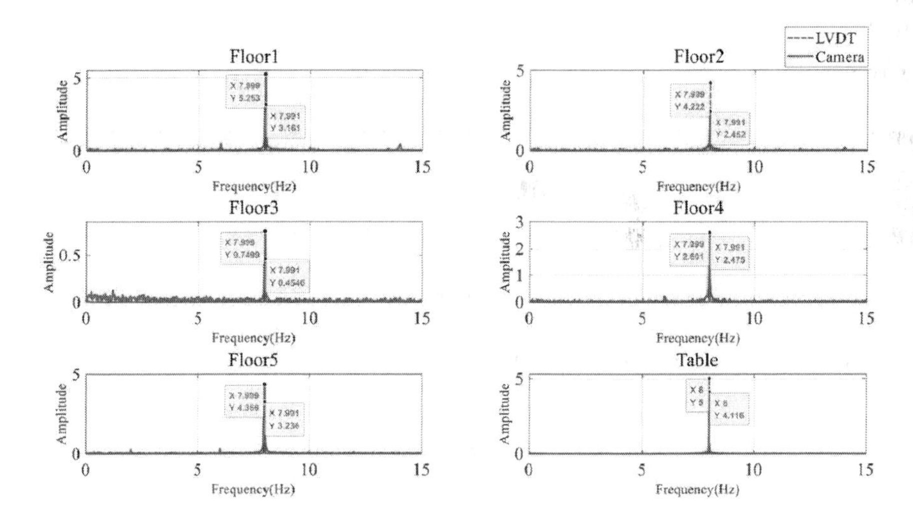

FIGURE 14.13 Comparisons of the power spectrum density.

model reaches 0.95 on the test set, while the intersection over union (IoU) of the second stage of the model achieves 0.48. Besides, after processing by the third stage of the method, the IoU of the test set reaches 0.70, which is 45.8% higher than the output of the second stage of the model. The visualization of the detection result is shown in Figure 14.14.

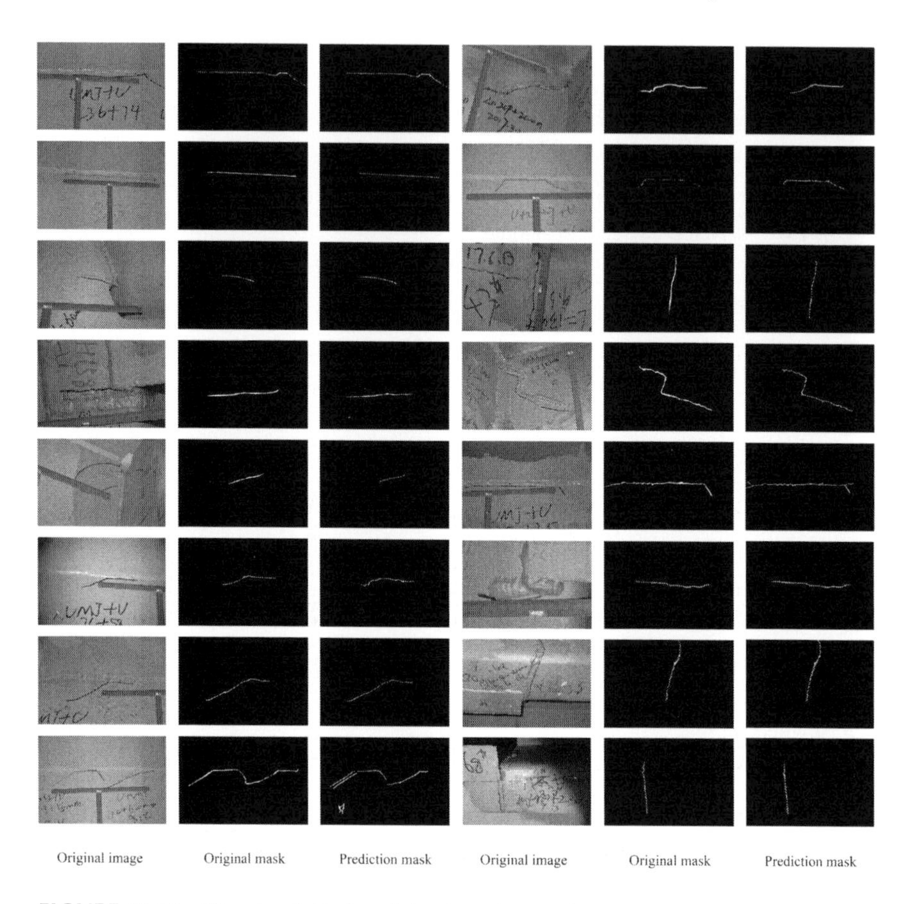

FIGURE 14.14 The visual display of the detection result.

14.5.3 Experiment of the UAV-Based Real-Time Crack Detection Method

The real-time high-precision crack detection method based on UAVs is verified on a two-story building and a shaking-table test building. The drone used in the experiment is a DJI M300 RTK equipped with a Jetson AGX Xavier and a Zed2 binocular camera, as shown in Figure 14.15. In an experiment on a two-story building, the flight strategy was planned through fixed-point navigation. First, 18 control points are roughly set around the building, and the distance between each control point and the building surface is about 3 m. When the drone detects a crack in the image, the drone's position at that time is recorded, and the algorithm is used to automatically control the drone using crack information to guide the drone toward the crack. When the high-precision crack segmentation algorithm is completed, the drone returns to the previously recorded position and moves to the next control point. Due to the complicated environment in this experiment and the errors in the result of the lightweight crack detection algorithm, the

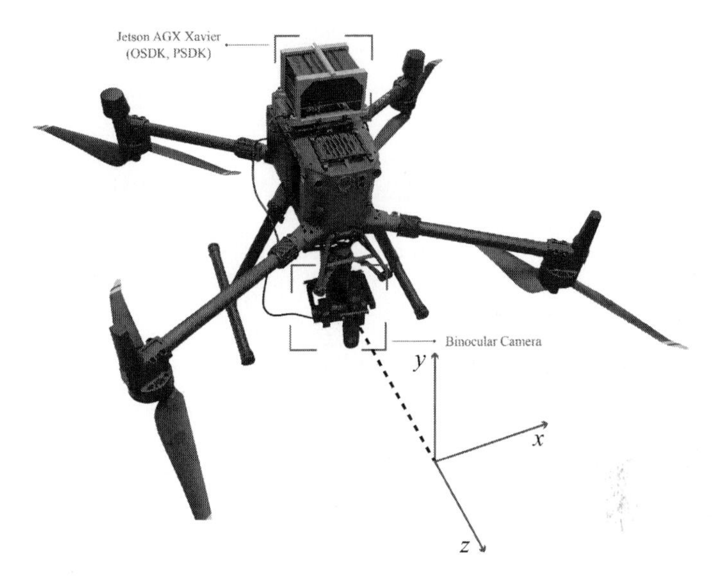

FIGURE 14.15 Components of the drone system.

crack information-assisted drone flight automatic control algorithm is only activated when more than five consecutive frames of images are identified as cracks. Furthermore, in this experiment, the values of c_{outmax}, c_{inmax}, Th_1, d_t, and α are set to 10, 50, 0.5 m, 0.5 m, and 0.9, respectively. The value of the sliding time window changes dynamically. When the number of keyframes is less than 10, n is equal to the number of keyframes, and when the number of keyframes is greater than 10, n is equal to 10. The overall drone path is shown in Figure 14.16 to demonstrate the effectiveness of the crack information-assisted drone flight automatic control algorithm. Figure 14.17 shows the visualization of the crack width measurement results, where the actual width of all cracks and the measured width in both detection cases are shown in Figure 14.17a, and Figure 14.17b shows the absolute error of all cracks in both detection cases. This experiment shows that the crack information-assisted drone flight automatic control algorithm introduced in this chapter can effectively drive the drone close to the cracks on the building surface and obtain significantly better results of crack geometry edge extraction as well as crack width measurement.

For the experiment on the shaking-table test building, the original structure has a total height of 161.5 m and the shaking-table test model adopts a scale of 1:10. In the fixed-point navigation, 37 control points are specified, and the distance between the control points and the building is 2 m. A schematic diagram of the specific location of the control points and the flight trajectory of the drone is shown in Figure 14.18. The position indicated by the red arrow in Figure 14.18 is the position where the drone automatically detected the crack. In this experiment, the parameters of the crack information-assisted drone flight automatic control algorithm are the same as those in the experiment on the two-story building.

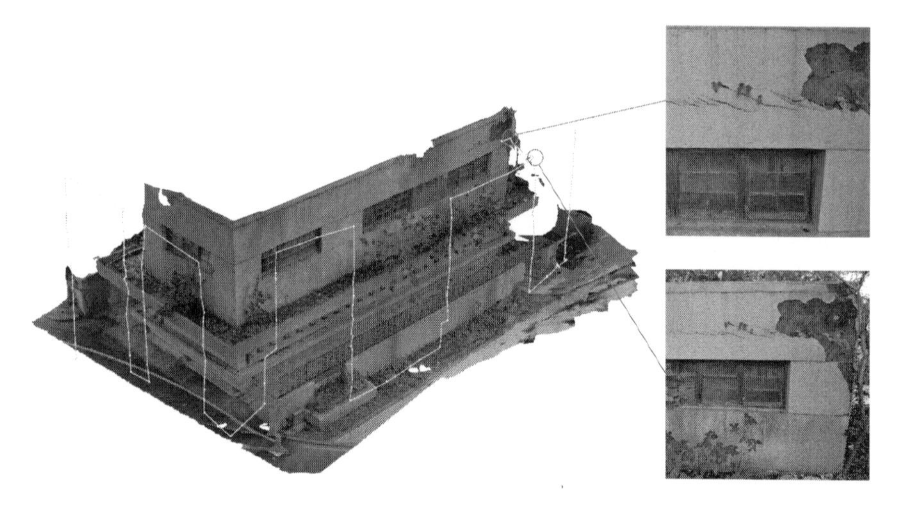

FIGURE 14.16 Experimental results of the automatic control algorithm. The yellow line shows the drone movement route. Located in the upper right section of the image, two distinctive red circles are noticeable. The right-side circle signifies the initial position of the drone prior to detecting the crack, while the left-side circle denotes the ultimate stable position achieved by the drone under the guidance of the automatic control algorithm. The two images show what the drone captured while in the corresponding positions.

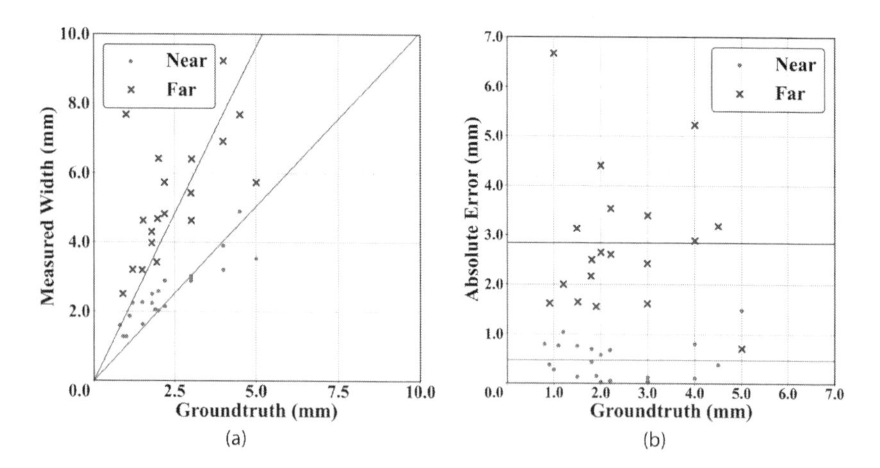

FIGURE 14.17 Visualization of the crack width measurement results: (a) The actual width of all cracks and measured width with and without the crack information-assisted drone flight automatic control algorithm, and (b) the absolute error of crack width measurement with and without the crack information-assisted drone flight automatic control algorithm. Among them, the "near" in the legend represents the detection result using the crack information-assisted drone flight automatic control algorithm, and the "far" means the detection result without using the algorithm.

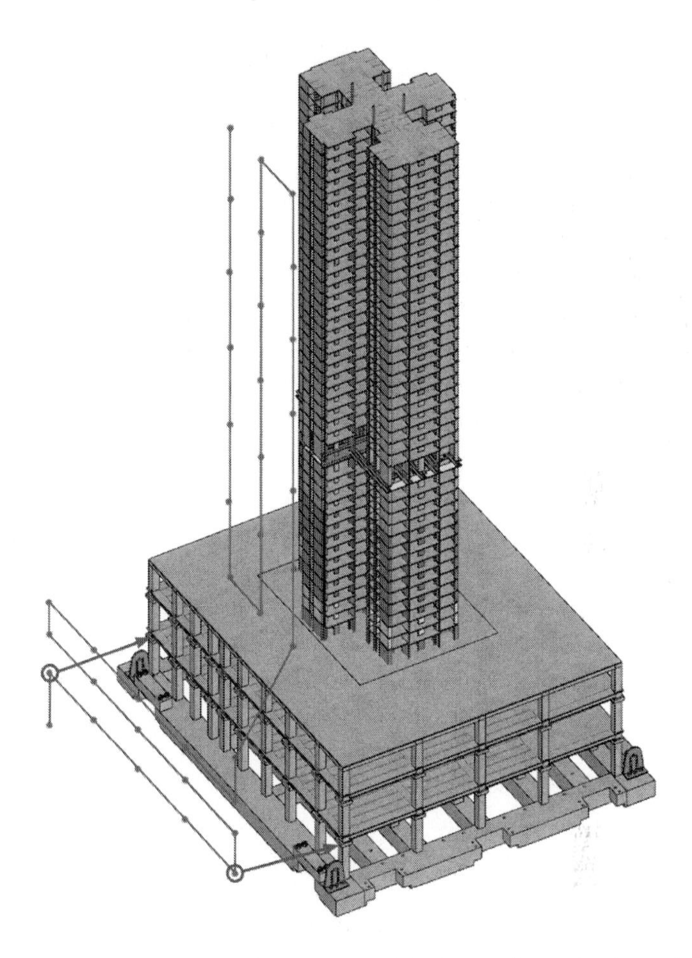

FIGURE 14.18 Visualization of the trajectory of the fixed-point navigation and the location of the crack detected by the drone.

All crack detection calculations are completed during the movement of the drone. In this experiment, the recall, precision, and mean IoU (MIoU) of crack edge extraction after the drone is close to the building are 63.26%, 59.11%, and 44.00%, respectively, and only 48.88%, 26.31%, and 20.63% when the crack information-assisted drone flight automatic control algorithm is not used. The experimental results show that the average absolute error of detecting the maximum crack width through the drone at 2 m from the building surface is 2.89 mm. When the drone is close to the building surface by implementing the crack information-assisted drone flight automatic control algorithm, the average absolute error of maximum crack width measurement is only 0.42 mm. This is because when the drone is far away from the building, the calculation error of the pixel resolution is larger, and the high-precision segmentation model has a lower detection accuracy for the extraction of the edge of the crack.

REFERENCES

1. Ribeiro, D., Calcada, R., Ferreira, J., Martins, T.: Non-contact measurement of the dynamic displacement of railway bridges using an advanced video-based system. Engineering Structures **75**, 164–180 (2014).
2. Feng, M.Q., Fukuda, Y., Feng, D., Mizuta, M.: Nontarget vision sensor for remote measurement of bridge dynamic response. Journal of Bridge Engineering **20**(12), 04015023 (2015).
3. Niassif, H.H., Gindy, M., Davis, J.: Comparison of laser doppler vibrometer with contact sensors for monitoring bridge deflection and vibration. NDT & E International **38**(3), 213–218 (2005).
4. Olawale, D.O., Kliewer, K., Okoye, A., Dickens, T.J., Uddin, M.J., Okoli, O.I.: Real time failure detection in unreinforced cementitious composites with triboluminescent sensor. Journal of Luminescence **147**, 235–241 (2014).
5. Zhou, Y., Zhang, L., Liu, T., Gong, S.: Structural system identification based on computer vision. China Civil Engineering Journal **51**(11), 17–23 (2018).
6. Liu, J., Yang, X., Li, L.: Vibronet: Recurrent neural networks with multi-target learning for image-based vibration frequency measurement. Journal of Sound and Vibration **457**, 51–66 (2019).
7. Zhou, Y., Liu, T.: Computer vision-based crack detection and measurement on concrete structure. Journal of Tongji University (Natural Science) **47**(9), 1277–1285 (2019).
8. Meng, S., Zhang, X., Qiao, S., Zhou, Y.: Research on grid optimized crack detection model based on deep learning. Journal of Building Structures **41**(S2), 404–410 (2020).
9. Yoon, H., Shin, J., Spencer Jr, B.F.: Structural displacement measurement using an unmanned aerial system. Computer-Aided Civil and Infrastructure Engineering **33**(3), 183–192 (2018).
10. Zhou, Y., Peng, S.-Y., Yan, L.-F., He, B.: A novel DNN tracking algorithm for structural system identification. Smart Structures and Systems. An International Journal **27**(5), 803–818 (2021).
11. Meng, S., Gao, Z., Zhou, Y., He, B., Kong, Q.: A three-stage deep-learning-based method for crack detection of high resolution steel box girder image. Smart Structures and Systems **29**(1), 29–39 (2022).
12. Meng, S., Gao, Z., Zhou, Y., He, B., Djerrad, A.: Real-time automatic crack detection method based on drone. Computer-Aided Civil and Infrastructure Engineering **38**(7), 849–872 (2022).
13. Bao, Y., Li, J., Nagayama, T., Xu, Y., Spencer Jr, B.F., Li, H.: The 1st international project competition for structural health monitoring (IPC-SHM, 2020): A summary and benchmark problem. Structural Health Monitoring **20**(4), 2229–2239 (2021).
14. Ma, N., Zhang, X., Zheng, H.-T., Sun, J.: Shufflenet v2: Practical guidelines for efficient CNN architecture design. In: Proceedings of the European Conference on Computer Vision (ECCV), pp. 116–131 (2018).
15. Yu, C., Gao, C., Wang, J., Yu, G., Shen, C., Sang, N.: BiSeNet V2: Bilateral Network with Guided Aggregation for Real-time Semantic Segmentation. arXiv:2004.02147 [cs] (2020).
16. Zhang, T.Y., Suen, C.Y.: A fast parallel algorithm for thinning digital patterns. Communications of the ACM **27**(3), 236–239 (1984).

Index

Note: Locators in *italics* represent figures and **bold** indicate tables in the text.

A

ABAQUS XFEM, 71
A-B-C triangle, 87
Acceleration mitigation, 209
Achillopoulou, D. V., 9
Acikgoz, S., 80
Acoustic emission (AE), 78, 89–91, *91*, *92*, 93, *94*, 95
Aeroelastic simulations, 122–124
Aero-servo-elastic simulations, 124
Agrawal, A. K., 220
Alavinasab, A., 204
Alexakis, H., 84, 89
Alkayem, N. F., 15, **35–36**
Aloisio, A., 9, **45**
Ambient vibration tests (AVTs), 259, 260
Amini, F., 204
Anh, P. H., **30**
Artificial intelligence (AI), 291–292
Artificial neural network (ANN), 226–228, *227*, 230, *232*, *233*, 238, 243–247, *246*
  algorithms, 247–248, **253**, **254**
  NN time series prediction, 248–249, *249*
Asadi, E., 9
ASTM C1702 Standard, 108
Attribute-based method, 286–287
Automatic control algorithm, 322–325, 331, *332*, 333
Autoregressive moving average (ARMA), 244
Azzara, R. M., 183, 188

B

Bakhary, N., 244
Barai, S., 243
Bassoli, E., **37–38**
Basu, B., 204
Bayesian filtering approach, 149
Bayesian hyperparameter identification, 122, 125
Bayesian method, 259–261; *see also* Cable-stayed road bridge
  for damage detection, 266–267
  for FE model input parameters, 262
  for inverse problem, 261
  modeling and measurement prediction error, 262–263
Beheshti Aval, S. B., 15
Benmouiza, K., 244
Bibliographic analysis, 4–8
Bibliometric analysis, 4–8
Bighamian, R., **22**
Blade element momentum (BEM) theory, 152–153
Bridge; *see also* Cable-stayed road bridge; CFM-5 bridge; Half-joint bridge; Kien bridge; Truss bridge
  Éric Tabarly, 245, 249–250, *250*, **251**, 254
  Ferrandet, 60, 66, 74
  Quisi, 60, 62, 74
Brilakis, I., 303
Brownjohn, J., 282
Building information model (BIM), 9
Bureerat, S., **33**, **39**, 46
Buza, E., 303

C

Cable-stayed road bridge; *see also* Éric Tabarly Bridge
  ambient vibration tests and operational modal analysis, 267–268, **268**, *268*
  damage detection, 273–274, *274*
  improved MCMC MH algorithm, 271, *272*
  initial FE model development, 268–269, **269**
  measurement uncertainties, 271, 273
  PC-based surrogate models, 269–271, *270*
  uncertain parameter PDF, 269, **270**
Cao, M., **35–36**
Cartesian coordinate system, 229
Cascardi, A., 244
Casciati, S., **20**
Catbas, F., 282
Cavalini, A. A. Jr., **26**
Celebi, M., 9

CFM-5 bridge, 85–86, *86*
analysis, 87–89, *88*
sensing network and installation, 85–87, *87*
Chang, C., 244
Cha, Y. -J., 220
Cheknane, A., 244
Chi-Chi earthquake, 209, **210**, 211–213, *212*
Chitholian, A. R., 301
Choi, K. M., 203
Chun, C., 305
Cimellaro, G. P., 4
Classical probabilistic approach, 259
Classic computer vision methods, 303–304
Climate Change Act 2008, 120
Clock Tower, 182–184, *183*, 187, **196**, 197, **197**, *198*
Clusters, 7
Cocking, S., 86
CO-LQG, 209, **210**, 211–214, *211–214*, **215**, *216*, 216–218, *217*
Complete pooling (CP), 172
Computer vision (CV)-based SHM systems, 280–281
artificial intelligence, 291
components, 285–286
importance, 284–285
limitations, 283–284
machine learning, 292
measurement, 281–282, *282*
quantification, 286–288
resilience, 284, 288–289, *290*
techniques, 282–283
technologies, 290–291
Conditional relational VAE, 132
Conditional relational variational auto-encoder (CRVAE) approach, 132–133, *133*, 135–138, *136*, *138*, *139*
Contact systems, 313
Continuous flight auger (CFA), 102
Convolutional neural networks (CNNs), 304
Covariance-driven stochastic subspace identification method (SSI/Cov), 183, 186, 187, **187**
Crack detection
automatic control algorithm, 322–325, 331, *332*, 333
drone problems for, 322
power spectrum density, 271, *282*, 327, *329*
shaking-table test building, **329**, 330, 331
three-stage deep-learning-based method for, 317–318, *318–321*, 320–322, 327, 329, *330*
UAV, application of, 322–326, *323*, 330–331, *331–333*, 333
vision and LVDT methods, 326–327, **327**, *327*, *328*, **329**, *329*
Crosshole sonic logging (CSL), 99
Cross pseudo-supervision, 306

D

Damage equivalent load (DEL), 121–124, *123*, 127–130, *128*, *129*, **130**, *131*, 145, 147, *147*
normalized mean squared error and, 130, *131*
wake-affected (downwind) turbine, 129–130
Darwin, C., 16
Das, D., 203
Data acquisition, 300–301, *301*
Datasets and benchmarks, 301–303, *302*, *303*
Decentralized output feedback polynomial control (DOPFC), 202
DeepLabv3+ architecture, 306
Deep learning methods, 304–307, *306*, *307*
Degrees of freedoms (DOFs), 16
De Oliveira Dias Prudente dos Santos, J. P., 244
De Schutter, G., 107, 111
Differential evolution algorithm (DEA), 16–17, *17*, 46, 48
application review, 19, **20–45**
crossover, 18–19
initialization, 17–18
multiple-step method, 48, *49*
mutation, 18
selection, 19
single-step method, 48, *49*
two-step method, 48, *49*
Digital video stabilization (DVS), 317
Ding, Z., **30**
Dinh-Cong, D., **32–34**
Discrete wavelet transform (DWT) analysis, 204
Displacement mitigation, 209
DJI M300 RTK, 330
Doebling, S. W., 2
Domaneschi, M., 4
Dong, C., 282
Downwind turbine, 121, *123*, 128, 129, *129*, 130
Dual Kalman filter (DKF), 149
Dutta, A., 243
DWM-FAST simulations, 127
Dyke, S. J., 202–203
Dynamical system formulation, 149–150
Dynamic wake meandering (DWM), 124

E

Earthquake
power loss, *218*, 218–219
Earthquake record, 213–214, *214*, **215**, *216*, 216–218, *217*
Earthquake time histories, 209, 211–213
Effective pile radius, 110
Efficient correlation-based index (ECBI), 46, 50
El Centro earthquake, 209, **210**, *211*, 218, *218*
Enhanced frequency domain decomposition (EFDD) method, 268, *268*, 271
Environmental and operational parameter (EOP) measurements, 130, *131*
Éric Tabarly bridge, 245, 249–250, *250*, **251**, 254
Evidence lower bound (ELBO), 133–135, **136**, 162
Expectation maximization (EM), 142, 163
Experimental modal analysis, 258–259

F

Fallahian, S., **37**
Fang, X., 244
Fatigue, 145
Fatigue, aerodynamics, structures, and turbulence (FAST), 124, 127
Fatigue strength curve test, 61–63, *63–66*, 65–66
Feature point/keypoint matching, 283
Federal Emergency Management Agency (FEMA), 213
Federal Highways Administration (FHWA) Order 5520, 284
Ferrandet bridge, 60, 66, 74
Ferrocarrils de la Generalitat Valenciana (FGV), 59, 65, 66, 67
Fiber Bragg grating (FBG) sensor, 61, 78, 80, 90, 91, 93
37NA3A4, 84, 85, *85*
37NA5A6, 81, *81*
37NA6A7, 84, *85*
arrays, 80–81, *81*
networks, 80, 81, 94
rosettes, 87, *88*, 88–89
Finite element (FE) model, 2, 15, 16; *see also* Lucca towers
1D, 100, 110, *111*, 112, 118
simulators, 145, **146**, *147*, 147–148
3D, 268–269, **269**
2D, 100–101, 114, 116–118
updating process, 264
Firouzi, B., **44**
5 MW NREL prototype turbine, 137
FLOw Redirection and Induction in Steady State (FLORIS), 136, *138*
Frangopol, D. M., 243
Fu, Y. M., **25**, 46
Fuzzy logic controller (FLC), 202, 203
Fuzzy sliding mode controller (FSMC), 202–203

G

Gamma-gamma logging (GGL), 99
Gaussian process (GP) model, 124, 127, 149, 150, 153, 155, 160
Gaussian process regression (GPR), 121–122, 124–129, **128**, *128*, *129*, 130–131, *131*, 159, 162
Gebze earthquake, 209, **210**, 211–213, *213*
General graph network (GN) model, 132, 134–135
Georgioudakis, M., **31**, **35**
Ghannadi, P., 16
Girardi, M., 190
GoPro action camera, 281
Graph neural network (GNN), 121, 132–139
Guclu, R., 202
Guedria, N. B., **42**
Guinigi Tower, 182–191, *183–194*, **187**, 194–198, **195–197**, *198*
Guzmán, R., 302

H

Half-joint bridge, 89
analysis, 91, 93, *93*, *94*
sensing network and installation, 89–91, *90–92*
Haria, K., 9
Heat-of-hydration test, 100
H-infinity control method, 202
Hsu–Nielsen source, 91
Hydration model, 107–109, **109**, *109*

I

Iannacone, L., 8
ICITECH laboratories of the Universitat Politècnica de València (ICITECH-UPV), 59, 66, *68*, 72, 74
ImageNet, 305
Immediate resilience, 4
Inception-ResNet-v2, 305
Inception-v2, 305
Input–state estimation, 151–152

Institute of Information Science and Technologies of the Italian National Research Council (ISTI-CNR), 186, 190
Intelligent structural damage detection, 312–313, *313*
Intersection over union (IoU), 329
Ismail, Z., 224
Italian National Research Council (ISTI-CNR), 186

J

Jahangiri, M., 15
Jansen, L. M., 204
Jena, P. K., **23**, **26**
Jetson AGX Xavier, 330

K

Kahya, V., 16
Kalman filter approach, 122, 152, 203–205
Kanade–Lucas–Tomasi method, 283
Kang, F., **21**, **22**
Kien bridge, *228*, 228–229
  vibration measurement, *229*, 229–230, *230*
  virtual data and evaluation, 231–232, *234*, 235, **235**, *235–238*, 238
Kim, N. -I., **40**
Kim, S., **36**
Koch, C., 303, 304
Kourehli, S. S., 16
Kripakaran, P., 244
Kromanis, R., 244
Kullback–Leibler (KL) divergence, 122, 126, 127
Kumar, A., 301
Kyoshin Network (K-NET), 213

L

Laier, J. E., **24**
Lambeth Group, 102
Levenberg–Marquardt backpropagation algorithm, 228, 247–248, 252
Lieu, Q. X., **41**
Li, H., 244
Likelihood function, 259
Limongelli, M. P., 8
Linear quadratic Gaussian (LQG) controller, 204–209, *206*, *207*
Linear quadratic regulator (LQR) controller, 203
Linear regression, 166–168
Linear time-invariant (LTI) system, 204–205
Linear variable differential transformer (LVDT), 63, *65*, 67, *69*, 73, 313, 326–327, *328*
Link, 6–7
Liu, K., 202, 243
Local coordinate system (LCS), 314
Long short-term memory (LSTM) approach, 244
Lucca towers
  Clock Tower, 182–184, *183*, 187, **196**, 197, **197**, *198*
  Guinigi Tower, 182–191, *183–194*, **187**, 194–198, **195–197**, *198*
  San Frediano Bell Tower, 182–184, *183*, 187, **196**, 197, **197**, *198*
Luxonis OAK-D camera, 302

M

MACEC code, 183, 232
Machine learning (ML), 292
Machine learning models (MLMs), 291–292
Maeda, H., 301
Magalhães, F., 243
Magnetorheological (MR) damper, 202–204, 219
Maia, N. M. M., 224
Majidifard, H., 303
Makhoul, N., 288
Manson, G., **20**
Marginal likelihood, 125
Markov chain Monte Carlo (MCMC) methods, 125, 172, 260, 264, 265; *see also* Cable-stayed road bridge
  Metropolis Hastings (MH) algorithm, 261, 264–267, 271, 274
Marsh Lane Viaduct, 78–79, *79*
  sensing network and installation, 78–81
  statistical shape analysis, 81–85
Martinelli, L., 4
Masihullah, S., 305
Masonry bridge, 77–78, 85, 86, 89, 95
Masonry towers, *182*, 182–183, *183*
  Clock Tower, 182–184, *183*, 187, **196**, 197, **197**, *198*
  Guinigi Tower, 182–191, *183–194*, **187**, 194–198, **195–197**, *198*
  San Frediano Bell Tower, 182–184, *183*, 187, **196**, 197, **197**, *198*
MATLAB software, 251, 254
Maximum a posteriori (MAP), 128
Mean-field assumption, 163
MEteorological Terminal Aviation Routine Weather Report (METAR), 245

Metropolis-Hastings (MH) scheme, 122, 127, **128**
algorithm, 261, 264–267, 271, 274
Mirdamadi, H. R., **22**
Moazzam, I., 304
MobileNet-v2, 305
Modal assurance criterion (MAC), 260, 262–264, 273
Modal phase collinearity (MPC), 235
Model updating, 259
Moderate assurance criterion (MAC), 235, 238, *238*, 239
Mohebian, P., 15
Mohr's circle of strain, 87, *87*
Montazer, M., **28**, **29**
Morandi bridge, 77, 280
Mosaic data augmentation, 306, *306*, 308
Multilevel/hierarchical models, 159, 166, 172–174, **173**, *173*, *174*
encoding domain expertise, 170–172, *171*
linear regression, 166–168
segmented linear power curves, 168–170, *169*, *170*
Multiple-input multiple-output (MIMO) systems, 202

N

Nakamura, Y., 205
Ndambi, J. -M., 224
Network visualization, 5–7, *6*
Neural network (NN), 244, 247–249, *249*
Ngamkhanong, C., 9
Nguyen-Thoi, T., **31**
Noacco, V., 191
Non-contact systems, 313
Nonlinear autoregressive (NAR) networks, 244–245, 248, 251–254, *252*, *253*
Nopour, M. H., **39–40**
Normalized mean squared error (NMSE), 130, *131*
NOSA-ITACA code, 190, 191, 195, **195**, 199

O

Offshore wind sector (OWS), 120
Operational modal analysis (OMA), 181, 183, 259
Optical flow, 283
Orcesi, A. D., 243
Ordinary Procrustes analysis (OPA), 82, 83, 85
Output-only (also operational) modal analysis (OMA), 232
Overlapping mixture of Gaussian processes (OMGP), 158–166, *161*, *165*
Overlay visualization, 5, *6*, 7, *8*
Ozgunalp, U., 302

P

Pahnabi, N., **40–41**, **44–45**
Pandey, P., 243
Parhi, D. R., **26**
Particle swarm optimization (PSO), 15, 204
and the τ_p^{max} approach, 204–207, *207*, *208*, 209, **210**, 211, *211–214*, 214, **215**, *216–218*, 218, 219
Pearson correlation coefficient, 174, *174*
Performance-based approach, 287
Phares, B. M., 243
Pholdee, N., **33**, **39**, 46
Pile foundations, 99
construction and instrumentation, *102*, 102–103, **103**, *103*
hydration model, 107–109, **109**, *109*
Stage 1 (S1) analysis, 110–114, *110–114*, 118
Stage 2 (S2) analysis, 114–117, *115*, *116*, 118
staged data interpretation framework, 100–101, *101*
temperature profiles, 103, *104–106*, 105–107
Plevris, V., **31**, **35**
Polynomial chaos (PC)-based surrogate models, 269–271, *270*
Population-based structural health monitoring (PBSHM), 155, *156*
Pothole detection, 298–299, *299*, 302
Power spectrum density (PSD), 271, *282*, 327, *329*
Pratt-type trusses, 59
Precise CutMix data augmentation, 308
Prediction error variances, 259
Price, K., 14, 16, 108
Probability density functions (PDFs), 259
Problem formulation, 152
Proportional-derivative (PD) controllers, 202
Proportional-integral-derivative (PID), 202

Q

Quantity of interest (QoI), 123
Quisi bridge, 60, 62, 74

R

Ramaswamy, A., 203
Ranieri, A., 302
Rao, A. R. M., **21**
Rapidity, 3
Rapidity, 288

Rath, S. R., 301
Red Green Blue-Depth (RGB-D) technology, 300, 302, *303*, 305, 308, 327
Redundancy, 3
Redundancy, 288
Reed, H., **22**
Region of interest (ROI), 281–283
ResNet-101, 305
Resourcefulness, 3, 288
Ricatti matrix differential equation, 207
Road damage, automatic detection of, 300
  classic computer vision methods, 303–304
  cracks, 299, *299*, 302
  data acquisition, 300–301, *301*
  datasets and benchmarks, 301–303, *302*, *303*
  deep learning methods, 304–307, *306*, *307*
  pothole detection, 298–299, *299*, 302
Robustness, 3
Robustness, 288
Runge–Kutta integration, 153
Ryu, S. K., 305

S

Saisi, A., 188
San Frediano Bell Tower, 182–184, *183*, 187, **196**, 197, **197**, *198*
Santos, L., 288
SAP2000 software, 268
Savoia, M., **28**, 46
Scale-invariant feature transform (SIFT), 317
Schindler, A. K., 107
Search origin, 116
Segmented linear power curves, 168–170, *169*, *170*
Semantic segmentation, 305
Seyedpoor, S., **27**, **28**, **29**, **34–35**, **37**, **39–40**, **40–41**, **44–45**
Shaking-table test building, **329**, 330, 331
Shape-based trackin, 283
SiamSDN algorithm, 314–315, *316*, 326–327, *328*
Simple passive control (SPC), 202
Single-task learning (STL), 172, 173
Slicing stage, 100–101, 114
Sliding mode controllers (SMCs), 202–203
Sobrinho, B. E., **40**
Sonic pulse echo (SE) testing, 99
Sousa, H., 288
Spatiotemporal process, 150–151
Stochastic gradient descent (SGD), 316
Stochastic subspace identification (SSI) algorithm, 232
Storn, R., 14, 16, 108
Strain sensors, 73, **73**
Strategic Road Network (SRN), 89
Structural health monitoring (SHM), 224, 266, 280; *see also* Computer vision (CV)-based SHM systems
  damage detection, 9
  damage intensity, 10
  damage localization, 9–10
  errors during, *225*, 225–226
  MCMC MH method for, 267
  phases of, 1–2
  prognosis, 10
  time series data, 226
  truss bridge, *see* Truss bridge
Structural resilience
  defined, 3
  dimensions of, 3–4
  on SHM, 8–10, *10*
Sum square error (SSE), 247
Sun, Q., 100, 101
Supervisory control and data acquisition (SCADA), 122, 135, **136**, *136*, 139, 155–156
Support vector machine (SVM), 83–84
Support vector machines (SVMs), 304
Su, Y., **43**
Systems-level approach, 155
  mean-field variational inference, 163–164, *164*
  multilevel modeling, *157*, 157–159, *158*
  OMGP, 159–166, *161*, *165*
  PBSHM Schema, 155, *156*
  SCADA data, 155–156

T

Taerwe, L., 107, 111
Talukdar, S., 243
Template matching, 282
Thanet Sand Formation, 102–103
Thermal integrity profiling (TIP), 99–100, 102, 105, 115, 117
Thompson, M., 302–303, 306, 307, 308
3D pile shape, 113–114, *114*
3D Shape Retrieval Contest (SHREC), 306, *307*
Time delay neural networks (TDNNs), 244
Time variable LQR (TVLQR) method, 204
Tomosawa, F., 107
Top-down dataset, 302
Truss bridge, 58–59
  dynamic tests, 67, 70
  fatigue strength curve test, 61–63, *63–66*, 65–66

fracture mechanics and failure patterns, 71, *71*
geometry of, 59, *60*
load hypothesis, 61, *62*
methodology, 60–61
real-scale span test, 66–68, *67–70*, 70–71
SHM recommendations, 71–74, *72*, **73**
2D video stabilization algorithm, 317

U

Unmanned aerial vehicle (UAV), 314, *315*, 317, 322–326, *323*, 326, *327*, **329**, 330–331, *331–333*, 333
Unsupervised local cluster-weighted bootstrap aggregation (ULC-BAG) method, 139–140, 142–145, *143*, **144**, *145*, 147–148
algorithm, 140–141
bagging algorithm, 144
Upwind turbine, 122, *123*, 127–130, *128*, *129*, **130**, *131*, 135, 137, *138*

V

Vaisala WXT530 series 6 weather station, 90
Variational autoencoder (VAE), 133
Variational Bayes (VB) approach, 132, 134, 139
Variational Bayesian Gaussian mixture (VBGM), 140, 148
Variational inference (VI), 142–143
Vertical scanning stage, 100, 110
Vibration-based methods, 15, 16, 46, 150, 224, 260, 273, 275; *see also* Differential evolution algorithm (DEA)
Villalba-Morales, J. D., **24**
Villamizar Mejía, R., **24**
Vincenzi, L., **23**, **28**, 46
Viren, 301
Virtual experimental responses, 261
Virtual-sensing task, 148–149, 154–155
dynamical system formulation, 149–150
example, 152–153, *153*, *154*
input–state estimation, 151–152
spatiotemporal filtering, 150–151
Vision-based displacement measurement system, 314, *315*
Vo-Duy, T., **29**, **32**, 46
VOSviewer software, 5, 7

W

Wahab, M. A., 244
Wake-affected (downwind) turbine, 129–130
Wang, C., 244
Wang, N., **43**
Wide-view dataset, 302
Wind farms, 120–121; *see also* multilevel/hierarchical models; systems-level approach; unsupervised local cluster-weighted bootstrap aggregation (ULC-BAG) method; virtual-sensing task
aeroelastic simulations, 122–124
DEL estimation, 121–124, *123*, 127–130, *128*, *129*, **130**, *131*, 145, 147, *147*
Gaussian process regression, 121–122, 124–129, **128**, *128*, *129*, 130–131, *131*, 159, 162
graph neural networks, 132–139
hyperparameter identification, 127
layout, 121–122
operational data, 134–135
SCADA measurements, 122, 135, **136**
simulated dataset, 136–138, *137–139*
Wind loads, 145
Wind turbine (WT), 134–137, *138*, 139, 145, 148, 149, 152, *153*, *169*, *171*, 176
Wöhler exponent, 128, 130, 145
Worden, K., **20**
World coordinate system (WCS), 314

X

Xia, Y., 224
Xu, Y., 282

Y

Yakut, O., 202
Yantra IIT Bombay, 301–302
Yazdanpanah, O., **27**
Yebes, J. J., 305
Yu, L., **25**, 46
Yun-Lai, Z., 244

Z

Zed2 binocular camera, 330
Zhang, Y., 9

Printed in the United States
by Baker & Taylor Publisher Services